Stimuli-Responsive Water Soluble and Amphiphilic Polymers

Stimuli-Responsive Water Soluble and Amphiphilic Polymers

Charles L. McCormick, EDITOR
The University of Southern Mississippi

American Chemical Society, Washington, DC

Library of Congress Cataloging-in-Publication Data

Stimuli-responsive water soluble and amphiphilic polymers / Charles L. McCormick, editor.

p. cm.—(ACS symposium series, ISSN 0097–6156 ; 780)

Includes bibliographical references and index.

ISBN 0–8412–3725–5

1. Water soluble polymers—Congresses. 2. Smart materials—Congresses.

I. McCormick, Charles L.., 1946–   . II. Series.

QD382.W3 S75   2000
547´.7045422—dc21   00–58619

The paper used in this publication meets the minimum requirements of American National Standard for Information Sciences—Permanence of Paper for Printed Library Materials, ANSI Z39.48–1984.

PRINTED IN THE UNITED STATES OF AMERICA

Foreword

The ACS Symposium Series was first published in 1974 to provide a mechanism for publishing symposia quickly in book form. The purpose of the series is to publish timely, comprehensive books developed from ACS sponsored symposia based on current scientific research. Occasionally, books are developed from symposia sponsored by other organizations when the topic is of keen interest to the chemistry audience.

Before agreeing to publish a book, the proposed table of contents is reviewed for appropriate and comprehensive coverage and for interest to the audience. Some papers may be excluded in order to better focus the book; others may be added to provide comprehensiveness. When appropriate, overview or introductory chapters are added. Drafts of chapters are peer-reviewed prior to final acceptance or rejection, and manuscripts are prepared in camera-ready format.

As a rule, only original research papers and original review papers are included in the volumes. Verbatim reproductions of previously published papers are not accepted.

ACS Books Department

Contents

Indexes

Preface

This book is based primarily on the Symposium on Stimuli-Responsive Water Soluble and Amphiphilic Polymers held during the 1999 National American Chemical Society (ACS) meeting in New Orleans under sponsorship of the ACS Division of Polymer Chemistry, Inc. and the Petroleum Research Fund. From the 56 papers presented, the editor invited principal investigators of the most outstanding work to submit chapters for publication. As a result, this volume includes exten-sive literature background, important concepts, and frontier research conducted by some of the most active and innovative research groups in the area of water-soluble polymers. The international scope of the re-search is attested to by contributions from the United States, the United Kingdom, Canada, France, Belgium, Germany, the Netherlands, and Japan.

Stimuli-Responsive water soluble and amphiphilic polymers repre-sent a rapidly growing area with enormous technological and commercial potential. Impetus for innovation has come from the need for environmentally safe, yet "smart" materials synthesized, dispersed, and delivered in aqueous media. In many cases, stabilization in water and subsequent assembly onto appropriate substrates can be controlled by microstructural elements responsive to stimuli such as pH, ionic strength, shear stress, light or temperature.

That publication of this book occurs 10 years after an earlier volume, *Water-Soluble Polymers: Synthesis, Solution Properties, and Applications*, ACS Symposium Series 467, seems appropriate. Readers of both volumes can readily trace the development of new synthetic and analytical procedures as well as key application technologies during this 10 year period. The projected seamless integration of chemistry, biochemistry, polymer science, molecular biology, and engineering into frontier research has become a reality in water-soluble polymer research and should continue to be a source of inspiration for future endeavors.

It is also notable that this volume is the first of the ACS Symposium Series to be written, reviewed, compiled, and transmitted electronically via the internet. Special acknowledgements are made to Velda Moore, Geoff Smith, and Andrew Lowe at the University of Southern Mississippi for

extensive manuscript editing and formatting. Finally, the editor thanks
Anne Wilson and Kelly Dennis of the ACS Books Department for their
efforts in providing the electronic format for authors and for final book
assembly.

CHARLES L. McCORMICK
Department of Polymer Science
The University of Southern Mississippi
Hattiesburg, MS 39406–0076

Chapter 1

Stimuli Responsive Water-Soluble and Amphiphilic (Co)polymers

Andrew B. Lowe and Charles L. McCormick

Department of Polymer Science, The University of Southern Mississippi,
Hattiesburg, MS 39406-0076

In recent years much interest has been focused on (co)polymer systems that undergo a conformational change or phase transition in response to an external stimulus. The external stimulus can include, but is not limited to, temperature, added electrolyte, changes in pH, light, other molecules or a combination of these. Such (co)polymers are of great scientific and technological importance. They serve as important additives in fields as diverse as water treatment, enhanced oil recovery, controlled drug release, the formulation of water-borne coatings and personal care products. In this chapter we will examine specific examples of water-soluble stimuli-responsive (co)polymer systems.

Introduction

Water-soluble and Amphiphilic copolymers possessing intra- and intermolecular modes of reversible association have been the subject of intensive study in both academic and industrial laboratories. Conformational changes that can be 'triggered' in response to external stimuli usually involve ionic interactions, hydrogen bonding, or hydrophobic effects. Often, combinations of these manifest themselves in segmental microphase transitions or other self-organization into electrolyte, shear, temperature, pH, or photo-responsive systems. A major emphasis in this research area has been to elucidate how microstructural features can be optimized for stimuli-responsivness in water or in thin films assembled from water at interfaces. Impetus for this research comes from the enormous potential range of commercial applications including: personal care, pharmaceutics, coatings, oil field chemicals, rheology modification, colloidal stabilization, and water remediation.

Electrolyte Responsive (Co)polymers

The addition of low molecular weight electrolyte to an aqueous solution of a (co)polymer can induce one of four responses. The electrolyte can cause chain contraction, chain expansion, aggregation through chelation (conformational transitions), or precipitation (phase transition). The exact nature of the response will depend on various factors such as chemical structure, concentration, *i.e.* molecular weight and composition of the (co)polymer and nature of the added electrolyte *i.e.* mono, di or trivalent ions.

In a simple linear polyelectrolyte, such as poly(methacrylic acid) (PMAA), neutralization by addition of base ionizes the acid groups to the carboxylate form resulting in repulsive interactions and thus an extended conformation of the macromolecular backbone. As a result the polymer coils occupy a large volume in solution (this will affect properties such as intrinsic viscosity and diffusion coefficient). The addition of a small amount of low molecular weight electrolyte, such as sodium chloride, to this solution will lead to a Debye-Hückel shielding effect and the polymer chains undergo a conformational transition, adopting a smaller, more entropically favored conformation (*1, 2*) (Figure 1).

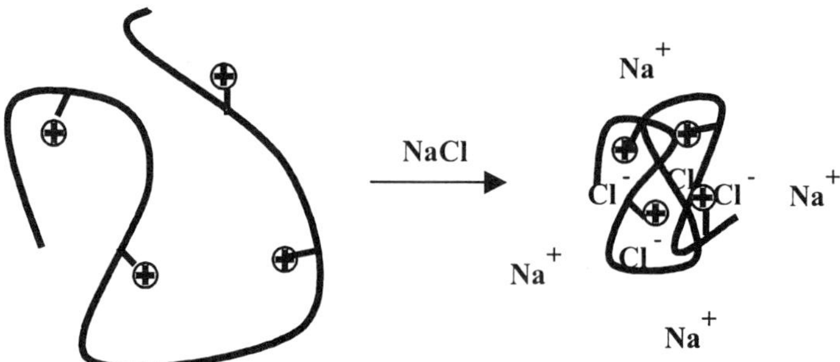

Figure 1. *The polyelectrolyte effect*

The opposite behavior, in which the addition of low molecular weight electrolyte causes chain expansion, is typical of aqueous solutions of polyzwitterions. These are (co)polymers that contain both anionic and cationic groups. These groups may be located on the same monomer unit (polybetaines) or on different monomer units (polyampholytes). When the ratio of anionic to cationic groups approaches unity the conformation adopted by the (co)polymer in aqueous solution is dictated by attractive electrostatic or dipolar interactions. This leads to a collapsed, globule-like conformation. Upon addition of low molecular weight electrolyte, the attractive interactions are screened and the (co)polymer adopts a more expanded conformation. This behavior, opposite to that exhibited by polyelectrolytes, is known as the *anti-polyelectrolyte effect* (*3*) (Figure 2).

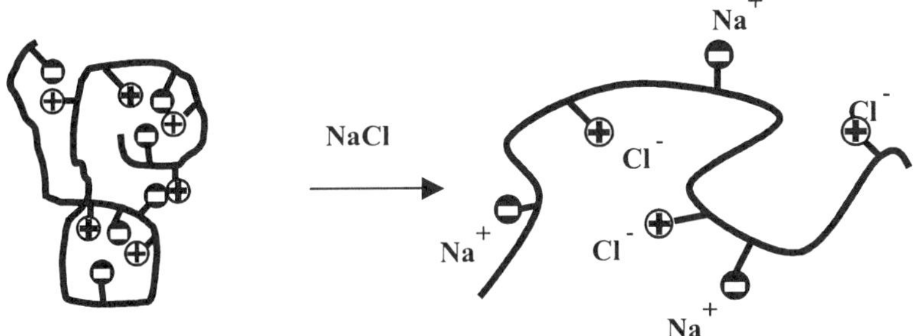

Figure 2. *The anti-polyelectrolyte effect*

Numerous workers have reported this type of behavior for polyzwitterions. (*4, 5, 6*) For example, Lowe and co-workers (*7, 8, 9*) reported the synthesis and characterization of near-monodisperse poly(sulfopropyl betaine) block copolymers based on 2-(dimethylamino)ethyl methacrylate (DMAEMA) with a hydrophobic comonomer such as methyl methacrylate (MMA). The DMAEMA residues of the precursor block copolymers were functionalized by reaction with 1,3-propanesultone yielding the corresponding sulfopropyl betaine block copolymers. In aqueous solution the hydrophilic-hydrophobic block copolymers aggregate into near-monodisperse micelles with the sulfopropyl betaine blocks forming the micelle corona and the hydrophobic comonomer residing in the micelle core. It has been shown that the hydrodynamic diameter of the micelles, as determined by dynamic light scattering, is a function of electrolyte concentration. Given the betaine nature of the micelles, the hydrodynamic diameter increases with increasing salt concentration (NaCl) from *ca.* 12-16 nm

4

up to maximum values of *ca.* 27 nm at around 0.1 M NaCl. This observed increase is due to the anti-polyelectrolyte effect exhibited by the betaine corona.

McCormick *et al.* (*10, 11, 12*) have studied the aqueous solution properties of acrylamido-based sulfobetaine and carboxybetaine polymers, either as statistical copolymers with acrylamide (*10, 11*) or as the sulfo/carboxy betaine copolymer. (*12*) For example, Kathmann and McCormick reported the aqueous solution properties of statistical copolymers of acrylamide with the betaine comonomer 6-(2-acrylamido-2-methylpropyldimethylammonio) exanimate. Anti-polyelectrolyte behavior was observed for the copolymers as determined by reduced viscosity measurements as a function of increasing NaCl concentration. The authors also demonstrated the effect of different electrolytes (NaSCN, CsCl and NaCl) on the aqueous solution behavior of the copolymers. The electrolytes with soft counterions (Cs^+ and SCN^-) show higher solution viscosities than the relatively hard Na^+/Cl^- counterions. This follows the well known Hoffmeister series and has been demonstrated previously by other authors for polybetaine systems. (*4, 13*)

It is well known that certain low molecular weight metal ions, such as Ca^{2+}, can induce aggregation in aqueous polymer solutions through the formation of polyion/metal complexes. In particular, Ca^{2+} is known to be an efficient binder for carboxylic acid groups.

Copolymers of acrylic acid with acrylamide (synthesized either by copolymerization of the two monomers, or by partial hydrolysis of the acrylamide homopolymer) represent one of the more widely studied systems. (*14, 15*) Recently, Peng and Wu carried out a light scattering study on the formation and structure of partially hydrolyzed polyacrylamide/Ca^{2+} complexes. (*16*) It was shown that complex formation could be controlled by both the degree of hydrolysis of the polyacrylamide homopolymer and the Ca^{2+} concentration. It was also shown that for a given sample of hydrolyzed polyacrylamide there exists a critical Ca^{2+} concentration at which point interchain complexation predominates over intrachain. Finally, for a fixed Ca^{2+} concentration, the size of the complexes increases with increasing degree of hydrolysis of the polyacrylamide.

Phase transitions may also be induced in some aqueous (co)polymer systems through the addition of low molecular weight electrolyte. A recent example of this is the unimer-to-micelle transition described by Bütün *et al.* (*17*) in block copolymers of 2-(diethylamino)ethyl methacrylate (DEAEMA) with 2-(*N*-morpholino)ethyl methacrylate (MEMA). MEMA is readily salted-out from aqueous solution on addition of electrolytes such as Na_2SO_4, whereas DEAEMA remains soluble. Addition of a critical concentration of electrolyte to an aqueous solution of the diblock copolymer results in the MEMA block undergoing a phase transition as the diblock associates to form well-defined micelles, with the MEMA forming the 'dehydrated' micelle core and the DEAEMA the solvated corona. This electrolyte-induced micellization is completely reversible (Figure 3). Removal of the added electrolyte by dialysis leads to molecular dissolution of the block copolymer.

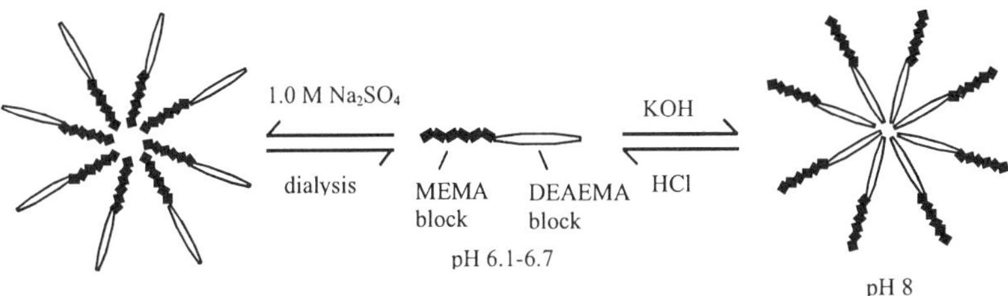

Figure 3. *Formation of micelles and reversible micelles for a DEAEMA-MEMA block copolymer at 20 °C*

Recently, Schöps *et al.* reported the salt-induced switching of microdomain morphology in an ionically functionalized styrene-*block*-isoprene copolymer. (*18*) In this block copolymer the ionic groups were introduced at the chain ends. Living homopolystyrene was end capped with a tertiary amine functional diphenylethylene, prior to the addition of isoprene, which was subsequently terminated by the addition of 1,3-propanesultone, thus introducing sulfonate functional groups (Figure 4).

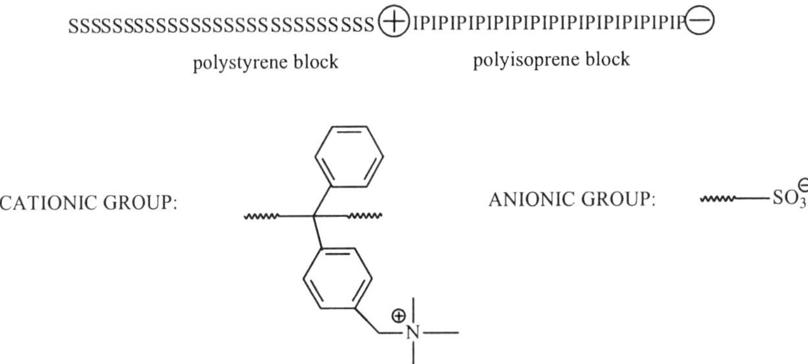

Figure 4. *Schematic representation of the cationic/anionic functionalized styrene-isoprene block copolymer.*

In the absence of salt the solid state morphology consists of hexagonally packed polystyrene cylinders in a polyisoprene matrix. This is a result of attractive ionic interactions between the cationic and anionic groups. Upon the addition of salt (6 equivalents of salt per chain) the ionic interactions are screened and the solid state morphology changes to one of a lamellar nature.

Temperature Responsive (Co)polymers.

The aqueous solubility of many water-soluble polymers (WSPs) depends on solution temperature. Most WSPs are soluble at room temperature, but will undergo a phase transition upon cooling, *e.g.* poly(acrylic acid) or upon heating *e.g.* poly(ethylene oxide). Phase separation as a function of increasing temperature is more common and is referred to as inverse temperature dependent water solubility. The temperature at which the WSP undergoes this phase transition is termed the *cloud point.* (*19*) The cloud point for any given homopolymer/copolymer system is dependent upon its intrinsic nature, *i.e.* the presence of ionic, or potentially ionizable species, hydrogen bonding character, tacticity effects etc. Also other factors such as the molecular weight and concentration of the polymer and the presence of other low molecular weight additives such as salts, (*20*) urea (*21*) or surfactants, (*22*) will all have some effect on the cloud point.

Recently, for example, Idziak *et al.* (*23*) reported on the thermosensitivity of aqueous solutions of poly(N,N-diethylacrylamide), and demonstrated the variation in the cloud point as a function of sodium chloride and sodium dodecyl sulfate concentrations. Similarly, Tong *et al.* (*24*) have reported cloud-points of aqueous solutions of poly(N-isopropylacrylamide) as a function of polymer concentration and molecular weight. The synthesis of hydrophilic-hydrophilic diblock copolymers with very different cloud points allows for the construction of reversible micelle forming species. For example, Topp *et al.* (*25*) have described the thermo-reversible micellization of ethylene glycol-N-isopropylacrylamide block copolymers for application as potential drug delivery systems. Poly(N-isopropylacrylamide) is a widely studied thermosensitive polymer due to the fact that it has a relatively low cloud-point (~ 31 °C). The ethylene glycol-N-isopropylacrylamide block copolymers exist as molecularly dissolved unimers below this lower critical solution temperature (LCST) as determined by aqueous size exclusion chromatography. Upon raising the solution temperature above the LCST, the N-isopropylacrylamide undergoes a phase transition and the block copolymer chains associate forming well-defined micelles with the ethylene glycol as the solvated corona and the phase-separated N-isopropylacrylamide as the micelle core. Micelle sizes ranging from ~ 60 – 120 nm, depending on molecular weight and block copolymer composition, were observed as determined by dynamic and static light scattering techniques. The aggregation behavior was completely reversible with the block copolymer becoming molecularly dissolved when the solution temperature was lowered below the LCST of the *N*-isopropylacrylamide block.

Similar thermoreversible aggregation has been demonstrated by Lowe *et al.* (*26, 27*) for methacrylate-based block copolymers of DMAEMA with methacrylic acid (MAA). The precursor block copolymers of DMAEMA and 2-tetrahydropyranyl methacrylate (THPMA) were synthesized by group transfer polymerization. Subsequent acid hydrolysis of the THPMA residues yielded the DMAEMA-MAA block copolymers (Figure 5).

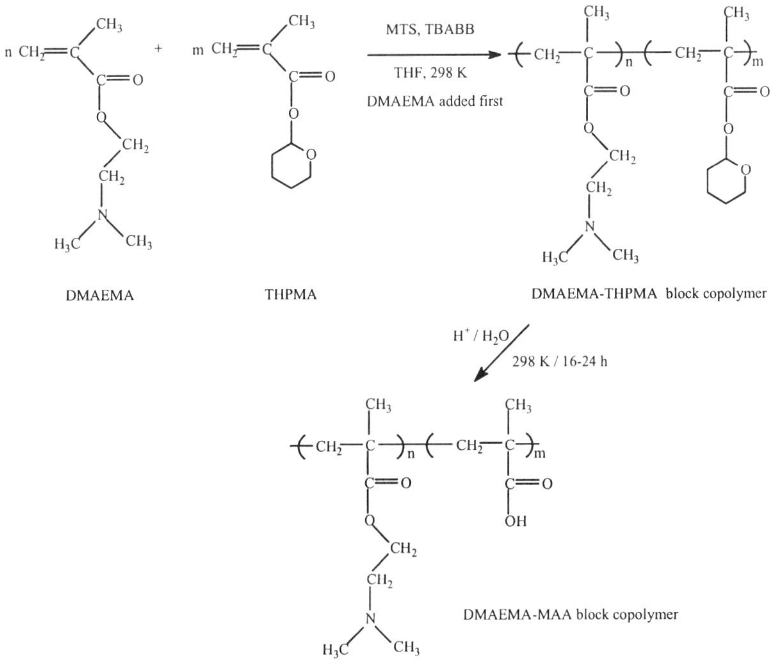

Figure 5. *Reaction scheme for the synthesis of DMAEMA-MAA zwitterionic block copolymers via group-transfer polymerization.*

Upon raising the temperature of a 1 w/w % aqueous solution at pH ~ 9.5 in the presence of 0.01 M NaCl to *ca.* 50 °C, the DMAEMA-MAA block copolymers micellize with the now hydrophobic DMAEMA residues forming the micelle core and the MAA residues forming the solvated corona. At *ca.* pH 9.5, the DMAEMA residues are almost completely deprotonated and, therefore, show inverse temperature water-solubility, whereas the MAA residues are fully ionized and remain soluble. Cooling the solution to room temperature resulted in the molecular dissolution of the block copolymer. Lowe *et al.* monitored the micellization/unimeric dissolution of the block copolymers by both dynamic light scattering and variable temperature ^{1}H NMR spectroscopy.

pH Responsive (Co)polymers.

Amphoteric (co)polymers may undergo conformational changes or phase transitions in response to changes in the aqueous solution pH. For example, DMAEMA homopolymer exhibits an LCST in aqueous solution in the range 32 - 46 °C depending on molecular weight (in its non-protonated form). (*28*) However, in a partially ionized (protonated) form DMAEMA homopolymer remains soluble in aqueous solution up to 100 °C. It is the ability to control such transitions that allows for the synthesis of novel (co)polymer structures. Recently a host of authors have reported the synthesis and applications of pH sensitive (co)polymer systems. Huang and Wu (*29*) have described the effect of pH (and other factors) on the phase transition in poly(*N*-isopropylacrylamide-*co*-MAA) nanoparticles. Both Lee *et al.* (*30*) and Martin *et al.* (*31*) have discussed the pH dependent micellization of tertiary amine containing block copolymers and McCormick and co-workers have studied both acrylamido and vinyl ether-based pH responsive copolymer systems. (*32, 33, 34*) For example, Hu *et al.* (*33*) reported pH responsive microdomains in labeled *n*-octylamide-substituted poly(sodium maleate-*alt*-ethyl vinyl ethers) (Figure 6), using nonradiative energy transfer and steady-state fluorescence techniques. It was shown at low pH (< pKa of the acid groups) that the octylamide modified poly(sodium maleate-*alt*-ethyl vinyl ethers) adopt highly collapsed conformations in aqueous media. At moderate pH values, the polymer coil becomes slightly more expanded, as indicated by reduced viscosity measurements. At 7.0 < pH < 8.0 the polymer adopts its most expanded chain conformation as determined by nonradiative energy transfer measurements. The observed behavior is a direct result of the increasing degree of ionization of the carboxylic acid functional groups along the polymer backbone. As the pH is raised, so too is the degree of ionization of the acid groups. This results in polyelectrolyte formation and chain expansion due to charge-charge repulsions. At some critical pH, the reduced viscosity again begins to fall (*ca.* pH 8.5). This is due to shielding of the anionic charges by the now relatively high concentration of small ions in solution (Figure 7).

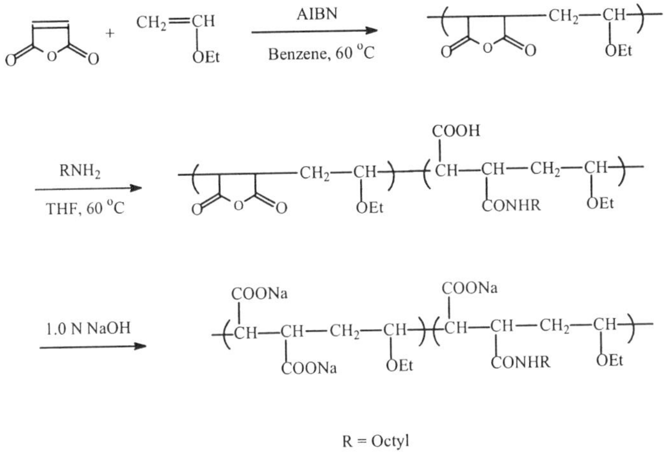

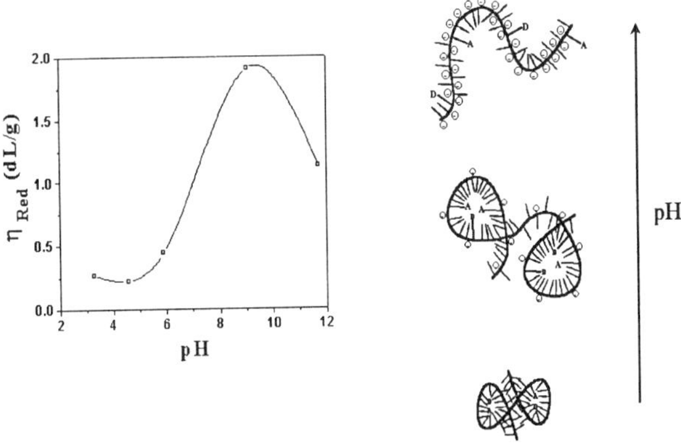

R = Octyl

Figure 6. *Synthetic pathway for the preparation of hydrophobically modified copoly(sodium maleate-alt-ethyl vinyl ethers).*

Figure 7. *Reduced viscosity as a function of pH for a 30 % n-octylacrylamide-substituted poly(sodium maleate-alt-ethyl vinyl ether) with a schematic showing chain expansion with increasing pH.*

10

Photo Responsive (Co)polymers.

Photochromic compounds undergo a conformational change, such as isomerization or dimerization, or a phase transition in response to exposure to light of an appropriate wavelength. The existence of two such states, which may be reversibly switched by means of an external light stimulus, can form the basis of molecular switches and/or optical storage data for example. The incorporation of photochromic molecules in (co)polymers yields photoactive materials of improved processibility and stability. The property changes of the chromophores that occur after irradiation have been utilized to induce conformational changes in the (co)polymers to which they have been bound. Common photoresponsive species that have been incorporated into synthetic (co)polymers include triphenylmethane derivatives, (*35, 36*) azobenzenes, (*37, 38*) spiropyrans (*39*) and coumarin (*40*). For example, Irie and Kunwatchakun have described the reversible swelling of polyacrylamide gels containing photoresponsive triphenylmethane derivatives. (*35*) Triphenylmethane derivatives dissociate when irradiated into ion pairs which thermally recombine. The process may be monitored visually since the tertiary cation produced upon irradiation is highly colored (Figure 8).

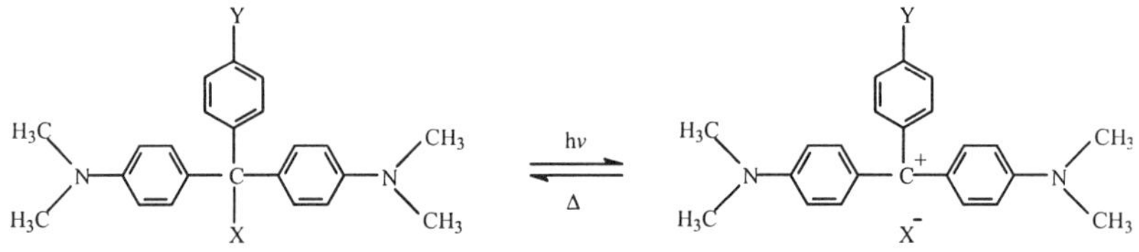

Figure 8. *The reversible, photo-induced dissociation of triphenylmethane derivatives.*

Upon irradiation of a polyacrylamide gel containing several mole percent of a triphenylmethane derivative the gel more than doubled its size, as a result of the ion-pair formation. The gel reverted to its original size when placed in the dark.

One of the more widely studied groups of (co)polymers is that comprised of poly(α-amino acids). (*41, 42, 43*) For example, Fissi and co-workers have studied spiropyran modified poly(L-lysine), (*39*) while Yamamoto and co-workers have studied coumarin modified poly(L-lysine). (*40*) In the latter system, the poly(L-lysine) undergoes photoinduced crosslinking – the coumarin species undergoes a cycloaddition reaction upon irradiation which occurs both inter- and intramolecularly. This results in a transparent gel with potential controlled release applications.

Molecule Responsive (Co)polymers

There are only a few reports in the literature of (co)polymers/hydrogels that show a response to the presence of a specific molecule. Examples include the work of Aoki *et al.* (*44*) and Watanabe *et al.* (*45*)   More recently, Miyata *et al.* (*46*) reported an acrylamido-based antigen sensitive hydrogel. The hydrogel was prepared from the radical copolymerization of acrylamide, *N,N'*-methylenebisacrylamide and an antigen-functionalized acrylamide monomer (antigen: Rabbit immunoglobulin G – Rabbit IgG) in the presence of Goat anti-rabbit IgG. The hydrogels were immersed in a phosphate buffer, and allowed to attain equilibrium. At this point free Rabbit IgG was added and the response of the hydrogel monitored by optical microscopy. An increase in the equilibrium swelling ratio was seen with increasing concentration of added Rabbit IgG (a control experiment with a crosslinked acrylamide hydrogel with no antibody actually showed a small decrease in swelling ratio). This indicates that the antigen-antibody hydrogel is Rabbit IgG sensitive. The hydrogel was also shown to be specifically responsive to Rabbit IgG, since the addition of free Goat IgG to the antigen-antibody hydrogel did not induce any swelling.

Conclusions

In this chapter we have highlighted some recently reported examples of water-soluble, stimuli responsive (co)polymer systems. We have focused on the more common modes of inducing conformational changes/phase transitions in these systems such as added electrolyte, changes in solution pH/temperature and irradiation.

Acknowledgements

The Department of Energy, Office of Naval Research, and Eveready Battery Company are kindly thanked for financial support of the McCormick research group.

References

1. Munk, P. *Introduction to Macromolecular Science* Wiley-Interscience 1989, p 59.
2. Billmeyer, F. W. *Textbook of Polymer Science* Wiley-Interscience 2nd Edn., 1971, p 95.
3. Kathmann, E. E. L.; Davis, D. D.; McCormick, C. L. *Macromolecules* **1994**, *27*, 3156.
4. Schulz, D. N.; Peiffer, D. G.; Agarwal, P. K.; Larabee, J.; Kaladas, J.; Sori, L.; Handwerker, B.; Garner, R.T. *Polymer* **1986**, *27*, 1734.
5. McCormick, C.L.; Kathmann, E. E. In *Polymeric Materials Encyclopedia*; Salamone, J. C., Ed.; CRC Press, Inc., 1996, Vol. 7, p 5462.
6. Huglin, M. B.; Radwan, M. A. *Makromol. Chem.* **1991**, *192*, 2433.
7. Lowe, A. B.; Billingham, N. C.; Armes, S. P. *Chem. Commun.* **1996**, 1555.
8. Tuzar, Z.; Popisil, H.; Plestil, J.; Lowe, A. B.; Baines, F. L.; Billingham, N. C.; Armes, S. P. *Macromolecules* **1997**, *30*, 2509.
9. Lowe, A. B.; Billingham, N. C.; Armes, S. P. *Macromolecules* **1999**, *32*, 2141.
10. Kathmann, E. E.; White, L. A.; McCormick, C. L. *Polymer* **1997**, *38(4)*, 879.
11. Kathmann, E. E. and McCormick, C. L. *J. Polym. Sci., Polym. Chem.* **1997**, *35*, 243.
12. Kathmann, E. E.; White, L. A.; McCormick, C. L. *Macromolecules* **1997**, *30*, 5297.
13. Liaw, D.; Lee, W.; Whung, Y.; Lin, M. *J. Appl. Polym. Sci.* **1987**, *34*, 999.
14. Muller, G.; Laine, J. P.; Fenyo, J. C. *J. Polym. Sci., Polym. Chem.* **1979**, *17*, 659.
15. Truong, D. N.; Galin, J. C.; Francois, J.; Pham, Q. T. *Polymer Comm.* **1984**, *25*, 208.
16. Peng, S.; Wu, C. *Macromolecules* **1999**, *32*, 585.
17. Bütün, V.; Billingham, N. C.; Armes, S. P. *J. Am. Chem. Soc.* **1998**, *120*, 11818.
18. Schops, M.; Leist, H.; DuChesne, A.; Wiesner, U. *Macromolecules* **1999**, *32*, 2806.
19. Molyneux, P. *Water-Soluble Synthetic Polymers: Properties and Behavior Vol 1*, CRC Press 1982.
20. Hahn, M.; Görnitz, E.; Dautzenberg, H. *Macromolecules* **1998**, *31*, 5616.
21. Platé, N. A.; Lebedeva, T. L.; Valuev, L. I. *Polymer Journal* **1999**, *31(1)*, 21.
22. Jehudah, E. *J. Appl. Polym. Sci.* **1978**, *22*, 873.

23. Idziak, I.; Avoce, D.; Lessard, D.; Gravel, D.; Zhu, X. X. *Macromolecules* **1999**, *32*, 1260.

24. Tong, Z.; Zeng, F.; Zheng, Xu. *Macromolecules* **1999**, *32*, 4488.

25. Topp, M. D. C.; Dijkstra, P. J.; Talsma, H.; Feijen, J. *Macromolecules* **1997**, *30*, 8518.

26. Lowe, A. B.; Billingham, N. C.; Armes, S. P. *Chem. Commun.* **1997**, 1035.

27. Lowe, A. B.; Billingham, N. C.; Armes, S. P. *Macromolecules* **1998**, *31*, 5991.

28. Bütün, V. D.Phil thesis, University of Sussex, Brighton, East Sussex, 1999.

29. Huang, J.; Wu, X. Y. *J. Polym. Sci., Polym. Chem.* **1999**, *37*, 2667.

30. Lee, A. S.; Gast, A. P.; Bütün, V.; Armes, S. P. *Macromolecules* **1999**, *32*, 4302.

31. Martin, T. J.; Procházka, K.; Munk, P.; Webber, S. E. *Macromolecules* **1996**, *29*, 6071.

32. Kathmann, E. E.; McCormick, C. L. *J. Polym. Sci., Polym. Chem.* **1997**, *35*, 231.

33. Hu, Y.; Smith, G. L.; Richardson, M. F.; McCormick, C. L. *Macromolecules* **1997**, *30*, 3526.

34. Richardson, M. F.; Armentrout, R. S.; McCormick, C. L. *J. Appl. Polym. Sci.* **1999**, *74*, 2290.

35. Irie, M.; Kunwatchakun, D. *Macromolecules* **1986**, *19*, 2476.

36. Mamada, A.; Tanaka, T.; Kungwatchakun, D.; Irie, M. *Macromolecules* **1990**, *23*, 1517.

37. Irie, M.; Iga, R. *Macromolecules* **1986**, *19*, 2480.

38. Yamamoto, H. *Macromolecules* **1986**, *19*, 2472.

39. Fissi, A.; Pieroni, O.; Ruggeri, G.; Ciardelli, F. *Macromolecules* **1995**, *28*, 302.

40. Yamamoto, H.; Kitsuki, T.; Nishida, A.; Asada, K.; Ohkawa, K. *Macromolecules* **1999**, *32*, 1055.

41. Ueno, A.; Adachi, K.; Makamura, J.; Osa, T. *J. Polym. Sci. Polym. Chem.* **1990**, *28*, 1161.

42. Menzel, H.; Hallensleben, M. L.; Schmidt, A.; Knoll, W.; Fischer, T.; Stumpe, J. *Macromolecules* **1993**, *26*, 3644.

43. Fissi, A.; Pieroni, O.; Ciardelli, F. *Biopolymers* **1987**, *26*, 1993.

44. Aoki, T.; Nagao, Y.; Sanui, K.; Ogata, N.; Kikuchi, A.; Sakurai, Y.; Kataoka, K.; Okano, T. *Polym. J.* **1996**, *28*, 371.

45. Watanabe, M.; Akahoshi, T.; Tabata, Y.; Nakayama, D. *J. Am. Chem. Soc.* **1998**, *120*, 5577.

46. Miyata, T.; Asami, N.; Uragami, T. *Macromolecules* **1999**, *32*, 2082.

Chapter 2

Stimuli-Responsive Associative Behavior of Polyelectrolyte-Bound Nonionic Surfactant Moieties in Aqueous Media

Akihito Hashidzume, Tetsuya Noda, and Yotaro Morishima

Department of Macromolecular Science, Graduate School of Science, Osaka University, Toyonaka, Osaka 560-0043, Japan

The self-association behavior in water of hydrophobically-modified polyelectrolytes can be controlled by macromolecular architecture. This chapter will discuss the stimuli-responsive associations of an amphiphilic polyelectrolyte derived from the micellization of polymer-bound amphiphiles. The first half of the chapter will review the associative properties of random copolymers of an electrolyte monomer and hydrophobic comonomers as a function of polymer architectures with an emphasis put upon effects of spacer between the hydrophobe and polymer backbone. The second half will focus on the effects of ionic strength and shear stress on solution properties of a polyelectrolyte-bound nonionic surfactant moiety having a poly(ethylene oxide) (PEO) spacer. The polymer-bound surfactant moieties can form micelles whose cores are formed by hydrophobes at an end of the PEO spacer chain. When the hydrophobes on the same polymer chain occupy different micelle cores, the micelles are bridged by polymer chains, and thus a network structure is formed. Micelle bridging depends not only on the polymer concentration but also on ionic strength and shear stress. Thus, aqueous solutions of these polymers exhibit unique stimuli-responsiveness.

The spontaneous structural self-organization of hydrophobically-modified water-soluble polymers in aqueous media has been a subject of extensive studies over the past decades because it is of importance for molecular understanding of many biological phenomena and also for designing of novel polymer-based water-borne systems for practical applications (*1–5*). The self-organization of amphiphilic polymers in water is derived mainly from hydrophobic associations, resulting in various types of micellelike nanostructures (*6–10*). Amphiphilic polyelectrolytes are among such systems that have been extensively studied in the past decade. These include block (*9,11–13*), graft (*14*), alternating (*15–17*), and random copolymers (*15,18–23*), consisting of electrolyte and hydrophobic monomer units. Some of these polymers show stimuli-responsiveness in their self-association behavior, responding to changes in external conditions such as ionic strength (added salt), shear stress, temperature, and pH. These stimuli-responsive systems have attracted considerable interest because of their potential ability to capture and deliver materials, which may find pharmaceutical or environmental applications (*10,24*).

Various random copolymers have been synthesized from various electrolyte monomers and hydrophobic comonomers possessing a hydrophobe in the side chain because of their ease of synthesis and a wide range of monomer selections as compared to the case of block copolymers. A wide variety of association phenomena can be expected for the random copolymers depending on their molecular architectures. In this article, we focus on the stimuli-responsive association behavior in water of random copolymers of sodium 2-(acrylamido)-2-methylpropanesulfonate (AMPS) and an associative surfactant macromonomer.

Self-Association of Random Copolymers of Electrolyte and Hydrophobic Monomers

The self-association and the type of nanostructures formed are strongly dependent on whether hydrophobic associations occur in an intra- or intermolecular fashion. One can think about the two extreme cases where hydrophobic associations are completely inter- or intramolecular. Figure 1a illustrates an extreme case where all polymer-bound hydrophobes undergo interpolymer association. When hydrophobe content is sufficiently low, the interpolymer hydrophobic association may result in a situation where several polymer chains are cross-linked. This hydrophobic cross-linking would cause a large increase in solution viscosity. As the hydrophobe content is increased, an infinite network may be formed, leading to gelation or bulk phase separation. Figure 1b illustrates another extreme case where all polymer-bound hydrophobes undergo completely intrapolymer associations. In this extreme case, hydrophobic associations lead to the formation of single-molecular self-assemblies or unimolecular micelles. When the content of hydrophobes in a polymer is sufficiently low, a "flower-like" unimolecular micelle may be formed, which consists of a hydrophobic core surrounded by hydrophilic loops shaped into "petals". This type of micelles has been theoretically predicted by de Gennes (*25*), Halperin (*26*), and Semenov et al. (*27*) for the self-organization of sequential multi-block copolymers in selective solvents. As the content of hydrophobes in a copolymer is increased, the flower-like micellar structure is expected to become unstable because a larger portion of the surface of the hydrophobic core is exposed to the water phase. This would lead to a further collapse into a more compact micelle with a third-order structure due to secondary association of hydrophobic cores.

In general, the association process in water for polymer-bound hydrophobes follows either a closed or open association model (*26–28*). In closed association, an increase in polymer concentration results in an increase in the number of polymer micelles of the same size and structure. In open association, in contrast, the size of polymer aggregates increases with increasing polymer concentration. The first and second extreme cases (Figures 1a and b, respectively) are typical of open and closed associations, respectively. These two extreme cases may not appear to be realistic because a large penalty of entropy as well as enthalpy should be imposed on the formation of such association structures.

A more realistic case is illustrated in Figure 1c where intrapolymer associations mainly occur but a portion of hydrophobes undergo interpolymer association. In this situation, intermolecularly bridged flower-like micelles would be formed. The extent of such micelle bridging may depend strongly on the content of hydrophobes in the polymer as well as the polymer concentration. As the content of hydrophobes in a polymer is increased, both the size of the micelle core and the number of micellar bridges would increase. As a result, a collapsed micelle network would eventually be formed, and the system may end up with gelation or bulk phase separation.

Effects of Molecular Architecture on the Self-Organization

It is well-documented that the associative properties of hydrophobically-modified polyelectrolytes strongly depend on macromolecular architectures. Some of the structural parameters that control intra- vs. interpolymer associations include the type of hydrophobes, their content in the polymer, the sequence distribution of electrolyte and hydrophobic monomer units along the polymer chain, and the type of spacer between hydrophobes and the main chain (*15–17,19–23,29–31*). Among these structural parameters, spacer bonding between the hydrophobe and polymer backbone is an important structural factor

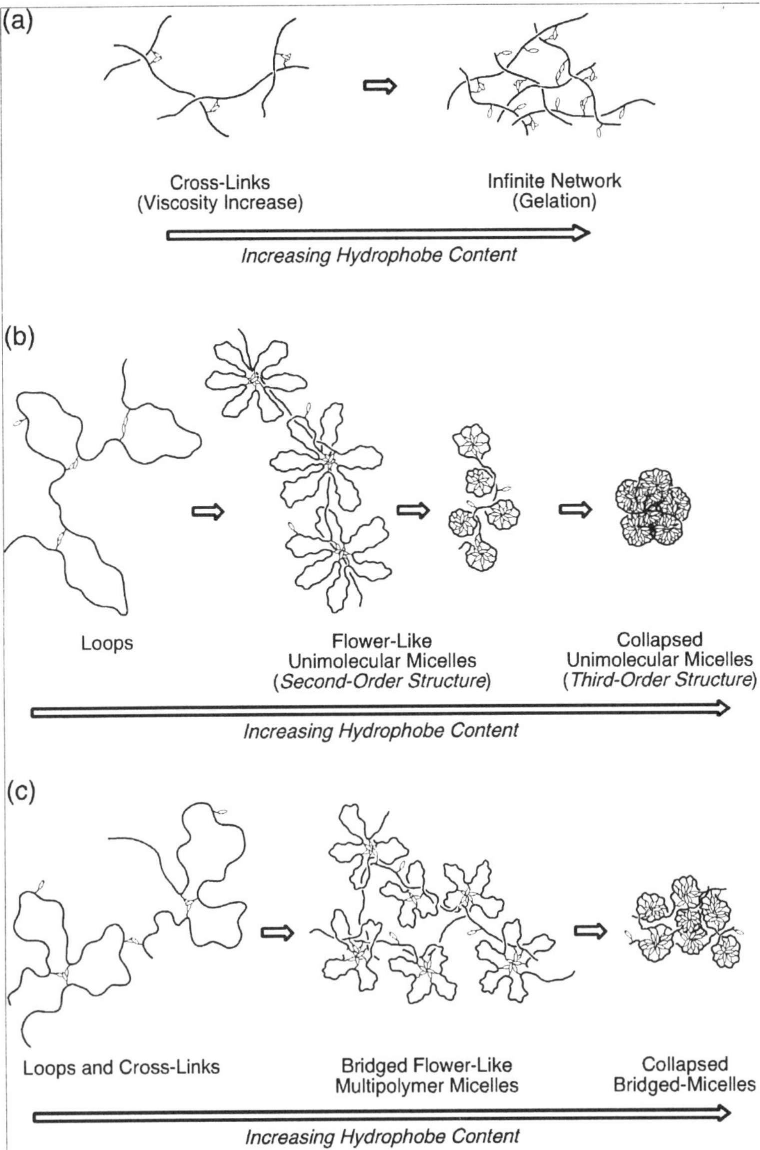

Figure 1. *Conceptual illustrations of hydrophobic associations of hydrophobically-modified water-soluble polymers; (a) interpolymer associations, (b) intrapolymer association, and (c) intra- plus interpolymer associations.*

18

that controls the associative behavior of polymer-bound hydrophobes (*29–31*). This is due to the fact that in the process of the self-association of polymer-bound hydrophobes, the polymer chain would impose steric constraints to the hydrophobes and affects the motional and hence geometrical freedom of the hydrophobes. Therefore, the length and flexibility of the spacer are expected to have a large effect on the association behavior of polymer-bound hydrophobes. Two possible effects of spacer have been discussed in the literature (*32*); (i) alleviation of an excluded-volume effect on interpolymer hydrophobic association with increased spacer lengths, and (ii) decoupling of motions of the polymer chain and hydrophobes linked to it. For interpolymer association, the excluded-volume effect is expected to become less important for longer spacer lengths. Moreover, the spacer is expected to have an effect on an entropy loss accompanied by association of hydrophobes into hydrophobic aggregates. The association of polymer-bound hydrophobes into hydrophobic aggregates can be regarded as an ordering process (*32*), and therefore the insertion of a long hydrophilic spacer would minimize this loss of entropy through decoupling the motions of the polymer chain and hydrophobe aggregates.

Not only the length but also the flexibility and hydrophilicity are important parameters for the spacer. The associative properties of hydrophobes directly linked to a polyelectrolyte chain via amide and ester bonds are remarkably different (*8,29,31,33,34*). Amide-linked hydrophobes exhibit a strong tendency for intrapolymer association whereas ester-linked hydrophobes show a propensity for interpolymer association along with concurrent intrapolymer association. Random copolymers of sodium 2-(acrylamido)-2-methylpropanesulfonate (AMPS) with *N*-dodecylmethacrylamide (DodMAm) (Chart 1) show a strong preference for intrapolymer self-association even in concentrated solutions, leading to the formation of unimolecular micelles (*8,29,33*). This AMPS-DodMAm copolymer with a DodMAm content of ca. 40 mol % forms a highly compact unimolecular micelle with a third-order structure (Figure 2). This is a typical example for the second extreme case discussed earlier (Figure 1b). In sharp contrast, random copolymers of AMPS and dodecyl methacrylate (DMA) (Chart 2) exhibit a tendency for interpolymer association,

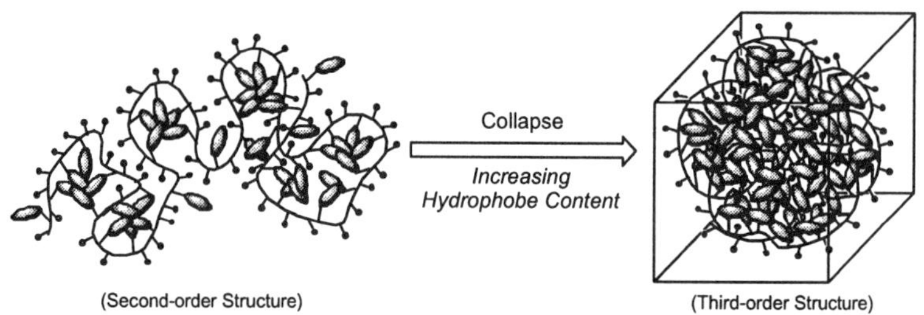

Figure 2. Conceptual illustration of unimolecular micelles of second- and third-order structures formed by AMPS-DodMAm copolymers.

Chart 1. AMPS-DodMAm Copolymer

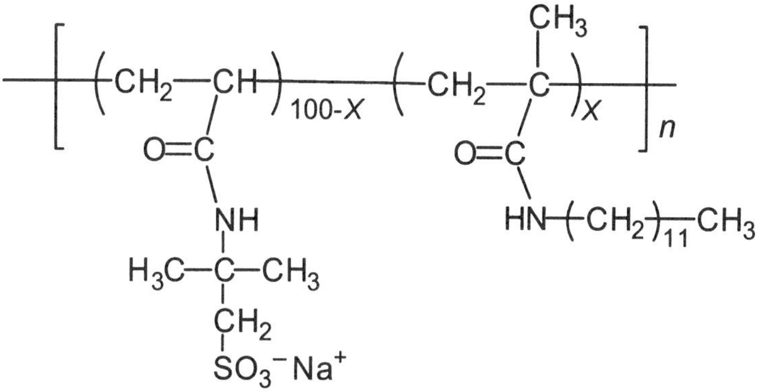

Chart 2. AMPS-DMA Copolymer

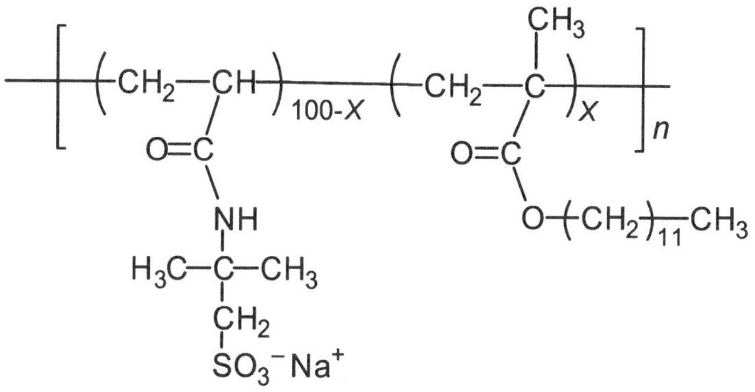

leading to the formation of multipolymer micelles (*31*). Furthermore, there is a great difference in the solubility in water between the AMPS copolymers possessing amide- and ester-linked hydrophobes. AMPS-DodMAm copolymers are soluble in water up to ca. 60 mol % hydrophobe content (*33,34*), whereas AMPS-DMA copolymers are soluble in water only when the hydrophobe content is as low as 15 mol % (*31*).

Copolymers of Electrolyte Monomer and Surfactant Macromonomer

There is a class of hydrophobically-modified water-soluble polymers that exhibit a greatly increased solution viscosity arising from formation of a transient network via interpolymer association of polymer-bound hydrophobic groups (*32,35–45*). Such polymers, known as hydrophobically associating thickening (AT) polymers, are designed, for a practical reason, such that a largest extent of interpolymer association, and hence a greatest enhancement of viscosity, can occur with a smallest extent of hydrophobic modification of a water-soluble polymer. Most of AT polymers, including their model polymers, reported so far are copolymers of acrylamide or (meth)acrylic acid with associative comonomers. Some of the associative monomers contain a $(CH_2CH_2O)_n$ chain as a spacer between a polymerizable moiety at one end and a terminal hydrophobe at the other end (*32,35–40*). The contents of the associative comonomers in AT copolymers are very low; ≈ 5 mol % at the highest and normally < 2 mol %. For this class of polymers, it has been reported that the preference of interpolymer association (i.e., thickening efficiency) is dependent on the length of the $(CH_2CH_2O)_n$ spacer in the comonomer (*32,40*). For example, in the case of copolymers of acrylamide and an acrylate substituted with a hydrophobe end-capped $(CH_2CH_2O)_n$ group with *n* ranging from 1 to 3, longer $(CH_2CH_2O)_n$ spacers result in more enhanced viscosities of copolymer, indicating that the longer flexible hydrophilic spacer is more effective in promoting interpolymer hydrophobic association (*32*). However, a maximum thickening efficiency seems to be achieved when *n* is more or less 10 for a terpolymer of methacrylic acid, ethyl acrylate, and the same type of the associative comonomers with a 2 mol % content, a model polymer for hydrophobically modified alkali swellable/soluble emulsions (HASE) (*40*). It has been reported that when the length of the $(CH_2CH_2O)_n$ spacer is increased from *n* = 10 to 40, a substantial decrease in solution viscosity is observed because the probability of forming interpolymer associations is decreased and in turn the probability of forming intrapolymer associations is increased (*40*).

The amphiphilic polyelectrolytes that we focus on in this chapter are the copolymers of AMPS and a methacrylate substituted with a nonionic surfactant $HO(CH_2CH_2O)_{25}C_{12}H_{25}$ ($C_{12}E_{25}$) with the content of the methacrylate-end-capped $C_{12}E_{25}$ macromonomer (DE25MA) unit (f_{DE25}) ranging from 10 to 30 mol % (Chart 3) (*46,47*). This macromonomer was synthesized by treating $C_{12}E_{25}$ with methacryloyl chloride in the presence of triethylamine in benzene. Copolymerizations of water-soluble monomers and surfactant macromonomers are often carried out in aqueous media in the presence or absence of non-polymerizable surfactants (*19,32,35,36,48*). In micelle or emulsion polymerization, copolymers with a blocky distribution of the surfactant comonomer are likely to be formed (*19,49,50*). To obtain copolymers with random distributions of the AMPS and DE25MA units, homogeneous solution

Chart 3. AMPS-DE25MA Copolymer

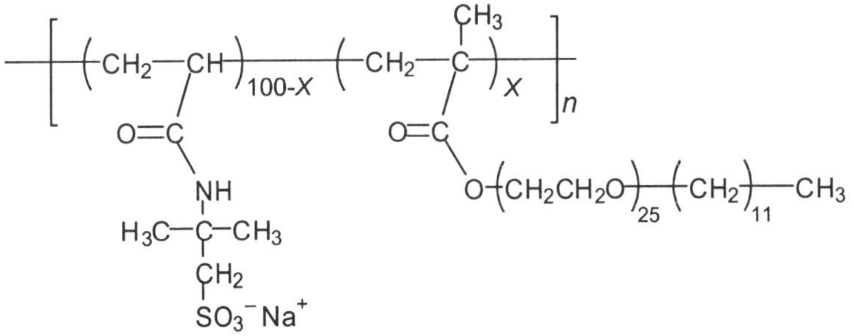

$$\left[\left(CH_2-CH\right)_{100-X}\left(CH_2-\underset{\underset{O=C}{|}}{\overset{\overset{CH_3}{|}}{C}}\right)_X\right]_n$$

AMPS unit: $O=C-NH-C(CH_3)_2-CH_2-SO_3^-\,Na^+$

DE25MA unit: $O=C-O(CH_2CH_2O)_{25}(CH_2)_{11}-CH_3$

polymerization was employed using *N*,*N*-dimethylformamide (DMF) as a solvent that can dissolve the two monomers and resulting copolymers (*46,47*). The copolymerization was performed in the presence of 2,2'-azobis(isobutyronitrile) in DMF at 60 °C for 21 h. The compositions of the copolymers were determined by ^{1}H NMR spectroscopy. The copolymer compositions were found to be practically the same as that in the monomer feed. The compositions and apparent molecular weights estimated by SEC are listed in Table 1.

These copolymers are different from the AT polymers described above in the amount of associative macromonomers incorporated in the polymer chain. Since we are interested in the self-organization of the polymer derived from micellization of polymer-bound hydrophobes, rather than in thickening efficiency due to interpolymer hydrophobic interactions, we intended to incorporate a large number of surfactant comonomer units into a polyelectrolyte chain. Furthermore, for that reason, we chose AMPS, a strong-electrolyte monomer, to render the copolymer water soluble up to a high loading amount of the associative macromonomer.

These copolymers can be visualized as a multi-surfactant system where a large number of surfactant fragments are linked to a polyelectrolyte backbone. It is expected that, since the $(CH_2CH_2O)_{25}$ spacer is a flexible chain, dodecyl groups at an end of the spacer chain can move freely and therefore they can easily associate with other hydrophobes on the same polymer chain as well as among different polymer chains. Furthermore, the polyAMPS main chain and the $(CH_2CH_2O)_{25}$ spacer chain are expected to be highly hydrated and thus the terminal dodecyl groups at an end of the spacer would be excluded from the hydrophilic macromolecular environment and forced to associate with one another.

Micelle Formation of PolyAMPS-Bound $C_{12}E_{25}$ Surfactant Moieties

The nonionic surfactant $C_{12}E_{25}$ molecules form micelles of an aggregation number (N_{agg}) of ca. 40 at surfactant concentrations above a critical micelle concentration (cmc) (ca. 3×10^{-4} M in 0.1 M NaCl) (46). As presented in Table 1, the numbers of dodecyl groups per polymer chain are 7, 13, 15, and 16 for the AMPS-DE25MA copolymers with f_{DE25} = 10, 15, 20, and 30 mol %, respectively. These numbers are smaller than the N_{agg} value found for the discrete $C_{12}E_{25}$ micelle. Therefore, if the polymer-bound $C_{12}E_{25}$ surfactant groups form micelles with an N_{agg} value similar to that for the discrete micelle, roughly 3 to 6 polymer molecules should associate to form a single micelle, depending on f_{DE25}.

Table 1. Characteristics of the Copolymers [a]

f_{DE25} [b] (mol %)	M_w [c] (10^4)	M_w/M_n [c]	Number of $C_{12}E_{25}$ units per polymer chain	Apparent cmc in $[C_{12}E_{25}$ unit] [d] (μM)	K_v [e] (10^5)	N_{agg} [f]
10	7.0	3.0	7	1.7	1.1	68
15	6.9	1.9	13	6.7	1.0	64
20	7.6	2.9	15	9.1	1.1	57
30	8.3	2.8	16	13	1.2	53

a. Apparent cmc and N_{agg} for nonionic surfactant $C_{12}E_{25}$ molecules in 0.1 M NaCl are 3×10^{-4} M and 40, respectively.

b. Mole percent content of DE25MA in the copolymer determined by ^{1}H NMR in D_2O.

c. Determined by GPC using a 0.1 M $LiClO_4$ methanol solution as an eluent. Standard poly(ethylene oxide) samples were used for the calibration of the molecular weight.

d. Determined from steady-state fluorescence excitation spectra of pyrene probes.

e. Partition coefficient for solubilization of pyrene. K_v value for the discrete $C_{12}E_{25}$ micelle is 1.7×10^5 *(46)*.

f. Averaged aggregation number of dodecyl groups in the polymer-bound micelle cores (see text). An N_{agg} value for the discrete $C_{12}E_{25}$ micelle is ca. 40 in 0.1 M NaCl *(46)*.

A fluorescence probe technique using pyrene as a probe is often employed for characterization of polymer micelles due to the fact that the steady-state fluorescence spectra and fluorescence lifetime of pyrene provide information about microenvironments where the probe molecule exists *(51)*. The formation of hydrophobic domains by polymer-bound $C_{12}E_{25}$ surfactant moieties was evidenced by fluorescence spectra of pyrene solubilized in the polymer phase in water. These hydrophobic domains can be regarded as micelle cores formed by dodecyl groups in the polymer-bound surfactant moieties. Aggregation numbers of dodecyl groups in the polymer-bound micelle cores were determined by a fluorescence technique based on the excimer formation of pyrene probes solubilized in the polymer-bound micelles *(52–55)*. When the concentration of solubilized pyrene is low such that each hydrophobic microdomain contains much less than one pyrene molecule on average, no excimer should be formed, exhibiting a single-exponential fluorescence decay. As the pyrene concentration is increased, probability of finding two or more pyrene molecules in the same micelle core increases and hence excimer can be formed in the same micelle, yielding to a fast decay component in the decay of monomeric pyrene fluorescence. Values of N_{agg} were estimated by fitting fluorescence decay data to the Infelta-Tachiya equation *(52–54)*;

$$\ln[I(t)/I(0)] = ([Q]_m/[M]) [\exp(- k_Q t) - 1] - k_0 t \tag{1}$$

where, $I(t)$ and $I(0)$ are the fluorescence intensities at time t and 0 following pulse-light excitation, respectively, k_Q is the pseudo-first-order rate constant for quenching of the excited probe, k_0 is the fluorescence decay rate constant for pyrene inside the micelle without excimer formation, $[Q]_m$ is the molar concentration of quencher inside micelle, and $[M]$ is the molar concentration of micelles. Because the quenching of pyrene monomeric fluorescence is due to the excimer formation of pyrene, $[Q]_m$ corresponds to the concentration of pyrene. The $[Q]_m /[M]$ ratio is determined from the best fit, from which N_{agg} can be calculated.

The core size (i.e., N_{agg}) of the polymer-bound $C_{12}E_{25}$ micelle is somewhat larger than that of the corresponding discrete $C_{12}E_{25}$ micelle, i.e., $N_{agg} = 53–68$ for the polymer-bound micelle whereas $N_{agg} \approx 40$ for the discrete micelle (Table 1). These aggregation numbers for the polymers are a little larger than that for the discrete micelle, and therefore it is obvious that the polymer-bound micelle is formed from the association of the $C_{12}E_{25}$ moieties not only on the same polymer chain but also on different polymer chains. In the case of f_{DE25} = 20 mol %, for example, one polymer chain possesses 15 hydrophobes while N_{agg} = 57 (Table 1). Therefore, one micelle unit is from formed at least 4 polymer chains assuming all the hydrophobes on the same polymer chain occupy the same micelle core. However, if some hydrophobes on the same polymer chain occupy different micelle cores, then the micelles are to be bridged by polymer chains.

Figure 3 shows a plot of N_{agg} for the copolymer with f_{DE25} = 20 mol % against the polymer concentration (C_p) at a concentration of added salt ([NaCl]) of 0.1 M. N_{agg} is virtually independent of C_p in the whole C_p range examined, suggesting that the polymer-bound micelle cores conserve their size over a significant range of C_p. On the other hand, N_{agg} increases with increasing salt concentration; it increases from 53 to 98 when [NaCl] is increased from 0.05 to 0.4 M (Figure 4).

Since the polymer-bound micelles are formed not only by intrapolymer associations but also by interpolymer association, an apparent cmc was observed. Apparent cmc values for the copolymers were estimated by analysis of fluorescence excitation spectra for pyrene probes solubilized in the micelle (*56*). This method is based on the fact that the pyrene 0–0 absorption band shifts towards longer wavelengths when pyrene is solubilized in a micelle core. The 0–0 absorption maximum for pyrene in water is at 334 nm whereas it shifts to 337 nm when pyrene is solubilized in the polymer-bound $C_{12}E_{25}$ micelle. The ratio of the intensity at 337 nm relative to that at 334 nm (I_{337}/I_{334}), estimated from excitation spectra, exhibits a sufficiently abrupt increase at a certain polymer concentration. From the plots of I_{337}/I_{334} against C_p, one can calculate the ratio of pyrene concentrations in the micellar phase and in the aqueous phase

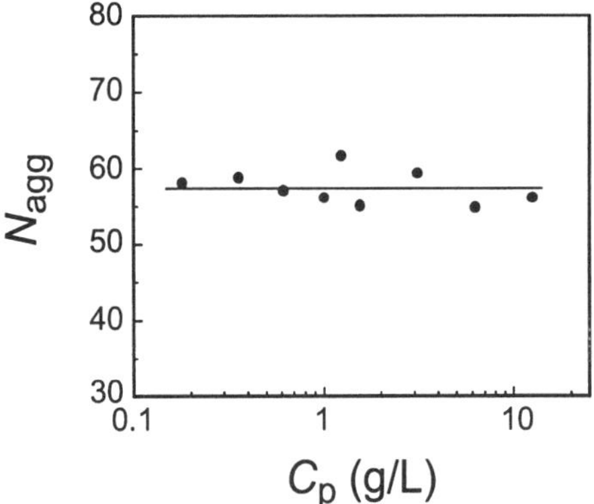

Figure 3. Plot of the aggregation number as a function of polymer concentration for the copolymer with f_{DE25} = 20 mol % in 0.1 M NaCl. A value of N_{agg} for the discrete $C_{12}E_{25}$ surfactant is indicated in the figure.

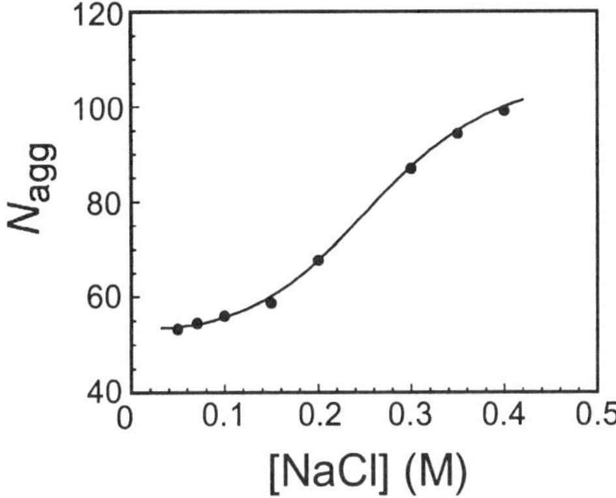

Figure 4. Dependence of the aggregation number on the salt concentration for the copolymer with f_{DE25} = 20 mol % at C_p = 1.56 g/L.

($[Py]_m/[Py]_w$). Plots of ($[Py]_m/[Py]_w$) as a function of C_p show a break at a certain C_p value followed by a linear relationship at higher C_p values. From the break and the slope of the linear line, one can estimate a cmc and partition coefficient (K_v) for pyrene solubilization, respectively (*13,31,56,57*). The results are listed in Table 1 in which cmc values are represented as the molar concentration of the $C_{12}E_{25}$ residue in the copolymer. The apparent cmc values for the copolymers are smaller than the cmc for the discrete $C_{12}E_{25}$ surfactant by more than an order of magnitude. The increase in the apparent cmc on going from f_{DE25} = 10 to 30 mol % is not well understood. The K_v values are fairly constant ($1.0 \sim 1.2 \times 10^5$) for the four copolymers (Table 1). These values are a slightly smaller than the K_v value for the discrete $C_{12}E_{25}$ (1.7×10^5). These results suggest that the polymer-bound $C_{12}E_{25}$ moieties form micelles that are somewhat similar to those formed by discrete $C_{12}E_{25}$ molecules.

Hydrodynamic Size of Polymer-Bridged Micelles

Figure 5 compares relaxation time distributions in quasielastic light scattering (QELS) observed at a scattering angle of 90° for varying C_p for the copolymer with f_{DE25} = 20 mol % in 0.1 M NaCl aqueous solutions. When $C_p \leq$ 18.8 g/L, the QELS relaxation time distributions are bimodal with a slow relaxation mode as a major component and a fast relaxation mode as a minor component. The slow-mode component is obviously due to polymer-bound $C_{12}E_{25}$ micelles bridged by polymer chains whereas the fast-mode component may be attributed to non-bridged micelles or an "oligomeric" micelles (an assembly of a small number of micelles that are bridged). When $C_p \geq 25$ g/L, the relaxation time distributions are unimodal. Apparent hydrodynamic radii (R_h) for the polymer aggregates (polymer-bridged micelles) were estimated from QELS data obtained at several different scattering angles.

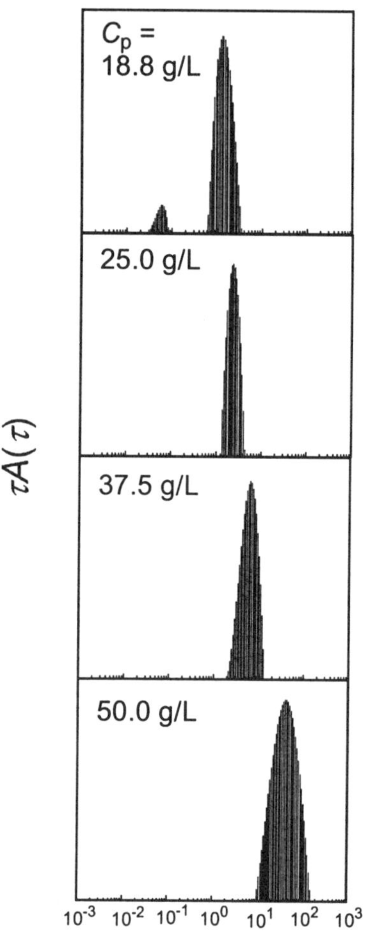

Figure 5. *Relaxation time distributions in QELS observed at a scattering angle of 90 ° for the copolymer with f_{DE25} = 20 mol % at varying polymer concentrations in 0.1 M NaCl.*

In Figure 6, R_h is plotted as a function of C_p. R_h increases from ca. 80 to ca. 120 nm when C_p is increased from 0.2 to 6 g/L. In the higher C_p region (i.e., $C_p \geq 20$ g/L), R_h increases markedly with increasing C_p, reaching on the order of 900 nm at $C_p = 37.5$ g/L. From the data shown in Figure 3, it is obvious that the polymer-bound $C_{12}E_{25}$ micelle cores conserve their size (i.e., N_{agg}) independent of the polymer concentration (at $C_p \leq 12.5$ g/L) whereas the hydrodynamic size of the polymer aggregate increases with increasing C_p. This means that the extent of micelle bridging by polymer chains increases with increasing C_p while the size of the polymer-bound $C_{12}E_{25}$ micelle core remains virtually the same. The micelle bridging occurs when some $C_{12}E_{25}$ moieties on different polymer chains occupy the same micelle. The sharp increase in R_h at C_p near 20 g/L is indicative of the formation of a larger network structure made up of an increased number of bridged micelles with increasing C_p.

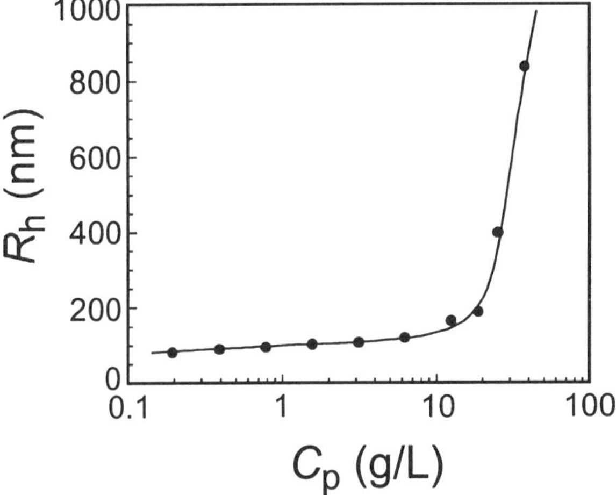

Figure 6. *Dependence of the apparent hydrodynamic radius on the polymer concentration for the copolymer with* f_{DE25} *= 20 mol % in 0.1 M NaCl. For* $C_p \leq$ *20 g/L,* R_h *values estimated for the slow mode are plotted.*

Hydrodynamic size decreases progressively with increasing salt concentration over a range of [NaCl] = 0.02–0.4 M (Figure 7). This trend is in contrast to that for the aggregation number, i.e., N_{agg} increases with increasing [NaCl] (Figure 4). These observations indicate that, with increasing [NaCl], the extent of micelle bridging decreases whereas the core size of the polymer-bound $C_{12}E_{25}$ micelle increases. The extent of micelle bridging is governed by the number of polymer chains that participate in the formation of the same micelle. As [NaCl] is increased, the polymer-bound micelles tend to be formed by a lower number of polymer chains although N_{agg} increases. Therefore, it can be concluded that a tendency for an intrapolymer association of polymer-bound surfactants increases as the salt concentration is increased.

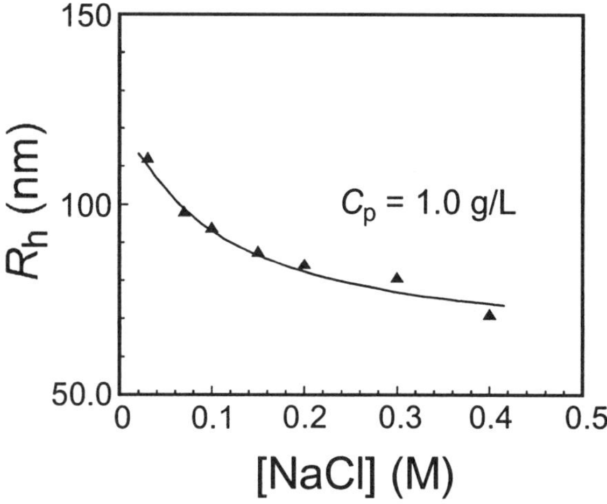

Figure 7. Dependence of the apparent hydrodynamic radius on the salt concentration for the copolymer with f$_{DE25}$ = 20 mol % *at* C$_p$ = 1.0 g/L.

Solution Viscosity of Polymer-Bridged Micelle Networks

The solution viscosity of the copolymer with f_{DE25} = 20 mol % at [NaCl] = 0.10 M is plotted in Figure 8 as a function of C_p. The viscosity was measured at a shear rate of 0.1 s^{-1}. Polymer solutions are Newtonian at this low shear rate and hence the viscosity is regarded as zero shear viscosity. The low-shear viscosity increases gradually with increasing C_p when C_p < ca. 10 g/L. In this low C_p region, the extent of the micelle bridging is relatively small. However, the viscosity starts to increase sharply at C_p near 10 g/L. A three-order increase in magnitude in the viscosity is observed when C_p is increased from 10 to 70 g/L. In this polymer concentration range, macroscopic network structures are formed due to a large number of micelle bridges, and the size of the network increases sharply with increasing C_p.

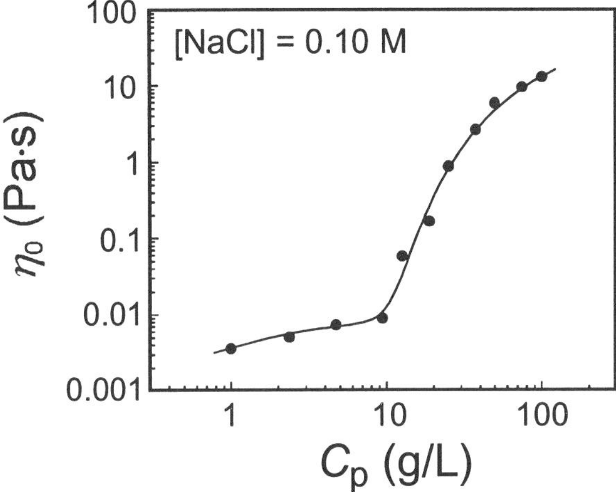

Figure 8. *Dependence of zero shear viscosity on the polymer concentration for the copolymer with* f_{DE25} = 20 *mol % in 0.1 M NaCl. Shear rate applied for the measurement is 0.1 s^{-1} at 25 °C.*

The extent of the micelle bridging is dependent on added salt. The low-shear viscosity (observed at a shear rate of 0.1 s^{-1}) is plotted in Figure 9 as a function of [NaCl] at C_p = 25 g/L. The viscosity increases with increasing salt concentration at [NaCl] < 0.15 M, reaching a maximum value at [NaCl] ≈ 0.15 M, and then decreases with a further increase in [NaCl]. The following considerations may explain these rather unusual observations. Micelle network formation is derived from interpolymer hydrophobic associations that occur in competition with interpolymer electrostatic repulsions. This process should be favored by added salt through an electrostatic shielding effect. As discussed in the previous section, when [NaCl] is increased, N_{agg} increases whereas R_h decreases, suggesting that interpolymer hydrophobic associations (i.e., micelle bridging) is disfavored at higher salt concentrations. Therefore, there are two competing salt effects on the polymer association behavior. When C_p is high enough for interpolymer associations to occur favorably, the extent of micelle bridging would increase with increasing [NaCl] in a low [NaCl] region due to electrostatic shielding of interpolymer repulsions. On the other hand, when [NaCl] is increased to a certain level, the polymer-bound micelles tend to be formed through intrapolymer associations and thus the extent of micelle bridging decreases with increasing [NaCl]. This tendency was observed in the increase in N_{agg} (Figure 4) and decrease in R_h (Figure 7) with increasing salt concentration for polymer solutions of lower concentrations (C_p = 1–1.56 g/L).

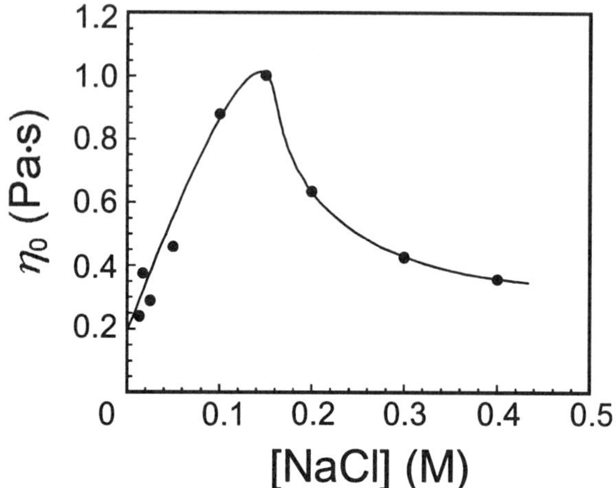

Figure 9. Dependence of zero shear viscosity on the salt concentration for the copolymer with f_{DE25} = 20 mol % at C_p = 25 g/L.

An important observation is that the viscosity depends strongly on shear stress. Figure 10 exhibits the shear-rate-dependent viscosity of polymer solutions at C_p = 25 g/L containing varying concentrations of added NaCl ranging from 0.05 to 0.40 M.

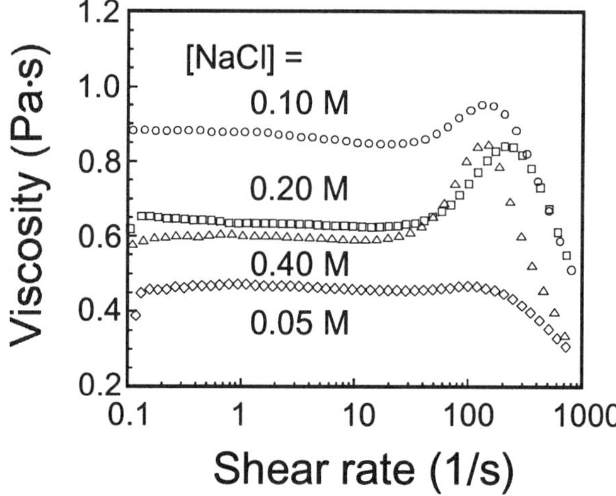

Figure 10. Effect of shear rate on the viscosity at varying salt concentrations for the copolymer with f_{DE25} = 20 mol % at C_p = 25 g/L.

All the solutions exhibit constant viscosity values up to an applied shear rate of ca. 30 s^{-1}. The viscosity value in this Newtonian region depends on the salt concentration, showing a maximum value at [NaCl] $\approx$ 0.1 M. This shear-rate-independent viscosity behavior is followed by shear thickening at intermediate shear rates, and the thickening behavior persists until ca. 100 to 200 s^{-1} shear rate. The shear-thickening behavior strongly depends on the concentration of added salt. Shear thickening can only be observed at [NaCl] $\geq$ 0.1 M, and the extent of the shear thickening increases with increasing [NaCl]. When the shear rate is further increased beyond ca. 100 to 200 s^{-1}, a large decrease in the viscosity (i.e., shear thinning) is observed.

Based on the fact that shear-thickening is derived from a shear-induced increase in the density of mechanically active chains, a plausible explanation for the observed shear thickening is a shear-induced increase in micelle bridging (*58,59*). When shear stress is applied to the polymer-bridged micelles beyond the Newtonian region, some of the polymer-bound surfactant moieties may be "pulled out" of the micelle and become available for associations with surfactant moieties on different polymer chains. This leads to an increase in the number of bridging polymer chains and hence a viscosity increase. A further increase in the shear rate may cause the interpolymer network junctions to fragment, leading to a decrease in the network structure in size and hence a decrease in the viscosity. An observation to be noted is that the viscosity restores instantly when applied shear stress is removed. This means that the shear-induced disruption and reformation of the micelle bridge occur reversibly, characteristic of reversible transient networks, as conceptually illustrated in Figure 11.

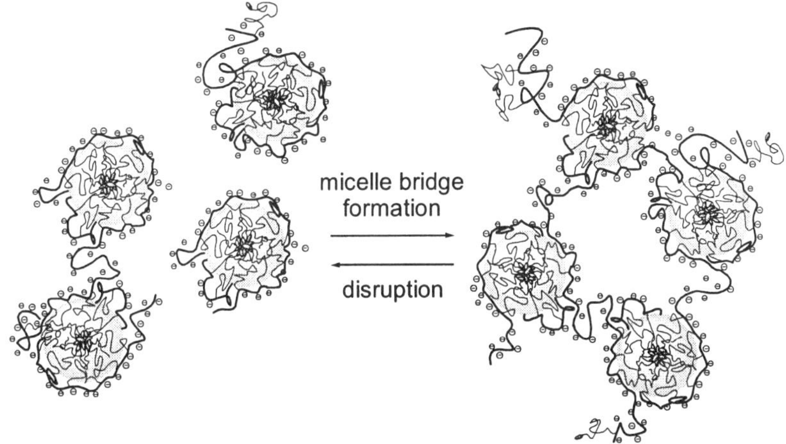

Figure 11. Hypothetical model for polymer-bound micelles reversibly bridged by polymer chains.

When the network is sheared at a high shear rate, micelle bridges are disrupted at a faster rate than the rate of their reformation, and this leads to a decrease in the bridge density and hence a drop in the steady shear viscosity.

Viscoelastic Behavior of Polymer-Bridged Micelle Networks

Figure 12 shows plots of storage modulus (G') and loss modulus (G'') as a function of angular frequency (ω) for the copolymer with f_{DE25} = 20 mol % in 0.1 M NaCl aqueous solutions at C_p = 25 g/L. The solution behaves as a viscoelastic liquid; the $G''(\omega)$ curve is proportional to ω while G' is proportional to ω^2, which is typical of a second-order fluid behavior. Both the $G'(\omega)$ and $G''(\omega)$ data in Figure 12 can be best-fitted to the simple single-element Maxwell model (60):

$$G'(\omega) = G_o\omega^2\lambda^2/(1 + \omega^2\lambda^2) \tag{2}$$

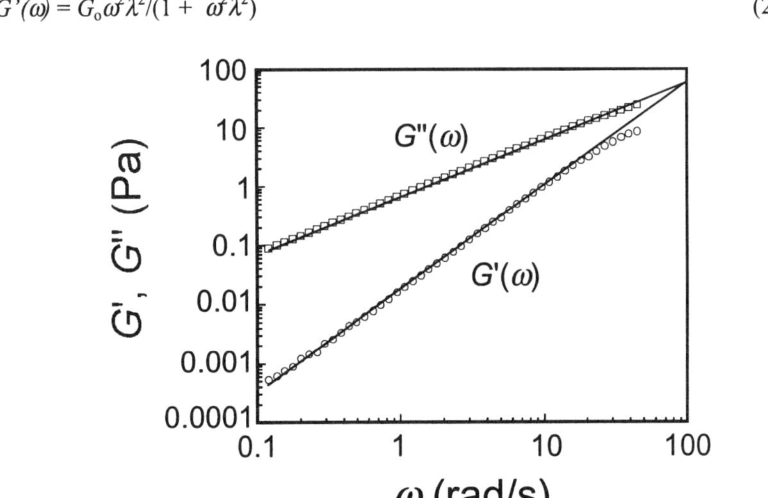

Figure 12. Plots of storage and loss moduli as a function of angular frequency for the copolymer with f_{DE25} = 20 mol % at C_p = 25 g/L in 0.1 M NaCl. Shear stress applied is 1.0 Pa at 25 °C.

$$G''(\omega) = G_o\omega\lambda/(1 + \omega^2\lambda^2) \tag{3}$$

The Maxwell model is a two-parameter model, consisting of a single elastic component connected in series with a viscous element, and describes $G'(\omega)$ and $G''(\omega)$ with a plateau modulus (G_0) and a terminal relaxation time (λ) at all ω values. Values of $G'(\omega)$ and $G''(\omega)$ were measured as a function of ω at varying concentrations of added salt at $C_p = 25$ g/L and the data were best-fitted to eqs 2 and 3, respectively. Values of G_0 and λ thus obtained are plotted in Figure 13 as a function of [NaCl]. As can be seen in Figure 13a, G_0 increases with increasing [NaCl] passing through a maximum at a salt concentration near 0.15 M and then decreases. On the other hand, λ is fairly constant over the range of [NaCl] investigated (Figure 13b). Given the zero shear viscosity can be expressed as $\eta_0 = G_0\lambda$, zero shear viscosity values thus calculated are plotted against [NaCl] in Figure 14. The dependence of the calculated zero shear viscosity on the salt concentration appears to be similar to that of G_0 on [NaCl]. Furthermore, the calculated viscosity values are in fair agreement with the experimental data (Figure 9) except for values near the viscosity maximum. Therefore, it can be concluded that the viscosity is practically governed by G_0 and not by λ.

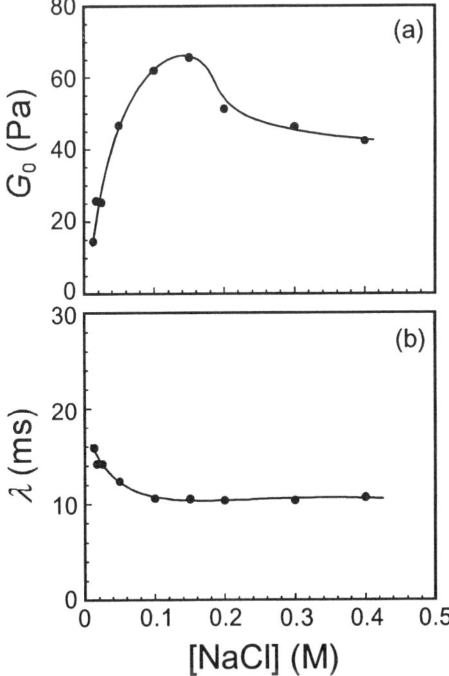

Figure 13. Dependence of (a) plateau modulus and (b) terminal relaxation time on salt concentration for the copolymer with $f_{DE25} = 20$ mol % at $C_p = 25$ g/L.

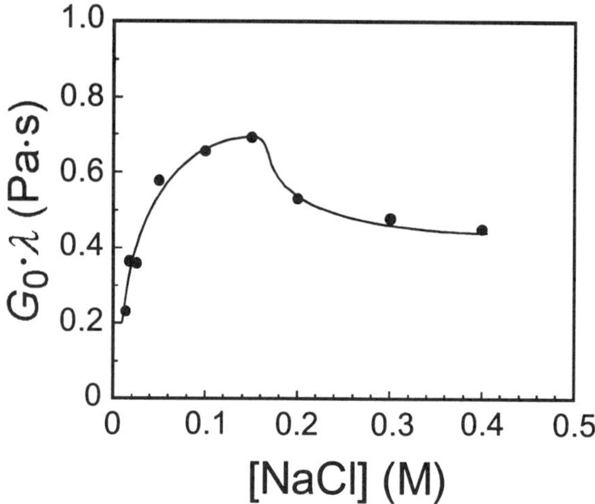

Figure 14. *Dependence of zero shear viscosity, calculated from the plateau modulus and terminal relaxation time, on the salt concentration.*

A simple theory of rubber elasticity extended to transient networks or reversible physical bonds (*61*), suggests that the magnitude of G_0 is proportional to the number density of mechanically active chains in the network. The micelle bridges (or cross-linked points) behave as transient junctions for a network structure, the junctions being in equilibrium of disruption and reformation (Figure 11). The lifetime of the micelle bridge depends on the residence time for a hydrophobe that occupies a micelle core (i.e., the reciprocal of the exit rate of the hydrophobe from the micelle bridge). The Maxwell terminal relaxation time obtained from the best-fits can be regarded as the lifetime of the micelle bridge. Consequently, the results in Figures 9 and 14 indicate that when [NaCl] < 0.15 M, the number of micelle bridges increases with increasing salt concentration whereas at [NaCl] > 0.15 M, the number of micelle bridges decreases with increasing [NaCl]. Moreover, the lifetime of the transient network (i.e., micelle bridging) remains virtually the same independent of the salt concentration (Figure 13b).

Summary

Aqueous solutions of random copolymers of sodium 2-(acrylamido)-2-methylpropanesulfonate (AMPS) and a methacrylate substituted with a nonionic surfactant $HO(CH_2CH_2O)_{25}C_{12}H_{25}$ ($C_{12}E_{25}$) with the content of the surfactant macromonomer ranging from 10 to 30 mol % exhibit unique stimuli-responsiveness. The polymer-bound $C_{12}E_{25}$ surfactant moieties form micelles with aggregation numbers somewhat similar to that of the discrete $C_{12}E_{25}$ micelles but they are bridged by polymer chains thus forming macroscopic network structures. The extent of the micelle bridging depends not only on the polymer concentration but also on ionic strength and shear stress. In addition, the micelle bridging is dynamic in nature, which makes the disruption and reformation of the micelle bridges occur reversibly in response to external stimuli. Thus, viscosity of aqueous solutions of these copolymers exhibits responsiveness to salt and shear stress.

References

1. Zang, Y. X.; Da, A. H.; Hogen-Esch, T. E.; Butler, G. B. In *Water Soluble Polymers: Synthesis, Solution Properties and Application*; Shalaby, S. W., McCormick, C. L., Butler, G. B., Eds.; ACS Symposium Series 467; American Chemical Society: Washington, DC, 1991; p 159.
2. Varadaraj, R.; Branham, K. D.; McCormick, C. L.; Bock, J. In *Macromolecular Complexes in Chemistry and Biology*; Dubin, P.; Bock, J.; Davis, R. M.; Schulz, D. N.; Thies, C., Eds., Springer-Verlag: Berlin and Heidelberg, 1994; p 15. And references cited therein.
3. Bock, J.; Varadaraj, R.; Schulz, D. N.; Maurer, J. J. In *Macromolecular Complexes in Chemistry and Biology*; Dubin, P.; Bock, J.; Davis, R. M.; Schulz, D. N.; Thies, C., Eds., Springer-Verlag: Berlin and Heidelberg, 1994; p 33.
4. Schmolka, I. R. *J. Am. Oil. Chem. Soc.* **1991**, *68*, 206.
5. Almgren, M.; Bahadur, P.; Jansson, M.; Li, P.; Brown, W.; Bahadur, A. *J. Colloid Interface Sci.* **1991**, *151*, 157.
6. McCormick, C. L.; Bock, J.; Schulz, D. N. "Water-Soluble Polymers". In *Encyclopedia of Polymer Science and Engineering;* 2nd ed.; Kroschwitz, J. I. Ed.; John Wiley: New York, NY, 1989; Vol. 11.
7. Laschewsky, A. *Adv. Polym. Sci.* **1995**, *124*, 1.
8. Morishima, Y. In *Solvents and Self-Organization of Polymers*, Webber, S. E.; Tuzar, D.; Munk, P., Eds., Kluwer Academic Publishers, Dordrecht, The Netherlands, 1996, P. 331.
9. Webber, S. E. *J. Phys. Chem. B.* **1998**, *102*, 2618.
10. McCormick, C. L.; Armentrout, R. S.; Cannon, G. C.; Martin, G. G. In *Molecular Interactions ant Time-Space Organization in Macromolecular Systems*, Morishima, Y.; Norisuye, T.; Tashiro, K., Eds., Springer-Verlag, Berlin, 1999, P. 125 and reference cited therein.

11. Zhang, L.; Shen, H.; Eisenberg, A. *Macromolecules* **1997**, *30*, 1001.
12. Zhang, L.; Eisenberg, A. *Science,* **1995**, *268*, 1728.
13. Astafeiva, I.; Zhong, Z. F.; Eisenberg, A. *Macromolecules* **1993**, *26*, 7339.
14. Cao, T.; Weiping, Y.; Webber, S. E. *Macromolecules* **1994**, *27*, 7459.
15. McCormick, C. L.; Chang, Y. *Macromolecules* **1994**, *27*, 2151.
16. Hu. Y.; Smith, G. L; Richardson, M. F.; McCormick, C. L. *Macromolecules* **1997**, *30*, 3526.
17. Hu. Y.; Armentrout, R. S.; McCormick, C. L. *Macromolecules* **1997**, *30*, 3538.
18. Morishima, Y. In *Solvents and Self-Organization of Polymers*, Webber, S. E.; Tuzar, D.; Munk, P., Eds., Kluwer Academic Publishers, Dordrecht, The Netherlands, 1996, P. 331 and references therein.
19. Chang, Y.; McCormick, C. L. *Macromolecules* **1993**, *26*, 6121.
20. Kramer, M. C.; Welch, C. G.; Steger, J. R.; McCormick, C. L. *Macromolecules* **1995**, *28*, 5248.
21. Hu, Y.; Kramer, M. C.; Boudreaux, C. J.; McCormick, C. L. *Macromolecules* **1995**, *28*, 7100.
22. Branham, K. D.; Snowden, H. S.; McCormick, C. L. *Macromolecules* **1996**, *29*, 254.
23. Kramer, M. C.; Steger, J. R.; Hu, Y.; McCormick, C. L. *Macromolecules* **1996**, *29*, 1992.
24. Harada, A.; Kataoka, K. *Macromolecules* **1995**, *28*, 5294.
25. de Gennes, P. G. *Israel. J. Chem.* **1995**, *35*, 33.
26. Halperin, A. *Macromolecules* **1991**, *24*, 1418.
27. Semenov, A. N.; Joanny, J.-F.; Khokhlov, A. R. *Macromolecules* **1995**, *28*, 1066.
28. Elias, H. G. *J. Macromol. Sci., Part A.* **1973**, *7*, 601.
29. Morishima, Y.; Nomura, S.; Ikeda, T.; Seki, M.; Kamachi, M. *Macromolecules* **1995**, *28*, 2874.
30. Yusa, S.; Kamachi, M.; Morishima, Y. *Langmuir* **1998**, *14*, 6059.
31. Noda, T.; Morishima, Y. *Macromolecules* **1999**, *32*, 4631.
32. Hwang, F. S.; Hogen-Esch, T. E. *Macromolecules* **1995**, *28*, 3328.
33. Yamamoto, H.; Mizusaki, M.; Yoda, K.; Morishima, Y. *Macromolecules* **1998**, *31*, 3588.
34. Yamamoto, H.; Morishima, Y. *Macromolecules* **1999**, *32*, in press.
35. Schultz, D. N.; Kaladas, J. J.; Maurer, J. J.; Bock, J.; Pace, S. J.; Schultz, W. W. *Polymer* **1987**, *28*, 2110.
36. Kumacheva, E.; Rharbi, Y.; Winnik, M. A.; Guo, L.; Tam, K. C.; Jenkins, R. D. *Langmuir* **1997**, *13*, 182.
37. Horiuchi, K.; Rharbi, Y.; Spiro, J. G.; Yekta, A.; Winnik, M. A.; Jenkins, R. D.; Bassett, D. R. *Langmuir* **1999**, *15*, 1644.

38. Tirtaatmadja, V.; Tam, K. C.; Jenkins, R. D. *Macromolecules* **1997**, *30*, 1426.
39. Tirtaatmadja, V.; Tam, K. C.; Jenkins, R. D. *Macromolecules* **1997**, *30*, 3271.
40. Tam, K. C.; Farmer, M. L.; Jenkins, R. D.; Bassett, D. R. *J. Polym. Sci., Part B: Polym. Phys.* **1998**, *36*, 2275.
41. McCormick, C. L.; Nonaka, T.; Johnson, C. B. *Polymer* **1988**, *29*, 731.
42. Biggs, S.; Selb, J.; Candau, F. *Polymer* **1993**, *34*, 580.
43. Volpert, E.; Selb, J.; Candau, F. *Macromolecules* **1996**, *29*, 1452.
44. Klucker, R.; Candau, F.; Schosseler, F. *Macromolecules* **1995**, *28*, 6416.
45. Branham, K. D.; Davis, D. L.; Middleton, J. C.; McCormick, C. L. *Polymer* **1994**, *35*, 4429.
46. Noda, T.; Hashidzume, A.; Morishima, Y. *Macromolecules*, in press.
47. Noda, T.; Hashidzume, A.; Morishima, Y. *Langmuir*, in press.
48. Ezzell, S. A.; Hoyle, C. E.; Greed, D.; McCormick, C. L. *Macromolecules* **1992**, *25*, 1887.
49. Hill, A.; Candau, F.; Selb, J. *Macromolecules* **1993**, *26*, 4521.
50. Dowling, K. C.; Thomas, J. K. *Macromolecules* **1990**, *23*, 1059.
51. Kalyanasundaram, K.; Thomas, J. K. *J. Am. Chem. Soc.* **1977**, *99*, 2039.
52. Infelta, P. P.; Grätzel, M.; Thomas, J. K. *J. Phys. Chem.* **1974**, *78*, 190.
53. Tachiya, A. M. *Chem. Phys. Lett.* **1975**, *33*, 289.
54. Infelta, P. P. *Chem. Phys. Lett.* **1979**, *61*, 88.
55. Yekta, A.; Aikawa, M.; Turro, N. J. *Chem. Phys. Lett.* **1979**, *63*, 543.
56. Wilhelm, M.; Zhao, C-L.; Wang, Y.; Xu, R.; Winnik, M. A.; Mura, J-L.; Riess, G.; Croucher, M. D. *Macromolecules* **1991**, *24*, 1033.
57. Vorobyova, O.; Yekta, A.; Winnik, M. A.; Lau, W. *Macromolecules* **1998**, *31*, 8998.
58. Tam, K. C.; Jenkins, R. D.; Winnik, M. A.; Bassett, D. R. *Macromolecules* **1998**, *31*, 4149.
59. Yekta, A.; Xu, B.; Duhamel, J.; Adiwidjaja, H.; Winnik, M. A. *Macromolecules* **1995**, *28*, 956.
60. Ferry, J. D. *Viscoelastic Properties of Polymers*; 3rd ed.; Wiley: New York, 1980.
61. Green, M. S.; Tobolsky, A. V. *J. Chem. Phys.* **1940**, *14*, 80.

Chapter 3

Micellar Polymerization for the Design of Responsive Amphiphilic Polymers

Geoffrey L. Smith & Charles L. McCormick

Department of Polymer Science, The University of Southern Mississippi Hattiesburg, MS 39406-0076

Micellar polymerization allows tailoring of responsive water-soluble amphiphilic polymers having a specified microstructural architecture. Hydrophobic groups can be incorporated in a microblocky fashion along the polymer backbone using this technique. In this manuscript, series of terpolymers containing acrylic acid, methacrylamide and a twin-tailed hydrophobic monomer have been synthesized using micellar polymerization methods. These polymer systems have been characterized using light scattering, viscometry, rheometry, and fluorescence methods. Viscosity studies indicate that the more nonpolar the hydrophobic monomer (longer chain length or twin tailed vs. single tailed), the more enhanced the solution viscosity. Energy transfer measurements indicate the onset of association at ~0.1 g/dl for the MAM/AA/DiC$_{10}$AM terpolymer. Changes in the energy transfer efficiency as a function of pH closely follow analogous viscosity behavior and are indicative of pH induced expansion and collapse.

INTRODUCTION

Precise microstructural design of stimuli-responsive amphiphilic polymers allows for direct correlation between specific microstructure variables and stimuli-responsiveness. Various polymerization techniques have been exploited to control polymer microstructure in amphiphilic systems including, but not limited to living anionic and cationic methods, TEMPO mediated polymerizations, ATRP, and more recently RAFT techniques. Another technique which allows for some measure of microstructual tailoring in amphiphilic systems is micellar polymerization. Although micellar polymerization does not allow for the level of control that these other methods provide, it is extremely useful for the tailoring of small hydrophobic blocks (microblocks) interspersed randomly along a water-soluble polymer backbone. The objective of this book chapter is to provide a short overview of this technique for the design of stimuli-responsive amphiphilic polymer systems as well as to discuss some recent advances made by our group in this area.

Micellar Polymerization

As mentioned above, micellar polymerization allows for the microstructural placement of hydrophobic microblocks randomly along a water-soluble polymer chain. This is somewhat troublesome, due to the heterogeneous nature of the water-soluble and hydrophobic monomers to be copolymerized, as well as the need to maintain control over the size of the hydrophobic microblocks (*1*). The utilization of micellar polymerization, which was first reported in the mid-1980s, overcomes these two difficulties and has attracted a great deal of attention since its introduction (*2,3*).

Micellar polymerization is classically defined as the copolymerization of a small amount of hydrophobic monomer with a water-soluble comonomer through the addition of a surfactant in excess of its critical micelle concentration (CMC). The surfactant solubilizes the hydrophobic monomer which allows for its polymerization in an aqueous environment(*1-4*). During the course of micellar polymerization, water-soluble initiator decomposes to produce free radicals, which initiate polymerization of the hydrophilic monomer. When the growing hydrophilic chain encounters a micelle containing a high local concentration of hydrophobic monomer, it is rapidly polymerized and incorporated on to the chain as a small block. It was initially postulated that this phenomenon would allow for the synthesis of a unique microblocky polymer architecture, yet this was not proven until several years after the introduction of the technique as will be discussed below (*5-7*).

Micellar polymerization is extremely versatile due to the number of monomers that can be utilized and the variety of reaction variables which can be adjusted to tailor specific polymer properties. These variables include hydrophilic monomer type, hydrophobic monomer size and structure, hydrophobic monomer content, number of hydrophobes per microblock (N_H), number of microblocks per polymer, and polymer molecular weight. Polymers with a diverse array of solution behaviors can be synthesized by adjusting any of these variables (*8-9*).

Some of the first polymers to be synthesized by micellar polymerization were hydrophobically modified polyacrylamides that contained small amounts of hydrophobic alkylacrylamide comonomers as reported by Turner *et al.*(*3*). McCormick and Johnson also reported acrylamide/N-(n-alkyl)acrylamide copolymers synthesized by a similar micellar procedure (*10*). The alkylacrylamide comonomer employed was insoluble in water; therefore, sodium dodecyl sulfate, an anionic surfactant, was added to the monomer solution to solubilize the hydrophobic comonomer. Alkylacrylamide was incorporated into monomer feed at levels of 0.25, 0.50, and 0.75 mole percent; n-octyl (C_8AM), n-decyl ($C_{10}AM$), and n-dodecyl ($C_{12}AM$) acrylamides were then copolymerized with acrylamide via free radical techniques.

Incorporation of small amounts of hydrophobic comonomer were found to dramatically alter the rheological characteristics. Apparent viscosity versus polymer concentration plots for poly(acrylamide) (PAM) and poly(acrylamide-*co*-N-(n-decyl)acrylamide) (PAM/$C_{10}AM$) modified with 0.75 mole percent $C_{10}AM$ in water are shown in Figure 1. Poly(acrylamide) displays an unremarkable, steady increase with polymer concentration, but the viscosity profile of the PAM/$C_{10}AM$ copolymer exhibits a pronounced upwards curvature at about 0.15 g/dL. Interpolymer hydrophobic associations drive this enhancement in the viscosity response.

As polymer concentration increases, so does the concentration of hydrophobic groups around which water structure ordering occurs. At a low degree of hydrophobic substitution and long intervening sequences, intrapolymer associations are less favorable than interpolymer associations, and hydrophobic moieties act as contact points or network junctions for polymer-polymer aggregation. Aggregation is absent at low concentration, but multimers begin to form as concentration increases. Network formation with transient hydrophobic cross-links occurs as polymer concentration increases further. This aggregation process manifests itself as a marked increase in the apparent viscosity response at concentrations lower than those corresponding to an onset of chain entanglement for unmodified polymer. Thus, hydrophobic substitution lowers the critical overlap concentration (C*) relative to unsubstituted polymer.

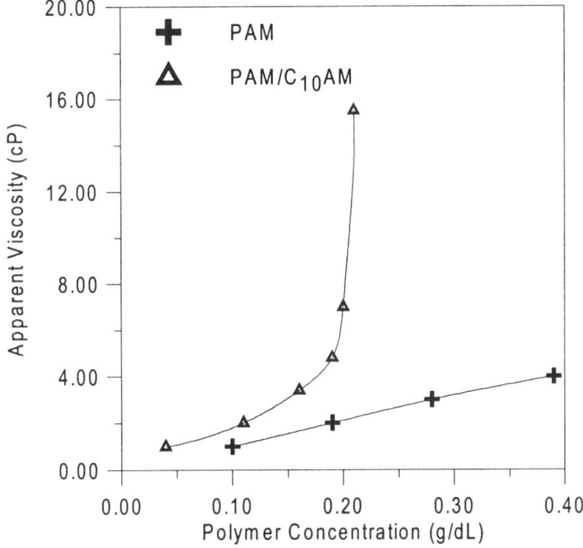

Figure 1. *Apparent Viscosity as a function of polymer concentration for poly(acrylamide) (PAM) and poly(acrylamide-*co*-N-(n-decyl)acrylamide) (PAM/C$_{10}$AM) in deionized water.(10)*

If an ionic group is incorporated into a hydrophobically modified acrylamide copolymer, charged group geometry, ionic strength and pH may have an effect on the extent and nature of associations. As electrostatic repulsive forces are shielded or eliminated between similarly charged ionic groups along the polymer chain, the conformational changes brought about may alter the degree of associative thickening.

Middleton and McCormick synthesized terpolymers of acrylamide (AM) with 0.5 mole percent of either N-(n-decyl)acrylamide (C$_{10}$AM) or N-(4-butyl)phenylacrylamide (BPAM) as the hydrophobic comonomer. The anionic monomers sodium acrylate (NaA), sodium 3-acrylamido-3-methylbutanoate (NaAMB), or sodium 2-acrylamido-2-methylpropanesulfonate (NaAMPS) were also incorporated in varying amounts (*11-12*) All terpolymers exhibited good associative thickening properties in deionized water, and carboxylate-based terpolymers exhibited a highly pronounced associative response relative to NaAMPS-containing polymers in high ionic strength environments. Terpolymers containing NaA tend to associate more strongly than NaAMB-containing polymers; charged group placement may account for this behavior. The anionic group in the NaA repeat unit is coupled directly to the polymer backbone, and the carboxylate group in the NaAMB unit is separated from the backbone by four atoms. This displacement obviously affects hydrophobic associations.

Similarly, Branham and McCormick examined an extensive list of hydrophobically modified terpolymers of AM, NaA, *N*-[(4-decyl)phenyl]acrylamide (DPAM), and *N*-[(4-hexyl)phenyl]acrylamide (HPAM) (*9,13-14*). In one study, the effect of the surfactant to monomer ratio (SMR) on polymer microstructure and solution properties was examined for a series of terpolymers (*9*). The feed ratio of the reaction was held constant (69.5% AM, 30.0% AA, and 0.5% DPAM), and only the SMR was varied (from 40 to 100). The polymers synthesized at low SMR showed the strongest associative behavior above C*, while those synthesized at higher SMRs behaved like unmodified polymers. This is one of several cases where the influence of N_H has been shown to be a key variable in the solution behavior of micellar polymers.

Another study by Branham and McCormick analyzed the effect of pH and polymer composition on the associative properties terpolymers of AM, AA, and HPAM (*13*). In these experiments, the hydrophobe content, sequence distribution, and molecular size were held constant, but the AA content of the terpolymers was varied. The solution behavior of the terpolymers was characterized using rheometry and fluorescence probe techniques. Terpolymers containing 9 and 21 mol % AA were shown to exhibit extensive intermolecular associations at low pH (4.0 – 5.0), as well as in the presence of NaCl at higher pH. A terpolymer containing 37 mol % AA showed no rheological evidence of intermolecular associations throughout the pH range examined in the study. Fluorescence experiments using pyrene probes confirmed that this polymer was forming intramolecular hydrophobic microdomains. Branham was able to show that by varying polymer composition, the association behavior of polymers could be controlled to respond in a specific manner to various environmental stimuli including the solution pH and electrolyte concentration.

As alluded to above, the microblocky architecture afforded by micellar polymerization was first suggested by Peer (*15*). However, Ezzell & McCormick were the first to provide evidence of this microblocky hydrophobic group placement in copolymers by performing fluorescence experiments on pyrenesulfonamide-labeled (APS) polymers (*6-7*). Copolymers having similar label content were synthesized by both micellar and solution polymerization. The copolymer polymerized by the micellar technique contained 0.25 mole% APS, while the solution polymerized (DMF/H$_2$O 57:43, v:v) polymer contained 0.35 mole%. Steady-state emission spectra of the two polymers in dilute solution indicated that I_E/I_M was higher for the micellar polymer even though it contained less of the pyrene-sulfonamide label. This was postulated to be indicative of a higher local label concentration in the micellar polymer if interpolymer aggregation in dilute solution is assumed to be minimal. Thus, a blocky copolymer microarchitecture is believed to result from micellar polymerization. However, these copolymers are not true block copolymers, but ones consisting of hydrophobe rich microblocks with long runs of intervening AM units.

Recently Candau *et al.* addressed the issue of compositional heterogeneity in micellar polymerized copolymers incorporating *N*-monosubstituted acrylamides (*16*). They observed an apparent decrease in the microblock size with conversion due to a stong propensity for the hydrophobic monomer to be consumed early in the polymerization. This behavior was found to result in compositional drift. However, when *N,N*-disubstituted acrylamides were utilized in the polymerization, no compositional drift was observed. Candau *et al.* explained this compositional drift behavior based on the polarity of the micellar phase. It had been shown by other researchers that the reactivity of *N*-monosubstituted acrylamides increased with decreasing solvent polarity, while the reactivity of *N,N*-disubstituted acrylamides was independent of solvent polarity (*16*). This behavior was attributed to hydrogen bonding of the amide proton on the *N*-monosubstituted acrylamides. However, when the amide is dialkylated, hydrogen bonding can no longer occur. Candau *et al.* reasoned that the observed compositional drifts were due to the nonpolar environment of the micelles, which enhanced the reactivity of the polar *N*-monosubstituted species in the micelle due to this hydrogen bonding phenomenon. For the *N,N*-disubstituted hydrophobic comonomers the nonpolarity of the micellar phase would not influence reactivity.

Candau *et al.* also demonstrated that *N,N*-disubstituted alkylacrylamido hydrophobes increase the associative behavior of hydrophobically modified polyacrylamides (*16*). Copolymers of AM and *N,N*-dihexylacrylamide displayed solution viscosities much higher than those of acrylamide and "single-tailed" hydrophobic comonomers. The stronger intermolecular association behavior was attributed to the greater hydrophobicity and increased microblocky character of the polymers. Thus, the use of *N,N*-disubstituted hydrophobic comonomers not only allowed for the synthesis of homogeneous polymer samples, but also offered a new strategy for increasing associative thickening efficiency of polymers produced using the micellar polymerization process.

Based on this work, our group recently has been exploring the effects of using more hydrophobic twin-tailed monomers (up to *N,N*-didecylacrylamide) to increase thickening efficiency and incorporating ionizable monomers to promote the solubility and pH-responsiveness of these more hydrophobic twin-tailed monomers. The main objective is to optimize an associative thickening system by incorporating twin-tailed hydrophobic monomers of varied lengths into a hydrophilic backbone in a microblocky fashion. The extent and reversibility of association of the stimuli-responsive microdomains are followed by rheological and photophysical (fluorescence) measurements.

EXPERIMENTAL

Materials and Monomers

All chemicals were purchased from Fisher Chemical Company or Aldrich Chemical Company at the highest purity available. Purification of VA-44 and

44

methacrylamide (MAM) were accomplished by recrystallization from methanol and acetone respectively. Acrylic acid (AA) was purified by vacuum distillation. The twin-tailed hydrophobic monomers, dihexylacrylamide (DiC$_6$AM), dioctylacrylamide (DiC$_8$AM), and didecylacrylamide (DiC$_{10}$AM) were prepared using a similar procedure to that described by Valint *et al.* (*17*). All solvents and other materials were used as received unless otherwise indicated.

Dansyl-2-aminocaprylic acid and succinic acid N-(1-naphthylmethyl) monoamide were chosen as model compounds. The former was purchased from Sigma and was recrystallized from methanol. The non-radiative energy transfer (NRET) donor used for labeling the copolymer backbone was 7-(1-naphthyl-methoxy)-heptylamine and synthesized using the methods of McCormick and Chang (18). The NRET acceptor, 8-dansyl octylamine, was synthesized as described by Shea *et al.* (19). The structures of these chromophores, are shown in Figure 2.

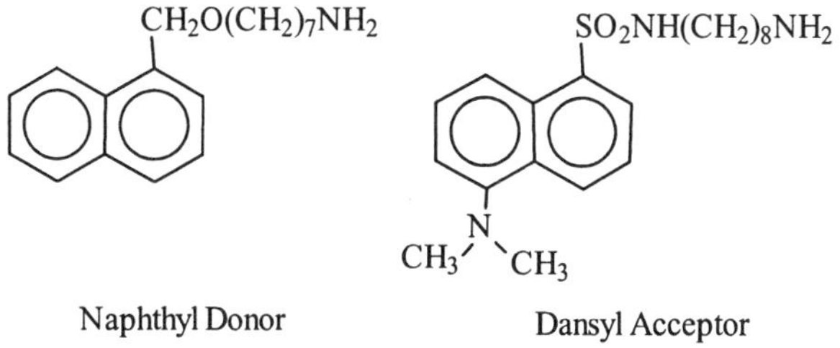

CH$_2$O(CH$_2$)$_7$NH$_2$
SO$_2$NH(CH$_2$)$_8$NH$_2$

N

CH$_3$ CH$_3$

Naphthyl Donor   Dansyl Acceptor

Figure 2. *Fluorescence Chromophores Utilized for Fluorescence Energy Transfer Experiments*

Micellar Polymerization

Co- and terpolymers were synthesized by micellar polymerization using SDS as the surfactant to solubilize the hydrophobic comonomer and VA-044 as the free-radical initiator. The total monomer concentration was held constant at 0.44 M and the [monomer]/[initiator] ratio at 3000. Also, the hydrophobic monomer content and the surfactant to hydrophobe ratio (SMR) were held constant at 1 mole% and 25, respectively. The reactions were performed for 3-6 hours at 50 ^{0}C. A typical micellar polymerization is provided below (Figure 3) for the synthesis of a terpolymer with MAM/AA/C$_6$AM at a feed ratio of 59/40/1.

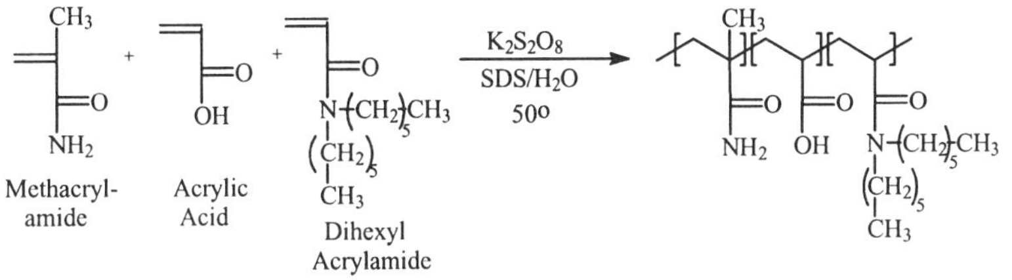

Figure 3. *Micellar polymerization of methacrylamide, acrylic acid and dihexyl acrylamide.*

Deionized water (600 mL) was sparged with N_2 for 30 minutes. SDS (18.0g) was added with stirring under N_2 purge. DiC_6AM (0.592g) was then added with continued stirring for approximately one hour or until the solution cleared. MAM (10.5g1) and AA (8.90g) were then dissolved in the reaction mixture. The pH of the reaction feed was measured and adjusted to assure that it was below the pKa of AA (~4.5). VA-044 (0.061g) was dissolved in 5 mL of deoxygenated, deionized water and injected into the polymerization container. The reaction was allowed to proceed under N_2 for 6 hours after which the terpolymer was precipitated into 1000 mL of methanol. The terpolymer was then washed with fresh methanol and dried overnight in a vacuum oven. Further purification was achieved by redissolving the terpolymer in water and dialyzing against deionized water in Spectra Por No. 4 dialysis tubing with a molecular weight cutoff of 12,000-14,000 for 5 days. The purified polymer samples were then freeze-dried to a constant weight. Synthetic parameters for the other co- and terpolymers made in this study appear below in Table I.

Fluorescence Labeling. The $MAM/AA/DiC_{10}AM$ terpolymer was dissolved in a 70/20 dioxane/formamide mixed solvent system. The appropriate fluorescence label was added to the solution along with the typical DCC/DMAP mixture and the solution was heated to 60 °C and allowed to react overnight. The details of this reaction are given elsewhere (*20*). The resulting solutions were then poured into Spectra Por No. 4 dialysis tubing and allowed to dialyze against deionized water for 15 days. The purified labeled polymer samples were then freeze-dried to a constant weight.

Instrumentation and Analysis

Light Scattering Measurements. Measurements of dn/dc were performed with a Chromatix KMX-16 Laser Differential Refractometer. MAM/AA terpolymer dn/dc values and molecular weights were determined in 0.5 M NaCl. Dust was removed from samples via centrifugation. Classical light scattering was performed using a Brookhaven Instruments BI-200SM automatic goniometer interfaced to a PC. Using standard Zimm analysis the molecular weights and radii of gyration were obtained. Prior to analysis all samples were cleaned by centrifuging for 5 mins. All dilution solvents were also cleaned by

Table I. Synthetic Parameters Utilized for Micellar Polymerizations

Sample	[SDS] mol/L	SMR	MAM	AA	Hydrophobe (1 mole %)
MAM/AA/DiC$_6$AM	0.10	25	49.0%	50.0%	DiHexAM
MAM/AA/DiC$_8$AM	0.10	25	49.0%	50.0%	DiOctAM
MAM/AA/DiC$_{10}$AM	0.10	25	49.0%	50.0%	DiDecAM

filtration though 0.45 μm filters to remove dust. Multiple analyses were performed to ensure reproducibility.

Viscosity Measurements. Viscosity measurements were performed with a Contraves LS-30 rheometer at 25 °C. All solutions were made at 1.0 g/dl in deionized water and diluted incrementally from there. Measurements at all concentrations were performed at a constant shear rate of 5.93 sec^{-1} unless otherwise noted. Solution pH values were adjusted with aqueous HCl or NaOH.

Fluorescence Quantum Yield. Prior to quantum yield determination, the absorbance of the dansyl labels at 330 nm for a polymer concentration of 0.50 g/L was determined. This then allowed the calculation of the fluorescence quantum yield using the equation:

$$\Phi_x = \Phi_{st} \frac{A_{st}}{A_x} \frac{I_x}{I_{st}} \frac{n_x^2}{n_{st}^2} \tag{1}$$

where, Φ is quantum yield, A is the absorbance at the excitation wavelength, I is the integral area of the corrected emission spectrum, and n is the reflective index at the excitation wavelength. The subscript x refers to the utilized chromophore and the subscript st refers to the standard compound.

Non-Radiative Energy Transfer. For NRET measurements, an Acton cut-off WG-305 optical filter was used at the excitation wavelength (282 nm) to prevent scattering of the excitation beam from the samples. The dansyl chromophores were excited at 330 nm to observe the dansyl emission spectra. Quantum yields (Φ) of the fluorescent labels were calculated by integrating the areas of the corrected emission spectra in reference to 2-amino pyridine in 0.10N H_2SO_4 as the standard (Φ=0.60 at 282 nm excitation) (*15*). Quantum yields (Φ) of the dansyl groups excited at 330 nm were calculated by integrating the areas of corrected emission spectra in reference to quinine bisulfate in 1.0 N H_2SO_4 as the standard (Φ = 0.55 at 330 nm excitation) (*16*). Beer's law corrections were applied for optical density changes at the excitation wavelength. Corrections were also made for refractive index differences.

The Förster distance, r_0 has been previously determined to be 23.45 Å for the naphthalene/dansyl donor/acceptor pair, (*23*) and the NRET quantum efficiency, χ, has been calculated using the method described by Guillet (*24*). In this case, the modified Guillet method (*25*) is used for calculating NRET quantum efficiency, χ, due to the minor absorbance of the dansyl chromophore when 282 nm is used as the excitation wavelength. The modified Guillet equation is given below:

$$\frac{\chi}{1-\chi} = \frac{\Phi_D^0 (I_A - I_A^0)}{\Phi_A^0 I_D} \tag{2}$$

in which Φ_D^0 is the fluorescence emission quantum yield of the donor in the absence of acceptor-labeled polymer excited at 282 nm and Φ_A^0 is the fluorescence emission quantum yield of the acceptor on the acceptor-labeled polymer. I_A and I_D are the integrated emission intensities of donor and acceptor, respectively, in the presence of both donor and acceptor-labeled polymer and I_A^0 is the integrated emission intensity of the acceptor in the absence of donor-labeled polymer.

RESULTS AND DISCUSSION

Synthesis of Twin-Tailed Associative Polymers

Initial attempts were made to synthesize copolymers consisting of methacrylamide (MAM) and one mole % of either the diC_6AM, diC_8AM, or $diC_{10}AM$ twin-tailed comonomer. Although the micellar copolymerization of these monomers was successful, attempts to resolubilize the diC_8 and diC_{10} based polymers following dialysis and freeze drying in aqueous solution were unsuccessful. As a result, terpolymers were synthesized incorporating acrylic acid (AA) to promote solubility in aqueous solution at high pH (pH > 6). The terpolymers synthesized consist of 49 mole % of MAM, 50 mole % of AA, and 1 mole % of either diC_6AM, diC_8AM, or $diC_{10}AM$ twin-tailed hydrophobic monomer. The polymer series synthesized are summarized in Table I.

Molecular Weight Determination By Light Scattering

Molecular weights and radii of gyration were determined for all polymer systems. Initially, formamide was used as a result of its excellent solvency and ease of dust removal. Formamide, however, proved not to be a suitable solvent. Accurate refractive index (dn/dc) values were unobtainable for these systems. This is often observed in low ionic strength media and is similar to findings by Branham and McCormick (*14*). Thus, for these systems, measurements were carried out in 0.5 M NaCl (pH 8) at 25 °C.

Classical light scattering with the standard Zimm analysis was utilized to determine weight average molecular weights (M_w), and the mean radii of gyration (R_g). The results obtained for all polymers synthesized are reported in Table II.

Table II. Molecular Weights and Radii of Gyration for Synthesized Terpolymers

Sample	Hydrophobe	R_G (nm)	MW (g/mol) x10^6
MAM/AA/DiC$_6$AM	DiHexAM	72.3	1.02
MAM/AA/DiC$_8$AM	DiOctAM	46.2	0.747
MAM/AA/DiC$_{10}$AM	DiDecAM	63.1	1.12

For the MAM/AA/DiC$_x$AM terpolymers, the molecular weights ranged between 0.75 – 1.12 x 10^6 g/mol. The difference in molecular weights are expected due to differences in comonomer reactivity and likelihood of chain transfer.

Viscometric Studies

The macroscopic solution properties of the aqueous soluble MAM/AA/DiC$_x$AM (x = 6, 8, 10) at pH 8 were determined utilizing low shear viscometry.

Effect of Hydrophobe Length On Solution Viscosity For MAM/AA Terpolymers

Apparent viscosities as a function of concentration in deionized water at pH 8 were determined for the three MAM/AA/DiC$_x$AM terpolymers as shown in Figure 4.

This study illustrates the effect of hydrophobe length (hydrophobicity) on the viscosification ability of the three similar polymer systems. The terpolymer incorporating the DiC$_{10}$AM hydrophobe shows the most pronounced increase in viscosity with concentration. Above ~ 0.6 g/dl, the viscosity is beyond the measurable value of the Contraves LS-30 rheometer. Similar trends are observed for the DiC$_8$AM terpolymer which reaches the measurable limit at ~ 0.8 g/dl and the DiC$_6$AM terpolymer which reaches this limit at 0.9 g/dl. Thus, the viscosification efficiency of these three analogous terpolymers closely follows the increasing hydrophobicity of the twin-tailed monomer. The smaller change in the viscosity profile between the DiC$_8$AM and the DiC$_6$AM terpolymers as compared to the DiC$_{10}$AM terpolymer, can be attributed to the lower molecular weight of the DiC$_8$AM terpolymer.

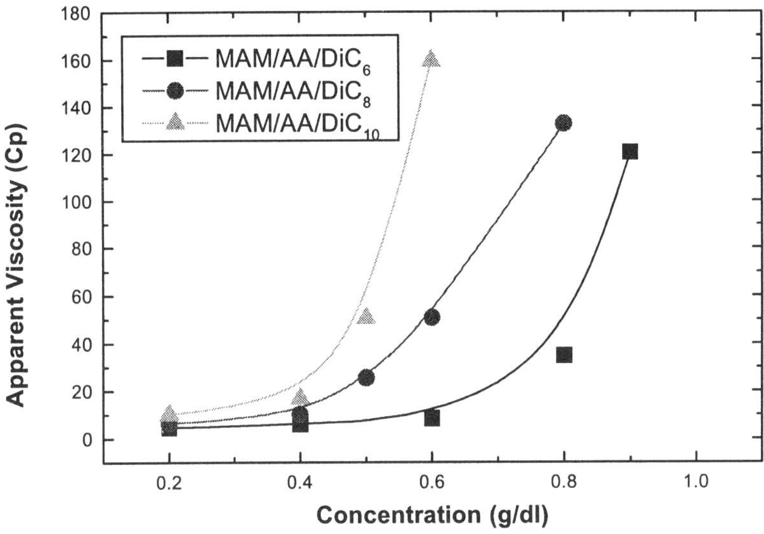

Figure 4. *Apparent viscosity vs. concentration for different MAM twin-tailed terpolymers.*

Single Tailed vs. Twin Tailed Hydrophobes

In order to evaluate the increased viscosification efficiency of the twin-tailed hydrophobe-based terpolymer vs. the single tail hydrophobe based terpolymers, viscosity measurements were performed for both systems. The results are illustrated in Figure 5, normalized for the concentration of hydrophobe.

Since the twin-tailed terpolymers have twice the amount of hydrophobe as the single-tailed terpolymers, normalization allows evaluation of other effects contributing to solution viscosity besides hydrophobe concentration. From Figure 4 it can be seen that the twin-tail based terpolymers show dramatically higher increases in viscosity than their single-tailed counterparts. This is likely due to an effect of the twin-tailed hydrophobic monomer on the conformation of the polymer. This effect can be best explained by analogy to small molecule surfactants. While single-tailed surfactants typically form micelles above their CMC, twin-tailed surfactants typically associate with one another forming lamellar type associates. Thus, the increased viscosification efficiency of the twin-tailed hydrophobes is likely a result of their natural propensity to form bilayer type associates with microblocks on other polymer chains due to geometry factors associated with twin-tailed amphiphiles, relative to those of single-tailed amphiphiles, which tend to more readily form intramolecular associates at low concentrations (*26*). Another suggested explanation for the enhanced viscosity of the twin-tailed polymer systems is that the CMC for twin-tailed amphiphiles is lower than for analogous single-tailed amphiphiles. The observed behavior likely results from a combination of these two phenomena.

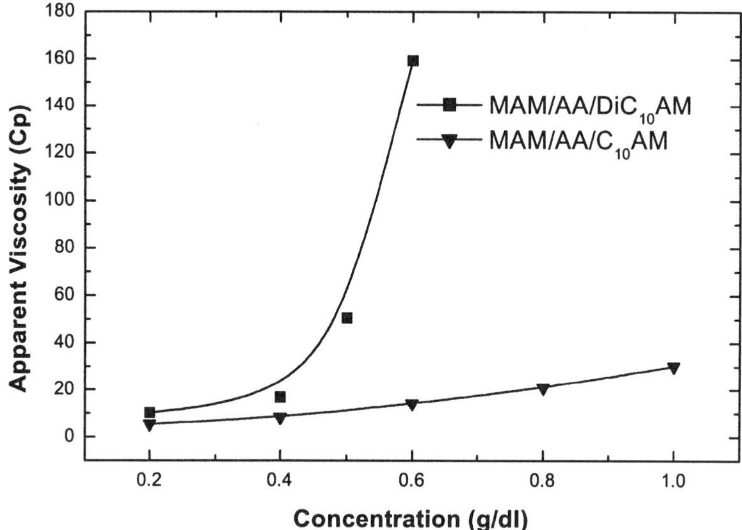

Figure 5. *Apparent viscosity vs. concentration comparing single and twin-tailed terpolymers.*

Effect of pH on Viscosity Behavior

As a result of the incorporation of acrylic acid moieties into the polymer backbone, the conformation of the polymer chain and its hydrodynamic volume in aqueous solution will be greatly affected by the solution pH. In order to evaluate this effect, the apparent viscosities for the MAM/AA/DiC$_{10}$AM terpolymers at a concentration of 0.1 g/dl vs. pH are reported in Figure 6.

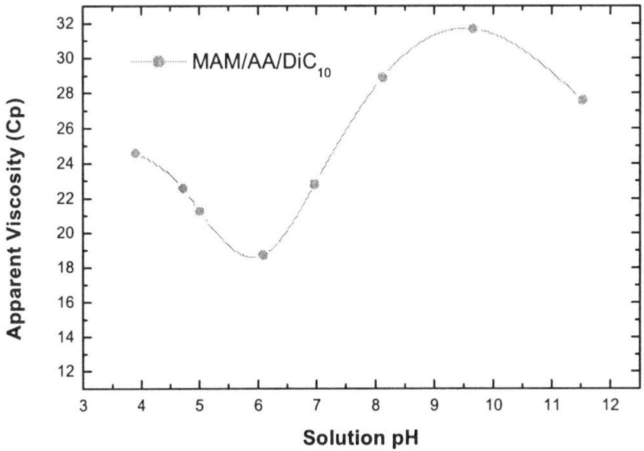

Figure 6. *Apparent viscosity vs. pH for MAM/AA twin-tailed terpolymers.*

The above graph indicates the pronounced effect that pH can have on the conformation of polyelectrolytes. For the MAM/AA based terpolymer an initial decrease is viscosity is observed over the pH range from 4.0 to 6.0. This behavior is typical for AA based polymers and results from the partial ionization of the polymer backbone, leading to hydrogen-bonding between ionized AA moieties and adjacent protonated ones and a subsequent decrease in hydrodynamic volume. At pH 6, the MAM terpolymer reaches a critical degree of ionization where further ionization beyond which AA moiety repulsion and chain expansion occurs. Finally, around ~ pH 9.5-10, a slight decrease in viscosity is observed due to the high ionic strength of the solution and a screening of the charge-charge repulsions.

Fluorescence Energy Transfer Measurements

In order to obtain a better understanding of the microscopic solution behavior exhibited by twin-tailed, hydrophobically modified acrylic acid-based polymers, the $DiC_{10}Am$ terpolymer, which showed the most pronounced associative thickening tendency, was labeled with either a naphthyl or a dansyl chromophore. Mixed solutions containing naphthyl donor and the dansyl acceptor on different chains were prepared and nonradiative energy transfer (NRET) measurements were performed, following a procedure previously reported by Hu *et al.* (*25*). The Förster distance for this energy transfer pair has previously been documented as 23.45 Å. This characteristic distance is indicative of the distance at which energy transfer efficiency is 50% between a chromophore pair. Thus, this method is very sensitive for indicating hydrophobic association and network formation.

Figure 7 illustrates the emission intensity vs. wavelength behavior for the mixed, individually labeled chains containing the dansyl and naphthyl chromophores when excited at 282 nm.

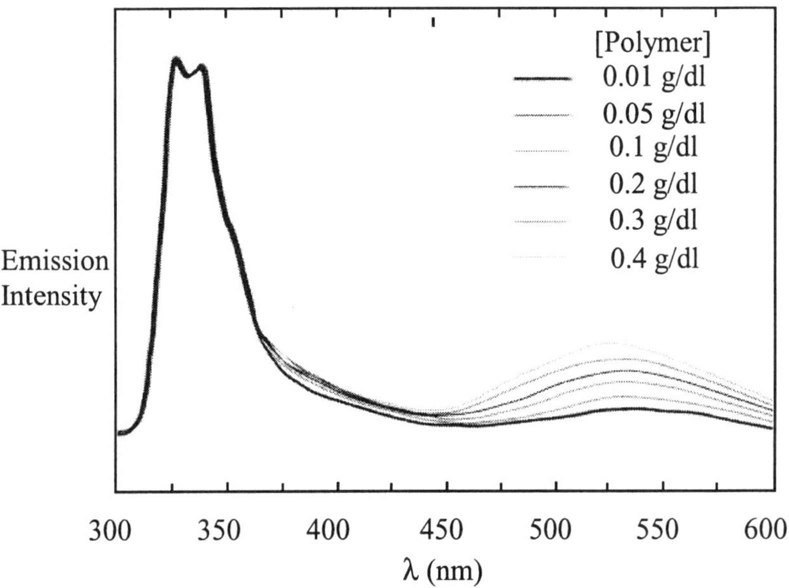

Figure 7. *Normalized Emission Spectra for MAM/AA/DiC$_{10}$AM Polymers Excited at 282 nm at pH 8.0*

As the polymer concentration is increased, an increase in the emission intensity is observed in the dansyl emission region between 450-580 nm. At 0.01 g/dl and 0.05 g/dl little emission in this region is observed. At concentrations near 0.1 g/dl, significant energy transfer occurs. While this is perhaps not so evident from direct examination of the emission spectra, a better picture concerning the amount of energy transferred may be obtained by plotting the NRET quantum efficiency vs. polymer concentration (Figure 8).

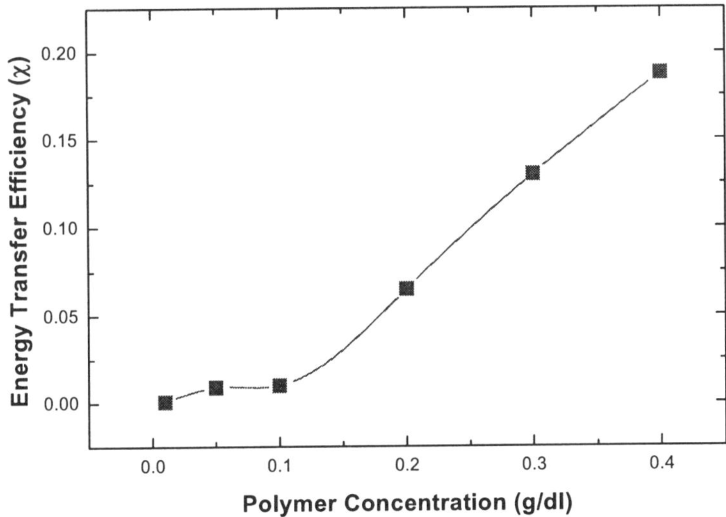

Figure 8. Energy Transfer Efficiency Plot for Labeled MAM/AA/DiC$_{10}$AM Terpolymer at pH 8.0 and 25 °C.

Due to the small amount of direct excitation of the dansyl chromophore at 282 nm, a correction must be made as discussed in the experimental section. In Figure 8 it can be seen that a significant increase occurs in NRET quantum efficiency at ~0.1 g/dl. This concentration is indicative of the microscopic overlap concentration at which network formation occurs. Below this concentration, most polymer chains likely exist as unimers in solution. As the polymer concentration is increased and the microscopic overlap concentration is reached, intermolecular association is observed. Comparison of the overlap concentration determined by NRET with that determined by viscosity, shows that the NRET based overlap concentration is slightly lower. This behavior however, is rather typical due to the enhanced sensitivity of fluorescence for indicating label interaction and overlap.

NRET experiments were also performed on mixed, individually labeled MAM/AA/DiC$_{10}$AM terpolymer at a fixed concentration of 0.3 g/dl as a function of solution pH. Following corrections for direct excitation, the Guillet equation was utilized to calculate the NRET quantum efficiency at seven pH values between 4 and 10. These data are presented in Figure 9.

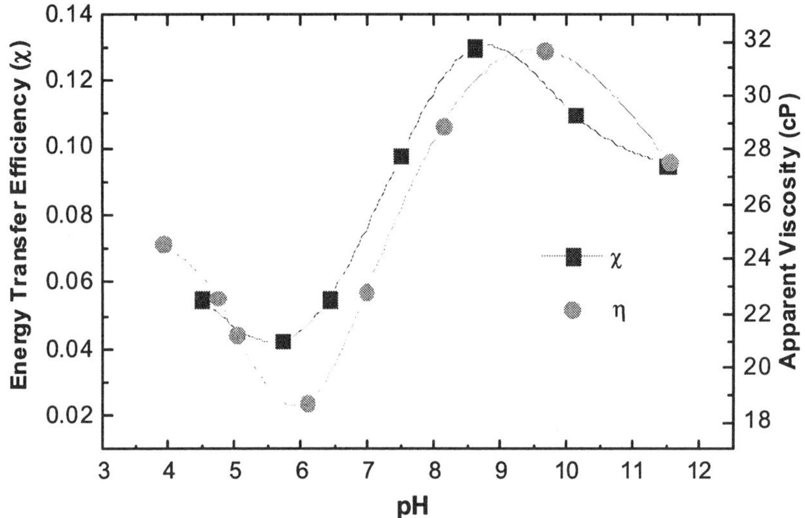

Figure 9. *Energy Transfer Efficiency Plot vs. pH for Labeled MAM/AA/DiC$_{10}$AM Terpolymer at a Fixed Concentration of 0.3 g/dl*

Although changes in the solution pH are expected to affect the population of reporting dansyl chromophores, calculation of the NRET quantum efficiency based on quantum yields corrects these data for concentration differences. An initial decrease in viscosity is observed from pH 4 to pH 5.5. This behavior is typical for AA based polymers and results from the partial ionization of the polymer backbone which leads to hydrogen-bonding between ionized AA moieties and adjacent protonated ones and a subsequent decrease in hydrodynamic volume. At pH 5.7, the MAM terpolymer reaches a critical degree of ionization beyond which AA moiety repulsion and chain expansion occur. Finally, around ~ pH 8.5-9, a slight decrease in viscosity is observed which results from the high ionic strength of the solution and a decrease in the Debye screening length   This behavior is modeled in Figure 10.

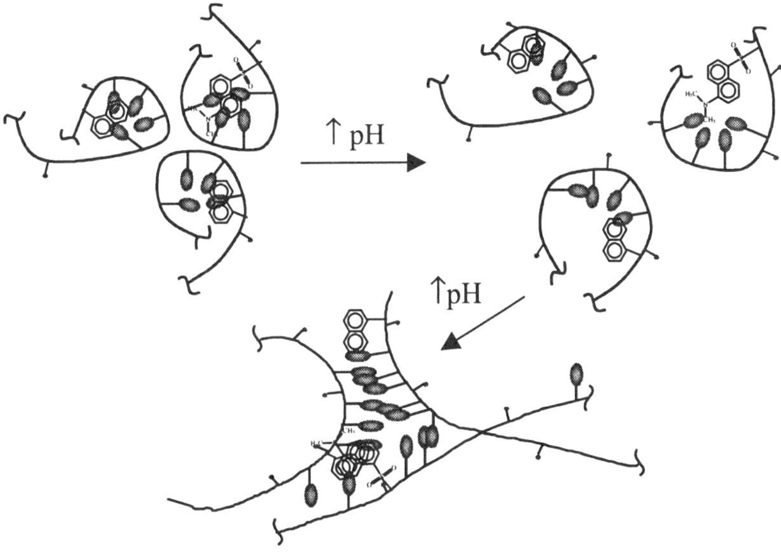

Figure 10. *Model Illustrating the Effect of Polymer Concentration on Nonradiative Energy Transfer Efficiency*

Also plotted in Figure 9 are the viscosity data from Figure 6. While both show the same trends, changes in chain expansion and label interaction indicated by NRET quantum efficiency occur at slightly lower pH values. This once again is believed to result from the greater sensitivity of fluorescence methods for indicating changes in labeled polymer chain interactions.

CONCLUSIONS

From the above results, it is apparent that the MAM/AA terpolymers exhibit solution properties conducive for associative thickening. Their tendencies toward intermolecular association closely follow the increase in hydrocarbon chain length (hydrophobicity) with the terpolymer containing the DiC$_{10}$Am hydrophobe exhibiting the most pronounced associative thickening. Also, twin-tail based terpolymers show dramatically higher increases in viscosity than their single-tailed counterparts. This is likely due to an effect of the twin-tailed hydrophobic monomer on the conformation of the polymer, which allows for enhanced associative junction formation. Energy transfer measurements indicate the onset of association occurring ~0.1 g/dl for the MAM/AA/DiC$_{10}$AM terpolymer. Changes in the energy transfer efficiency as a function of pH closely follow analogous viscosity behavior and are indicative of pH induced expansion and collapse caused by hydrogen bonding, ionic repulsion, and ionic shielding effects.

Acknowledgments. Financial support for this research from ICI, the Department of Energy, and the Office of Naval Research is gratefully acknowledged.

References

1. Schulz, D. N.; Bock, J.; Valint, Jr., P. L. In *Macromolecular Complexes in Chemistry and Biology*; Dubin, P.; Bock, J.; Davies, R. M.; Schulz, D. N.; Thies, C., Eds.; Springer-Verlag: Berlin, 1994; pp 3-12.
2. Evani, S. U.S. Patent 4,432,881, 1984.
3. Turner, S. R.; Siano, D. B.; Bock, J.; U.S. Patent 4,528,348, 1985.
4. McCormick, C. L.; Bock, J.; Schulz, D. N. In *Encyclopedia of Polymer Science and Engineering*; Mark, H. F.; Bikales, N. M.; Overberger, C. G.; Menges, G.; Kroschiwitz, J. I., Eds. John Wiley & Sons, Inc.: New York, 1989; Vol. 17, pp 730-784.
5. Ezzell, S. A.; McCormick, C. L.; In *Water Soluble Polymers. Synthesis, Solution Properties, and Applications*; Shalaby, S.; McCormick, C. L.; Butler, G. B., Eds. ACS Symposium Series No. 467, American Chemical Society: Washington, DC, 1991; pp 130-150.
6. Ezzell, S. A.; McCormick, C. L.; *Macromolecules*, **1992**, *25*, 1881-1886.
7. Ezzell, S. A.; Hoyle, C. E.; Creed, D.; McCormick, C. L.; *Macromolecules*, **1992**, *25*, 1887-1895.
8. Regalado, E. J., Selb, J; Candau, F.; *Macromolecules*, **1999**, *32*(25), 8580-8588.
9. Branham, K. D.; Davis, D. L.; Middleton, J. C.; McCormick, C. L.; *Polymer*, **1994**, *35*(20), 4429-4436.
10. McCormick, C. L.; Nonaka, T.; Johnson, C. B.; *Polymer*, **1988**, *29*, 731-739.

11. McCormick, C. L.; Middelton, J. C.; Grady, C. E.; *Polymer*, **1992**, *33*, 4184-4191.

12. McCormick, C. L.; Middelton, J. C.; Cummins, D. F.; *Macromolecules*, **1992**, *25*, 1201.

13. Branham, K. D.; Snowden, S. L.; McCormick, C. L.; *Macromolecules*, **1996**, *29*, 254-262.

14. Branham, K. D. Ph.D. Thesis, University of Southern Mississippi, Hattiesburg, MS, 1995.

15. Peer, W., *Polymers in Aqueous Media*, Glass, J. E., Ed., Advances in Chemistry Series No, 223; American Chemical Scoiety, Washington D. C., 1989, 381.

16. Volpert, E.; Selb, J.; Candau, F.; *Macromolecules*, **1996**, *29*, 1452-1463.

17. Valint, P. L., Bock, J., Schultz, D. N., *Polym. Mater. Sci. Eng.* 1987, **57**, 482.

18. McCormick, C. L; Chang, Y. *Macromolecules* **1994**, 27, 2151.

19. Shea, K. J.; Stoddard, G. J.; Shavelle, D. M.; Wakui, F.; Choate, R. M. *Macromolecules* **1990**, 23, 4497.

20. Boudreaux, C.J., Bunyard, W.C., and McCormick, C.L. *Journal of Controlled Release*, **1996**, 40, 223.

21. Rusakowicz, R.; Testa, A. C. *J. Phys. Chem.* **1968**, 72, 2680.

22. Eastmen, J. W. *Photochem. Photobiol.* **1967**, 6, 55.

23. Lakowicz, J. R.; Wiczk, W.; Gryczynski, I.; Fishman, M.; Johnson, M. *Macromolecules* **1993**, 26, 349.

24. Guillet, J. E.; Rendall, W. A. *Macromolecules* **1986**, 19, 224.

25. Hu, Y.; Kramer, M. C.; Boudreaux, C. J.; McCormick, C. L. *Macromolecules* **1995**, 28, 7100.

26. Warr, G. G. and Patrick, H. N. *Specialist Surfactants.* Blackie Academic & Professional, London, **1997**.

Chapter 4

Macromolecular Assemblies Generated by Inclusion Complexes between Amphipathic Polymers and β-Cyclodextrin Polymers in Aqueous Media

Catherine Amiel,[1,*] Laurence Moine,[1] Agnès Sandier,[1,2] Wyn Brown,[2] Cristelle David,[1] Frederique Hauss,[1] Estelle Renard,[1] Martine Gosselet,[1] Bernard Sébille.[1]

[1] Laboratoire de Recherche sur les Polymères, UMR C7581 CNRS, 2-8 rue H. Dunant, 94320 Thiais France
[2] Department of Physical Chemistry, University of Uppsala, Box 532, S-751 21 Uppsala, Sweden.

Addition of a water soluble β-cyclodextrin polymer to an aqueous solution of an amphiphilic polymer has given the birth to a new class of associating polymer systems. The β-cyclodextrin polymers are used to promote the associations between amphipathic polymer chains by formation of inclusion complexes between the hydrophobic moieties and the β-cyclodextrin cavities. In this study, the water-soluble β-cyclodextrin polymers were formed by polycondensation of the monomer with epichlorohydrin. The amphipathic polymers are hydrophilic chains which have been modified by introducing hydrophobic moieties. The latter can be either rigid (adamantyl groups) or flexible structures (alkyl chains with more than 12 carbon atoms) but they must possess a strong affinity for β-cyclodextrin cavities. The influence of the chemical nature and the architecture of the amphipathic copolymer has been studied in order to elucidate the mechanisms of associations between these polymers. Different architectures have been studied: comblike polymers (hydrophobically modified dextran) and star polymers (hydrophobically end-capped star polyethyleneoxide). Polyelectrolytes have also been synthesized (hydrophobically modified degradable copolyesters). In this case, the associations show high sensitivity to pH and ionic strength. The hydrophobic molar ratio of the amphipathic polymer, as well as the respective concentrations of each polymer play a central role in determining the associating properties of the system. The associations have been studied by viscosity, rheology and dynamic light scattering measurements. Polymer networks are obtained above critical values of the number of hydrophobic groups per chain and are physical gels possessing original rheological properties because the microscopic association is reversible. Another alternative provided by the attractive interactions which generate the inclusion complexes is the possibility of associative phase separation. Phase separation takes place, in the case of neutral systems, when the number of hydrophobic moieties per chain is larger than 3.

Cyclodextrins are cyclic oligosaccharides (the structure of β-cyclodextrin is shown schematically in Figure1) which are well known for their ability to form inclusion complexes with a large variety of hydrophobic molecules *(1)*. Because of this property, great interest is attached to cyclodextrin compounds in food, drug and agricultural industries *(2-3)*. In addition to these applications, inclusion complexes allow creation of supramolecular structures due to the highly specific non-covalent binding between the cyclodextrin host and the guest molecule. The use of water-soluble polymers in combination with β-cyclodextrin gives rise to two different kinds of structures: a) when β-cyclodextrin interacts directly with the backbone units of the polymer, necklace structures called polyrotaxanes (or pseudopolyrotaxanes), in which many cyclic molecules are threaded along the linear polymeric chain, are obtained *(4)* and b) β-cyclodextrin acts as a structure breaker when added to amphiphilic copolymers, binding directly to the hydrophobic side groups and breaking up the hydrophobic microdomains *(5)*. A variety of structures can be obtained by interacting a β-cyclodextrin polymer (a polymer incorporating several β-cyclodextrin units) and an amphiphilic copolymer in aqueous media *(6-10)*. Association between these two polymers is shown schematically in Figure 2. This class of supramolecular structures is similar to those obtained with associating polymers. For instance, transient networks may be formed when the density of bonds is high enough. However, the originality of the present host-guest associating system comes from the nature of the physical bond itself: a mechanism of molecular recognition is involved in the interaction between host β-cyclodextrin and guest hydrophobic moiety.

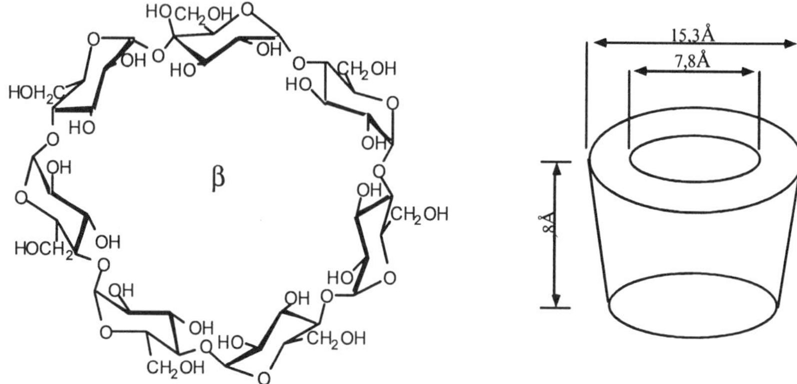

Figure 1. *Structure of β-cyclodextrin.*

In this study, the water-soluble β-cyclodextrin polymer is formed by polycondensation of the monomer with epichlorohydrin under strongly alkaline conditions. The amphiphilic polymers are hydrophilic chains which have been modified with suitable hydrophobic groups. The selected moieties may either be rigid (adamantyl groups) or flexible structures (alkyl chains with more than 12 carbon atoms) but must exhibit a strong affinity for the β-cyclodextrin cavity.

The influence of the chemical nature and the architecture of the amphiphilic copolymer has been studied in order to understand the mechanisms of association in the ternary system. Different architectures have been studied: comb-like polymers (hydrophobically modified dextran) or star polymers (hydrophobically end-capped star polyethyleneoxides). Polyelectrolytes have also been synthesized (hydrophobically modified degradable copolyesters). In this case, the association shows a high sensitivity to both pH and ionic strength.

The paper is organized as follows. The hydrophobically modified water soluble polymers are presented in a first section (guest structures). The β-cyclodextrin polymers and the thermodynamic parameters describing their association with guest polymers, determined by a spectroscopic method (fluorescence) and dialysis, are given in a second section. The influence of the nature of the hydrophobic moiety, as well as the polymer structures and sizes are discussed. The presence of reversible crosslinks between the host and guest polymers increases their effective interaction and reduces their affinities for the solvent. Increasing the interaction strength or the number of associating groups per chain may lead to associative phase separation. This is shown in a third section by phase diagrams determined as a function of the number of hydrophobic groups per chain. Macroscopic properties due to the aggregation phenomenon are described in a following section based on dynamic light scattering, steady shear viscosity and rheology measurements.

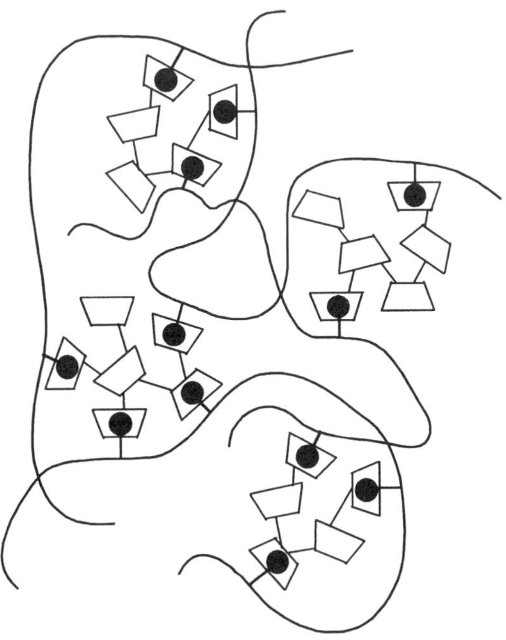

Figure 2. Schematic representation of the association between the hydrophobically modified polymer and the β-cyclodextrin polymer.

Hydrophobically Modified Water-Soluble Polymers

The water-soluble polymers bearing hydrophobic moieties distributed along the main chain or grafted at the end of the chains have been obtained either by chemical modification of commercial polymers or by synthesis of copolymers. The hydrophobic moieties, shown schematically in Figure 3, have been selected for their affinity for the β-cyclodextrin cavities. The adamantane group was selected since it fits precisely into the slightly apolar cavity of β-cyclodextrin *(11)*. The 1-1 complexes formed with β-cyclodextrin are among the most stable. Alkyl groups are more commonly used in associating polymers *(12)*. Stable complexes with β-cyclodextrin are formed when the number of carbon atoms is greater than 10. The stoichiometry of the complex may involve in some cases more than one guest or host unit *(13)*.

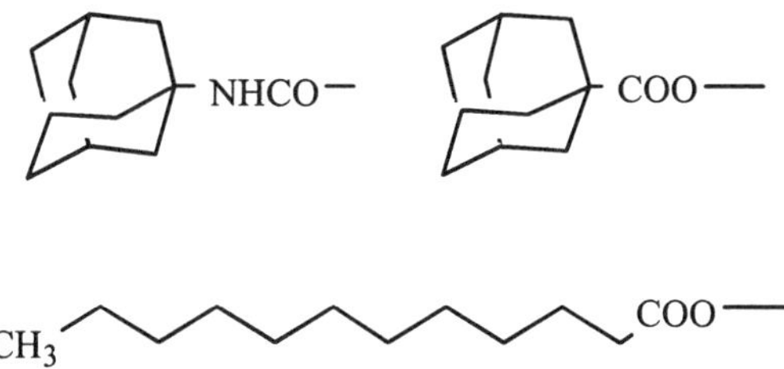

Figure 3. Structure of the hydrophobic groups: adamantane group linked via an amide function (to PEO) or an ester function (to dextran) and an alkyl group linked via an ester function.

Telechelic Polymers: Hydrophobically End-Capped PEO

Hydrophobically modified PEO was obtained by reacting the OH terminal functions of the polymer with isocyanate groups, 1-adamantyl isocyanate, to give PEO-Ad *(6)*. The hydrophobic molar ratio was varied either by using linear PEO of different molecular weights (6000 to 35000) or by using multi-arm branched PEO of constant arm size (5000) but variable number of arms (3 to 8). The latter compounds were obtained by ethoxylation of polyols (Shearwater Polymers, Inc., USA).

Comb-like Polymers: Hydrophobically Modified Dextran.

Dextran polymers (Pharmacia, France), molecular weights (10^4 to 10^6 g/mol), have been hydrophobically modified by reacting a small proportion of the hydroxyl functions with an acyl chloride: 1-adamantanecarbonyl chloride and lauroyl chloride (Aldrich, France). The number of hydrophobic groups per chain ranging from 3 to 140 were obtained by varying the hydrophobic molar ratio (1 to 7%).

Hydrophobically Modified Polyelectrolyte: Poly Malic Acid (PMLA) derivatives.

Poly(malic acid) is a degradable polyester. Due to its carboxylic acid functions, it has a polyelectrolyte character, with a pKa around 4. The amphiphilic copolymers, poly(β-malic acid-co-ethyladamantyl β-malate) (PMLA-Ad), were synthesized by copolymerization with benzyl malolactonate, which is conducive after deprotection to malic acid units, and ethyladamantyl malolactonate *(14)*. Its structural formula is presented in Figure 4. Molecular weights and hydrophobic molar ratio of the water soluble copolymers are reported on Table I.

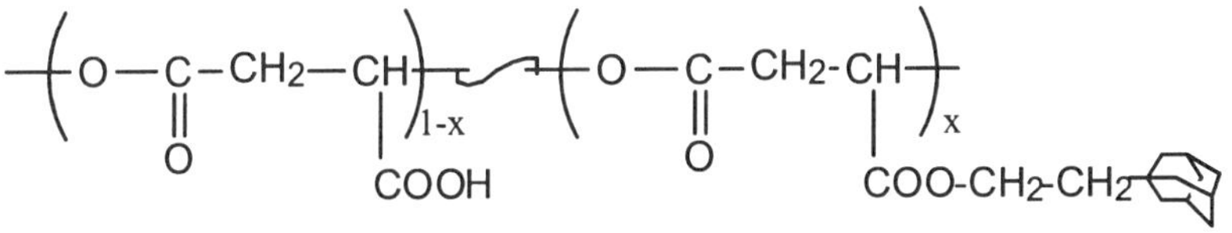

Figure 4. Structural formula of poly(β-malic acid-co-ethyladamantyl β-malate) (PMLA-Ad).

Table I. Characterization of the hydrophobically modified PMLA.

Polymers	Mw	Hydrophobic molar ratio
PMLA4Ad	18000	3.5
PMLA7Ad	74000	7.5
PMLA11Ad	58000	11.5

Associations of the Hydrophobically Modified Polymers

The hydrophobically modified polymers used here have similar characteristics to associating polymers: i.e they are water soluble polymers containing a small proportion of hydrophobic groups. It is necessary to characterize their possible auto-association properties because the incorporation of the hydrophobic groups into hydrophobic microdomains may compete with the inclusion complex interaction when the two polymers are mixed. Depending on the nature of the hydrophobic moieties, alkyl or adamantane, different behavior is observed:

a) Polymers modified with adamantane groups do not show auto-association properties at low and moderate concentrations (0 to 10% w/w solutions). The viscosities of the modified polymers are always comparable to those of the precursors, even at concentrations higher than the overlap concentration (C*) of the chains. Intrinsic viscosities given in Table II show that there is a little variation between precursor and modified polymers for the PEO chains. There is a slight difference between precursor and modified dextran which may be more related to a change in solvent quality (adding increasing amounts of hydrophobic groups along the chains) rather than to auto-association of the hydrophobic groups.

b) Dextran polymers modified with alkyl groups behave as associating polymers described in the literature *(12)*, showing lower intrinsic viscosities than the precursor and thickening properties at concentration above C*. This is due to the well known incorporation of alkyl moieties into the hydrophobic microdomains resulting in compact structures of the chains at low concentration and intermolecular associations in the semi-dilute range.

Table II. Intrinsic viscosities of precursor and modified polymers (dextran and PEO).

Polymers	*Molecular weight*	*Hydrophobic molar ratio*	*[η] precursor (l/g)*	*[η] modified polymer (l/g)*
PEO-Ad linear	6.10^3	1.5 %	0.033	0.033
PEO-4Ad 4 arms	2.10^4	0.9 %	0.028	0.028
Dext-Ad	2.10^6	1.1 %	0.065	0.062
Dext-Ad	5.10^5	2.4 %	0.047	0.043
Dext-Ad	10^4	4.7 %	0.0085	0.0080
Dext-Ad	4.10^4	7 %	0.018	0.016
Dext-Alk	4.10^4	2.8 %	0.018	0.0104
Dext-Alk	4.10^4	5.1 %	0.018	0.0056

β-Cyclodextrin Polymers and Thermodynamic Parameters describing their Interactions with Amphiphilic Copolymers.

β-Cyclodextrin Polymers.

Polymers have been obtained by polycondensation of β-cyclodextrin and epichlorohydrin under strongly alkaline conditions. The synthesis has been described in detail earlier *(15)*. Due to the multifonctionality of β-cyclodextrin (21 hydroxyl groups), branched architectures and highly compact coils are obtained *(16)*. Depending on the conditions of the synthesis (epichlorohydrin/ β-cyclodextrin ratio and reaction time), low or high molecular weights are obtained. In order to tailor the interactions between the β-cyclodextrin polymers and the amphiphilic copolymers, it was relevant to control the number of β-cyclodextrin units per guest polymer. Thus, three different kinds of samples have been synthesized for this study: oligomers, containing an average of 3 - 4 β-cyclodextrin units per chain, low molecular weight polymers with an average of 15 β-cyclodextrin units per chain and high molecular weight polymers containing more than 1000 β-cyclodextrin units per chain. The characteristics of these samples are given in Table III.

Table III. Characteristics of the β-cyclodextrin polymer samples.

β-CD/EP copolymers	*Molecular weights (Mn - Mw)*	*β-CD content (% w/w)*	*[η] (l/g)*
β-CD/EP oligomer	2.10^3 - 8.10^3	67 %	0.0038
β-CD/EP LM	26.10^3 - 36.10^3	41 %	0. 0063
β-CD/EP HM	2.10^5 - 1.10^6	87 %	0. 013

Thermodynamic parameters describing the interaction between β-cyclodextrin polymers and amphiphilic copolymers.

The hydrophobic moieties may form inclusion complexes with the β-cyclodextrin cavities. We assume to a first approximation that each group belonging to a multifunctional polymer, behaves independently in the same way as a small molecule. This allows us to write the following equilibrium:

$$H + CD \Leftrightarrow HCD \; , \qquad K_c = \frac{[HCD]}{[H][CD]} \qquad (1)$$

[H], [CD] and [HCD] are respectively the molar concentrations of free hydrophobic groups, free CD and bound CD. K_c is the complexation constant in $l \cdot mol^{-1}$. The consistency of the results obained by different methods justify this approximation in most of the cases.

Thermodynamic parameters describing the interaction between host and guest molecules have been studied by two methods. The fluorescence method involves a fluorescent probe, 1-anilino 8-naphthalene sulfonic acid (1-8 ANS) which has been used as a competitive inhibitor *(13)*. The conditions of the experiments are detailed in a previous work *(6)*. The dialysis method uses dialysis membranes of high molecular weight cut-off which are only permeable to the host polymers (PEO derivatives). The molecular weight of β-cyclodextrin polymer used in this study (β-CD/EP HM) was higher than the membrane molecular weight cut-off. Measurement of PEO concentration at equilibrium in the outer compartment allows detetmination of the complexation constant *(7)*.

Table IV. Complexation constants between linear PEO end-capped with adamantyl groups and β-cyclodextrin polymers.

Sample	M(PEO)	K ($l \cdot mol^{-1}$)
PEO-Ad	5000	3600
PEO-Ad2	6000	3200
PEO-Ad2	20000	2300
PEO-Ad2	35000	2000

Note: PEO-Ad has a methoxy group at one end and an adamantyl group at the other end.

The mechanism of association is not only related to the chain length but also to the architecture of the amphiphilic macromolecules. This is illustrated in Figure 5 where complexation constants (determined by the dialysis method) are shown for polymers of different architectures. For telechelic PEO, where the number of hydrophobic groups per chain has been varied from 2 to 8 using star PEO, the complexation constant decreases as a function of the molecular weight as for linear PEO (Table IV) when the number of hydrophobic groups is lower than or equal to 4. On the other hand, an increase of K by a factor greater than 2 is observed for PEO containing 8 adamantyl groups per chain. This effect can be attributed to a cooperative binding of the adamantyl groups which should be closer together when the number of arms of the PEO star molecule is increased. In the same way, dextran containing a higher hydrophobic molar ratio (3%) than the PEO derivatives of this study (less than 1.5%) show an even higher complexation constant. The cooperativity of the association seems to be

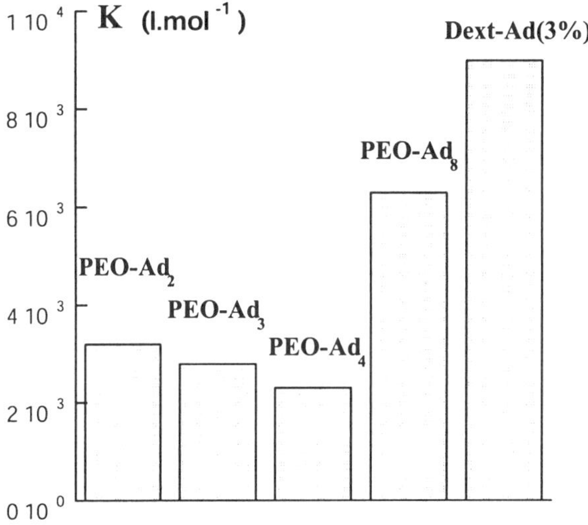

Figure 5. *Influence of the architecture of the chains on the complexation constants. The amphiphilic polymers are linear PEO-Ad2, branched PEO-Ad3, PEO-Ad4, PEO-Ad8 and Dext-Ad3%.*

controlled by the proximity of the hydrophobic moieties in the amphiphilic polymer coil: crowded adamantyl groups on a star PEO molecule or a sufficiently high molar ratio on comb-like dextran structure.

Associative phase separations

Ternary mixtures polymer1-polymer2-solvent generally display segregative phase separations, due to the low entropy of mixing of polymers *(17)*. This is the case for PEO (unmodified) and β-cyclodextrin polymers in aqueous solutions *(7)*. However, biphasic samples in which the two polymers are segregated in each phase are obtained at relatively high concentrations (above 10% w/w). Segregative phase separation is also observed for the pair PMLA (polyelectrolyte) and β-cyclodextrin polymer but the biphasic domain appears at much lower concentrations (≈1%), independent of the pH.

Associative phase separations are obtained in some cases with the neutral systems containing PEO, dextran or polydimethylacrylamide derivatives *(7, 8, 18)*. Attractive interactions yielding inclusion complexes, which are at the origin of the macrostructures formed in solution (see next section), also induce associative phase separation when the strength of these interactions is high enough. Different interaction mechanisms are known to induce associative phase separation in ternary mixtures: electrostatic interactions between polymers of opposite charges *(19)* and hydrogen-bonding interactions *(20)*. This family of polymer systems allows one to demonstrate that interactions of inclusion complexes may also induce associative phase separations. The strength of the interaction between the two polymers will depend not only on the affinity between the hydrophobic moiety for the β-cyclodextrin cavity but also on the number of interacting centers per polymer chain. In the following, we discuss the influence of the number of hydrophobic groups (N_H) per amphiphilic copolymer while maintaining constant at a high value the number of β-cyclodextrin cavities (N_{CD}) per β-cyclodextrin polymer. As these two numbers which characterize the functionality of each polymer play a reciprocating role, one expects that the same trend will be observed when the number of cavities per β-cyclodextrin polymer is varied while the number of hydrophobic groups per chain is kept at a high value.

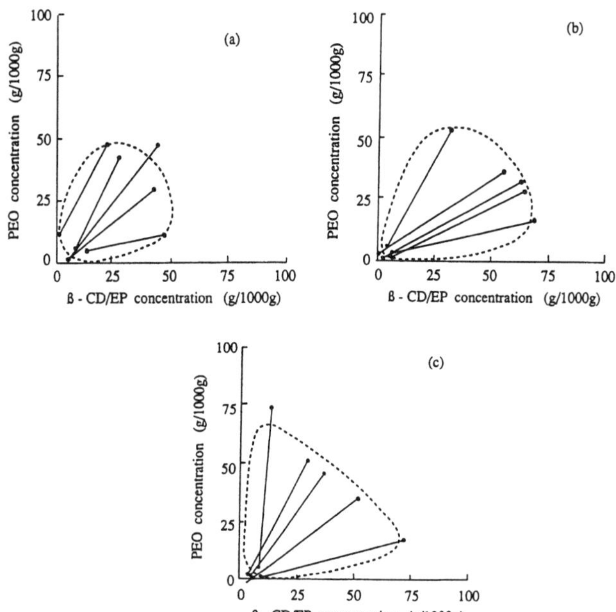

Figure 6. *Associative phase separations in aqueous media of β-cyclodextrin polymer and PEO-Ad3 (a), PEO-Ad4 (b) and PEO-Ad8 (c).*

For the telechelic PEO derivatives, N_H has been varied from 2 to 8 by using linear or star PEO with a varied number of arms. In this way, we have observed a transition from segregative to associative phase separation. As mentioned above, unmodified PEO (0 hydrophobic groups) and the β-cyclodextrin polymer are incompatible while linear PEO-Ad2 with 2 adamantyl groups per chain makes homogeneous phases when mixed with β-cyclodextrin polymer. PEO-Ad3 with 3 adamantyl groups per chain shows associative phase separation with β-cyclodextrin polymer. The phase diagrams obtained with PEO-Ad3, PEO-Ad4 and PEO-Ad8 are represented in Figure 6. Contrary to what might be expected, the upper limits of the biphasic domain are little influenced by the number of adamantyl groups per chain. However, these limits come closer to the axis when this number is increased from 3 to 8.

The same general behavior is observed with the modified dextran: associative phase separation only occurs when the average number of hydrophobic groups per chain is greater than 3. The nature of the hydrophobic moiety, adamantane or alkyl group (C12), does not appear to influence the mechanism of phase separation. The phase diagrams of two samples of modified

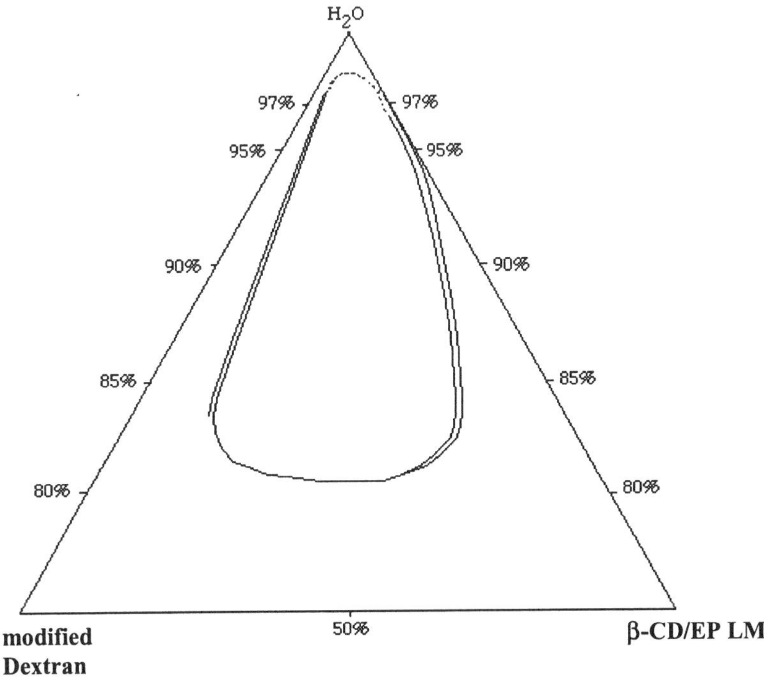

Figure 7. Associative phase separations in aqueous media of β-cyclodextrin polymer and modified dextran (40000): dext-Ad (<14> adamantane per chain) and dext-Alk (<18> alkyl per chain).

dextran (40000) bearing approximately the same average number of hydrophobic groups per chain but differing in nature (<14> adamantane and <18> alkyl) are compared in the same graph (Figure 7) and show a similar biphasic domain. However, the phase diagrams are significantly influenced when the average number of hydrophobic groups per chain is increased from 13 to 70: the two phases domains are extended toward higher concentrations (Figure 8).

In the systems containing the PMLA derivatives, no associative phase separation is observed, independent of the number of hydrophobic moieties per chain (5 - 50). The high "solubility" of the polymer-polymer complexes is explained by the polyelectrolyte nature of PMLA. It is well known that a high entropy of mixing of counter-ions does not favor phase separation *(19)*.

Aggregation in solution.

Inclusion complex interactions lead to macrostructures in solution which depend on polymer concentrations and also on the interaction strengths between the host and guest polymers. For amphiphilic copolymers of given N_H (number

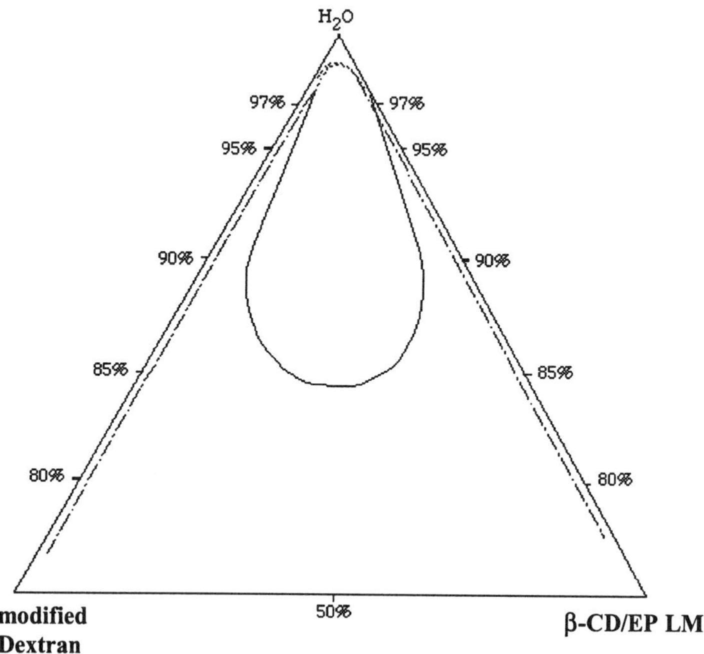

Figure 8. Associative phase separations in aqueous media of β-cyclodextrin polymer and dext-Ad : plain line diagram corresponds to dext-Ad 7.10⁴ (<13> adamantane per chain) and dotted line diagram corresponds to dext-Ad (<70> adamantane per chain).

of hydrophobic groups per chain), we were able to tailor the interaction strength in the polymer pair by using β-cyclodextrin polymers containing variable amounts of N_{CD} (number of β-cyclodextrin cavities per chain) ranging from 3 to 1000 (Table II). For instance, mixtures with βCD/EP oligomer ($N_{CD} \approx 3$) give monophasic samples over the whole concentration range, allowing studies at low concentration. Figure 9 shows specific viscosities of mixture Dextran-Adamantan (N_H =75) with βCD/EP oligomer as a function of the total concentration, at a given composition 50/50 (w/w) of the two components. At low concentrations, the association leads to compact aggregates as shown by the low intrinsic viscosities of the mixture. With the conditions in Figure 9, $[\eta]$= 0.1 dl/g, which is much lower than the intrinsic viscosity of the corresponding mixtures containing precursor dextran $[\eta]$= 0.25 dl/g. At concentrations of the order of chain overlap, association leads to intermolecular aggregates yielding moderate thickening properties: at a concentration of 50 g/l, the specific viscosity of the mixture is 10 times higher than the specific viscosity of the equivalent mixture containing precursor polymer.

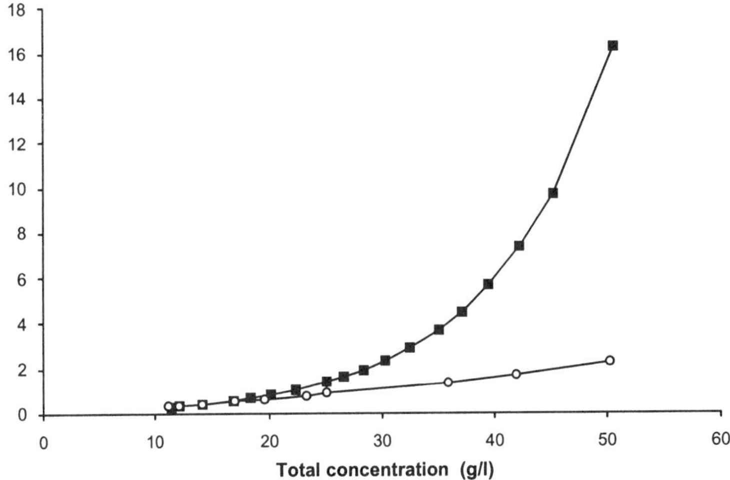

Figure 9. Specific viscosities of mixtures (50/50) Dext-Ad (5.10^5, 2.4%Ad)+ βCD/EP oligomer as a function of the total polymer concentration (■). Specific viscosity of the equivalent mixture containing precursor Dextran is given in comparison (O).

Monophasic phase diagrams can also be obtained with β-cyclodextrin polymer of higher N_{CD} (βCD/EP LM, N_{CD}= 13) when the amphiphilic copolymer added has a low N_H (lower than 3). Addition of increasing amounts of linear PEO-Ad (N_H =2) to a solution of βCD/EP LM shows increasing values of the average relaxation times measured by dynamic light scattering and these correlate with the viscosity enhancement (Figure 10) *(21)*. Associations between the two polymers lead to aggregates of increasing size.

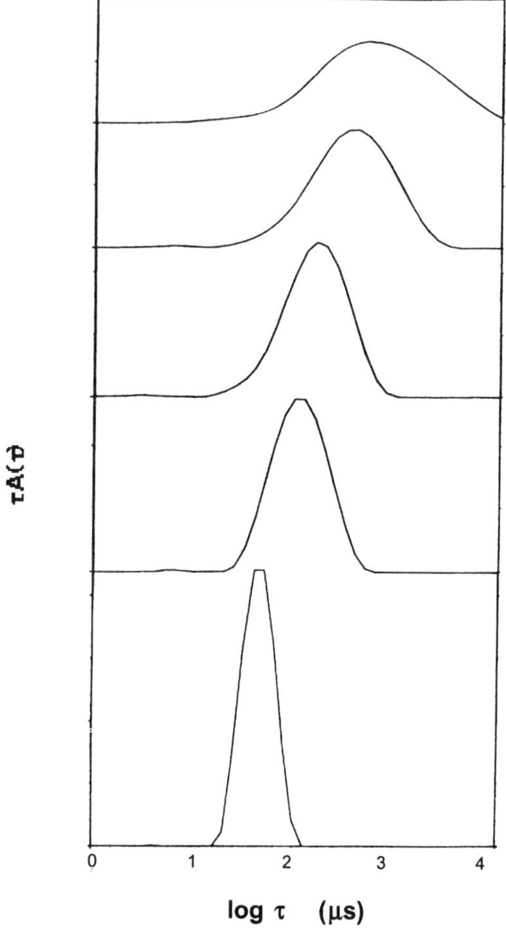

Figure 10. *Relaxation times distribution (DLS) for, bottom: βCD/EP LM c=1%*
and respectively the complex formed on addition of 0.25; 0.5; 1.0; 2.0 % PEO-
Ad linear, Mw=10000.

Depending on the connectivity of the polymer system and on the stoichiometry of the components, aggregates with different structures can be obtained. For instance, in the limit of infinite dilution of β-cyclodextrin polymer, the associated structure probably corresponds to an isolated β-cyclodextrin polymer coil "decorated" by PEO-Ad2 linked to it by one or its two ends. Figure

11 shows the hydrodynamic radius (R_h) determined from DLS measurements extrapolated to infinite dilution of the β-cyclodextrin polymer, as a function of the PEO-Ad concentration. The strong increase of R_h with PEO-Ad2 concentration can be attributed to an increase of the grafting density of PEO-Ad2 on the β-cyclodextrin particle (linking by only one end) as more PEO-Ad2 chain compete for complexation with the β-cyclodextrin cavities.

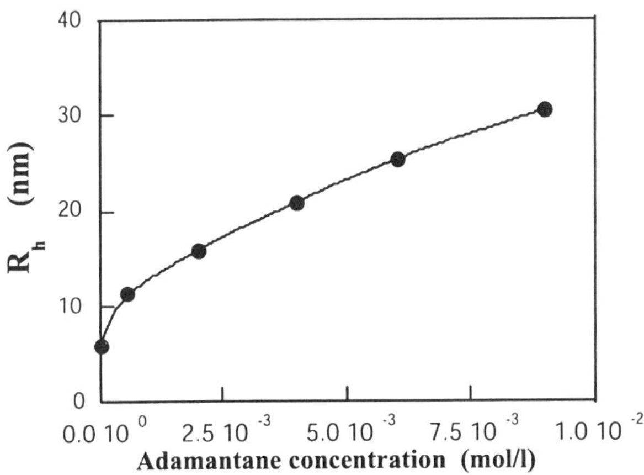

Figure 11. *Hydrodynamic radius (R_h) of β-cyclodextrin polymer coils (β-CD/EPLM) "decorated" with PEO-Ad2 (10^4), as a function of the PEO-Ad2 concentration.*

Polymer pairs which interact more strongly ($N_{CD} \approx 1000$ and $N_H = 4$ for instance) do not only show associative phase diagrams (see previous section) but also exhibit increased thickening properties, in the monophasic domain, than less interacting pairs. Figure 12 shows the viscosities of homogeneous mixtures of PEO-Ad4 (4 arms) mixed with βCD/EP HM. PEO-Ad4 concentration is kept constant (70 g/l) while βCD/EP concentration is varied. The sharp increase of viscosity with concentration is again ascribable to associating clusters. The viscosity enhancement, which can be as high as 6000 (in relative values), is more than one decade higher than those obtained with linear PEO-Ad of the same molecular weight *(7)*. Similar behavior has also been observed in associating polymers made of PEO *(22)* where enhanced thickening was observed when the number of alkyl groups per chain was increased from 2 to 3. This was interpreted as an increase of the percentage of intermolecular over intramolecular links. In our case, the density of active junctions in the aggregates is increased when N_H is increased.

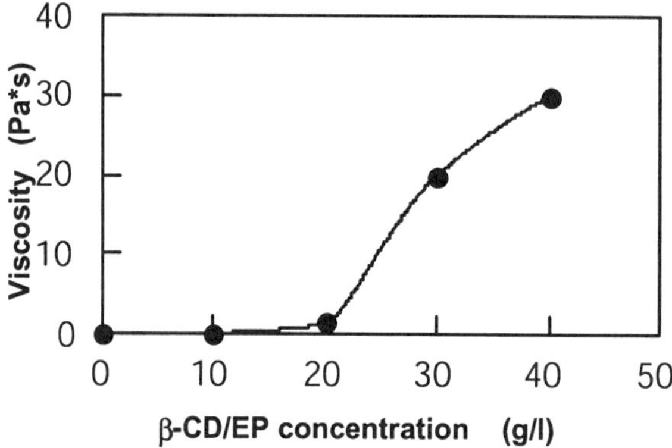

Figure 12. Viscosities of PEO-Ad4 + βCD/EP HM mixtures at constant PEO-Ad4 concentration (70 g/l).

Temperature influence

Inclusion complex interactions are reversible since no permanent links are involved. The reversibility can be studied by adding a competitor, which may either be a small molecule containing a hydrophobic moiety or a β-cyclodextrin cavity. Another simple way to dissociate the complexes is to increase the temperature. The temperature dependance of the specific viscosity is shown in Figure 13 for linear PEO-Ad + β-CD/EP LM   (concentrations 2% and 1% respectively). As expected, the viscosity decreases with temperature, following an Arrhenius relationship:

$$\frac{\eta}{\eta_0} = Ae^{\frac{E_a}{kT}}$$, η_0 is the solvent viscosity at temperature T and E_a is the activation energy.

A value of E_a=41kJ/mol gives the best fit to the previous relationship. As energies determined from Arrhenius plots represent activation enthalpies *(23)*, E_a is compared to the enthalpies of inclusion complexes of adamantane compounds with β-cyclodextrin described in the literature. These are 20 to 40 kJ/mol *(11)*, depending on substitutions on the adamantane or β-cyclodextrin moieties. The good agreement obtained supports again the mechanism of association due to inclusion complex interactions between adamantyl groups and β-cyclodextrin cavities.

The same qualitative result has been obtained when dextran-Ad is involved instead of PEO-Ad. In this case, dextran, M_w= 10^4, contained a low number of adamantyl groups per chain: N_H=3. The lower activation energy obtained, E_a=19.4 kJ/mol, can be explained by the different functions which link adamantane moieties to the chain in the two cases (an amide bond for PEO-Ad and an ester bond for dextran-Ad, see Figure 3).

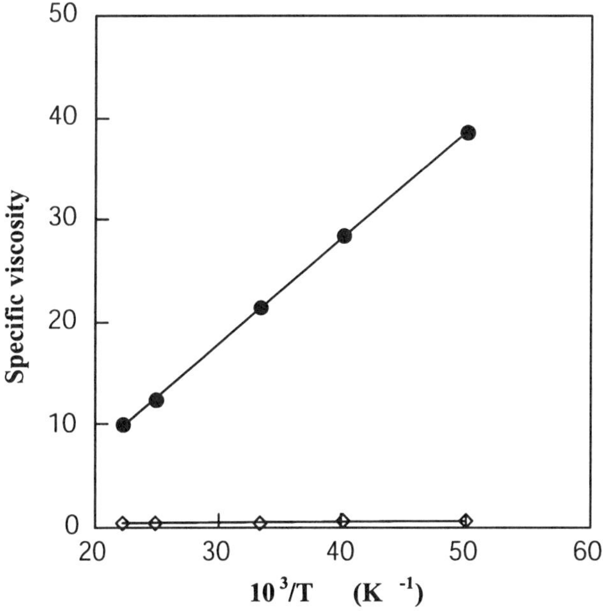

Figure 13. *Specific viscosities as a function of 1/T for a mixture containing 1% β-CD/EP LM and 2% linear PEO-Ad (●), and for 2% PEO-Ad (◊) solution.*

pH responsive system.

Use of a polyelectrolyte like PMLA-Ad allows the elaboration of systems which are responsive to pH and ionic strength. Thus mixtures of hydrophobically modified PMLA with polymers of β-cyclodextrin are highly sensitive to pH variation, as shown in Figure 14 where the viscosity increases sharply between pH 2 (uncharged PMLA) and pH 5 (high charge density of PMLA). This effect cannot be attributed to the copolyester alone at the same concentration as in the mixture: its viscosity varies between 1 and 3 cP in the pH range studied *(9)*. As demonstrated with the neutral systems, association by formation of inclusion complexes between the two polymers are again at the origin of the viscosity enhancements. The thickening properties are pH dependent because of the conformational changes experienced by the copolyester: from compact coils at low pH (low charge density) to extended conformations at pH around pKa (high charge density). This conformational change may influence the mechanism of association in two ways:

a) the interaction strength between the two polymers may be dependent on how compact the copolyester coil is, since the adamantyl groups will be less accessible for complexation when they belong to a compact coil than to an extended coil.

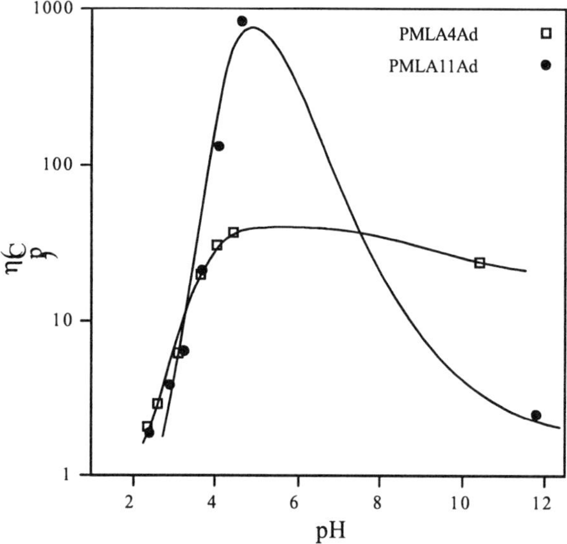

Figure 14. *Viscosity as a function of pH for 50/50 mixtures of β-CD/EP HM and PMLA4Ad (4% Ad) (□), and with PMLA11Ad (11%Ad) (●). The total concentrations are 30 g/l.*

b) for a given affinity between the two polymers, the connectivity of the polymer
system will be increased when the pH is increased due to the swelling of the
PMLA-Ad chains, leading to aggregates of greater size.

Independent of the weighting of these two effects, viscosity enhancement
with increasing charge density is expected in both cases.

As for the neutral systems, higher viscosities are obtained with the
copolymers of higher N_H: the viscosity at the optimum pH is more than 10 times
larger for the PMLA11Ad mixture (N_H=49) than for the PMLA4Ad (N_H=5).

No measurements are obtained in the range of pH comprised between 5 and
9 since addition of small amounts of NaOH solution induces large pH variation
in this domain. At pH larger than 9, Figure 14 shows viscosity drops which are
more pronounced for the mixture with the more hydrophobically modified
PMLA (PMLA11Ad). This is attributed to the increased ionic strength of the
medium, due to addition of greater volumes of NaOH solution in order to
increase the pH. Electrostatic interactions between the charged groups on the
PMLA-Ad chains are thus partially screened, leading to contraction of the coils.

Viscoelastic properties.

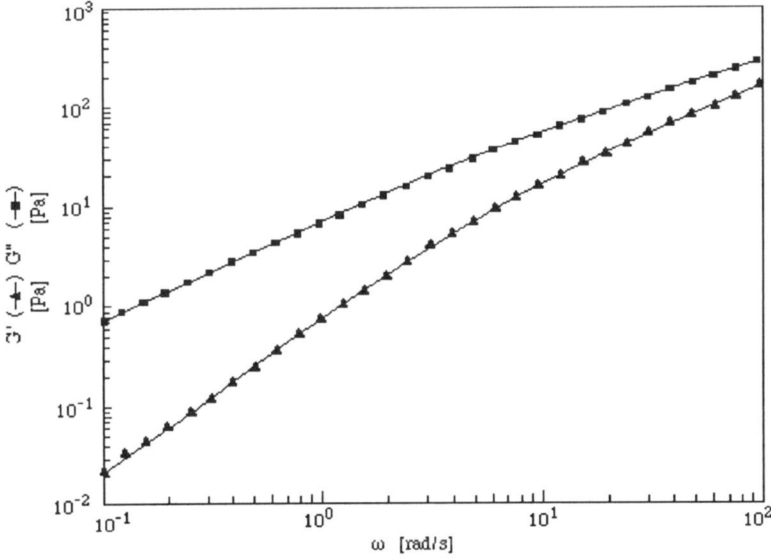

***Figure 15**. G' and G" as a function of frequency (w) for PMLA4Ad / β-CD/EP,
40/60, pH 4, c_t=80 g/l.*

78

Depending on the polymer pairs studied, two different kinds of viscoelastic behaviors are observed corresponding to media with unentangled aggregates at any concentrations and media in which a transition to percolating structures is observed as the concentration is increased.

When N_H is lower than 10, all the mixtures studied show a viscoelastic liquid behavior, with G' lower than G" over the frequency range (10^{-1} to 10^2 rad/s), and independent of the concentration. This is illustrated in Figure 15 for a mixture of PMLA4Ad (N_H=5) with β-CD/EP HM, at a total concentration of 80 g/l. The same qualitative features are observed with star PEO-Ad of N_H=3, 4, 8 *(7)*. Instead of generating a network, the inclusion complex interaction generates only aggregates of finite size.

Different behavior is obtained with the PMLA chains of higher grafting density. Figure 16 shows weak physical gel behavior for PMLA7Ad (N_H=42) mixed with β-CD/EP LM at a concentration of 30 g/l: G' is higher than G" over the frequency range (10^{-1} to 10^2 rad/s). At lower concentrations (lower than the overlap concentration of the chains), liquid behavior is obtained as in Figure 15. Experiments with dextran-Ad and dextran-alkyl show also the formation of reversible networks for N_H=70 at concentrations higher than overlap concentrations.

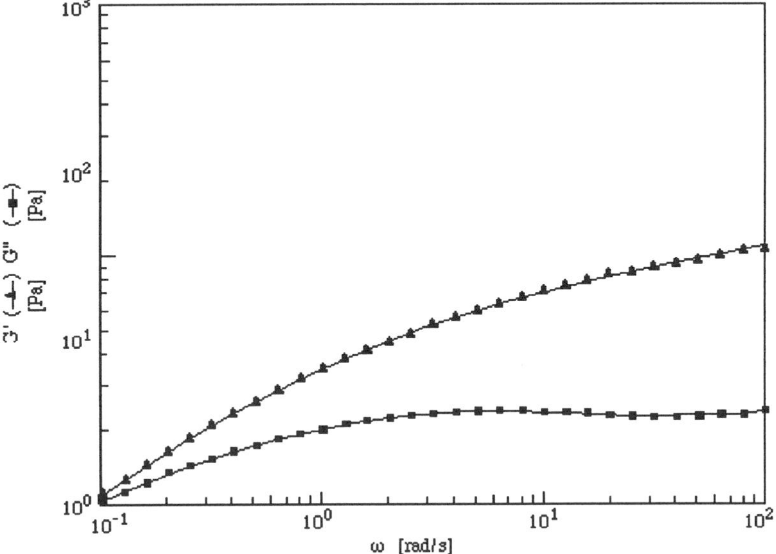

Figure 16. G' and G" as a function of frequency (w) for PMLA7Ad / β-CD/EP HM , 40/60, pH 4, c_t=30 g/l.

The viscoelastic properties of the mixtures appear to be controlled by the number of hydrophobic groups per chain N_H (the number of cavities per β-cyclodextrin polymer N_{CD} being higher than 10). For N_H comprised between 3 and 10, the mixtures show very good thickening properties but, in contrast to what is observed with associative polymers *(12)*, the viscosity is only weakly influenced by the shear rates. This can be explained by the structures of the aggregates which only grow to a limiting size. At higher N_H (the limit lies between 10 and 40 since no experiments have been performed in the intermediate range), the association between the two polymers can lead to weak physical gels or reversible networks. An association mechanism can be deduced from these viscoelastic properties: it seems that a high proportion of the interaction centers between the host and guest polymers are non-active junctions, i.e. they link less than 3 molecules. The density of active junctions, which is expected to be an increasing function of N_H , should reach a percolation threshold when N_H is above a critical value lying between 10 an 40. A closed association model, similar to that proposed for the association of alkyl end-capped PEO *(23, 24)*, where the individual chains associate in well defined structure can be considered in the case of low N_H. At higher N_H , the structures of the aggregates can be described by an open association model *(25)* since an increasing number of active junctions are involved.

Conclusion

The investigations show that supra"macromolecular" structures can be generated by mixing an amphiphilic polymer with a β-cyclodextrin polymer in aqueous solutions. The driving interaction mechanism is formation of inclusion complexes involving hydrophobic moieties of the amphiphilic copolymer and cavities of the β-cyclodextrin polymer. Depending on the interaction strength between the two polymers, different physical phenomena are observed:

a) aggregation in solution, leading in some cases to transient percolating structures, and
b) associative phase separations.

Use of amphiphilic copolymers of different architectures and chemical natures, and of β-cyclodextrin polymers of different molecular weights has demonstrated the role of the main parameters controlling the associations: N_H (number of hydrophobic groups per amphiphilic chains), and N_{CD} (number of cavities per β-cyclodextrin polymer).

In the case of neutral systems and for N_{CD} larger than 10, associative phase separation occurs for N_H greater than or equal to 3 whereas reversible networks are obtained when N_H is higher than a limit lying between 10 and 40. Associative phase separation is not observed when polyelectrolytes are used, owing to the high solubilities of these compounds.

Despite the clear influence of the parameters N_H and N_{CD} on the association properties, the influence of the other microstructural parameters, such as molecular weights, the nature of the hydrophobic moiety and hydrophobic molar ratio must also to be taken into account in order to understand completely the mechanisms of interactions. For instance the mechanism of autoassociation of the hydrophobic moieties, whose interaction strength is related to the hydrophobic molar ratio, competes with the mechanism of inclusion complexes between the polymer pair. On the other hand, the cooperativity of the inclusion complex interaction is also related to the hydrophobic molar ratio and to the architecture of the amphiphilic copolymer.

In view of the potential applications, this family of polymer systems combines two different kinds of properties:

a) the same properties as for associating polymers *(26),* i.e. the thickening properties which are a consequence of the aggregation phenomenon and the associative phase separations occuring over a range of (N_H, N_{CD}) values. Design of media with desired properties can be achieved by choosing the parameters (N_H, N_{CD}) and the polymer architectures, from viscoelastic liquids to reversible networks. For instance, for N_H comprised between 3 and 10, mixtures with high viscosities (more than 1 Pa.s) but low sensitivity to shear rates can be obtained, since aggregates of limiting size are formed in the medium.

b) specific binding properties due to the presence of β-cyclodextrin units. A high proportion of the β-cyclodextrin units in the medium are available for complexation with an additional guest without disrupting the aggregates *(27).* The guest can be a drug, as a large variety of drugs are known to make stable inclusion complexes with the β-cyclodextrin compounds *(2).* Combination of these two properties makes these systems very good candidates for drug delivery.

References

1. Wenz, G. ; *Angew. Chem* **1994.**, *106*, 851 .
2. Saenger, W.; *Angew. Chem. Int. Ed. Engl.* **1980**, *19*, 344-362.
3. Allegre, M.; Deratani, A.; *Agro-Food-Industry Hi-Tech* **1994**, *1*, 9-17.
4. Harada, A.; *Coord. Chem. Rev.* **1996**, *148*,115.
5. Zhang, H.; Hogen-Esch, T. E. ; Boschet, F.; Margaillan, A.; *Langmuir* **1998**, *14*, 4972-4977.
6. Amiel, C.; Sebille, B.; *J. Inclus. Phenomena* **1996**, 25, 61.

7. Amiel, C.; Sebille, B.; *Advances in Colloid and Interface Science* **1999**, *79*, 105.

8. Gosselet, N. M.; Borie, C.; Amiel, C.; Sebille, B.; *Journal of Dispersion Science and Technology* **1998**, *19*, 805.

9. a) Moine, L.; Cammas, S.; Amiel, C.; Renard, E.; Sebille, B.; Guerin, P.; *Macromol. Symp.* 1998, *130*, 45.
b) Moine, L.; Amiel, C.; Brown, W.; submitted to Langmuir.

10. Weickenmeier, M.; Wenz, G.; Huff, J.; *Macromol. Rapid Commun* **1997.**, *18*, 1117.

11. Eftink, M. R.; Andy, M. L. ; Bystrom, K.; Perlmutter, H. D. ; Kristol, D. S.; *J. Am. Chem. Soc.* **1989**, *111*, 6765.

12. *Hydrophilic Polymers: Performance with Environmental Acceptance*, Glass, J. E., Ed., ACS Adv. Chem. Ser. 248, American Chemical Society: Washington DC, 1996.

13. Park, J. W.; Song, H. J.; *J. Phys. Chem.* **1989**, *93*, 6454.

14. Moine, L.; Cammas, S.; Amiel, C.; Guerin, P.; Sebille, B.; *Polymer communications* **1997**, *38*, 12.

15. Renard, E.; Deratani, A.; Volet, G.; Sebille, B.; *Eur. Polym. J.* **1997**, *33*, 49.

16. Renard, E.; Barnathan, G.; Deratani, A.; Sebille, B.;,*Macromol. Symp.* **1997**, 122, 229.

17. Albertsson, P. A.; *Partition of Cell Particles and Macromolecules*, 3rd ed., Wiley & Sons: New York NY, 1986.

18. Amiel, C.; David, C.; Renard, E.; Sebille, B.; *Polymer Preprint* **1999**, *40 (2)*, 207.

19. Piculell, L; Lindman, B.; *Adv. Colloid Interface Sci.* **1992**, *41*, 149.

20. Neel, J.; Sebille, B.; *C. R. A. S.* **1960**, 250, 1052.

21. Sandier, A., Brown, W., Mays, H.; Amiel, C.; *Langmuir* **2000,** *16,* 1634.

22. Xu, B.; Li, L.; Zhang, K.; Macdonald, P. M.; Winnik, M. A.; *Langmuir* **1997**,13, 6896.

23. Annable, T.; Buscall, R.; Ettelaie, R.; Whittlestone, D.; *J. Rheol.* **1993**, *37 (4),* 695.

24. Yekta, A.; Xu, B.; Duhamel, J.; Adiwidjaja, H.; Winnik, M.; *Macromolecules* **1995**, *28*, 956.

25. Maechling-Strasser, C.; François, J.; Clouet, F.; Tripette, C.; *Polymer* **1992**, *33*, 1021.

26. Rubinstein, M.; Dobrynin, A. V.; *TRIP* **1997**,*5 (6),* 18.

27. Amiel, C.; David, C.; to be published.

Chapter 5

Evidence for Photoresponsive Cross-links in Solutions of Azobenzene Modified Amphiphilic Polymers

Iolanda Porcar[1], Philippe Sergot[2] and Christophe Tribet[1]

[1] Laboratoire de Physico-Chimie des Polymères
[2] Laboratoire de Physico-Chimie Structurale et Macromoleculaire
UMR CNRS 7615, ESPCI, Université Paris VI,
10 rue Vauquelin, 75005 Paris, France.

A series of hydrophobically modified polyacrylates grafted by octadecylamine and/or azobenzene groups were synthesized to compare macromolecules having the same length and varying percentage of hydrophobic comonomer groups. As expected for associating polymers, the viscosity of these modified macromolecules in pure water dramatically increased above a critical polymer concentration. Similar enhancement of viscosity was achieved at lower polymer concentration by supplementation with bovine serum albumin or sodium dodecylsulfate. These properties are likely to be due to the formation of reversible cross-links. The sensitivity to irradiation by light of these viscous solutions was studied by viscometry and spectroscopy. In all the mixtures, the azobenzene groups underwent rapid and reversible photoconversion from the *trans* to the *cis* isomer upon irradiation by a near-UV light. The viscosity of these samples was sensitive to the irradiation when inter-polymer association takes place. On the contrary, this coupling between the photo-isomerisation of the grafted azobenzene and the viscosity was not observed in the absence of cross-links between the chains.

A mechanism for the origin of the photosensitivity involving cross-links *via* protein or micellar surfactant "connectors", and independent on the extension of the polymer chain, is proposed.

Introduction

The self-association of hydrophobically modified polymers has been shown to provide an efficient control of the flow of aqueous solutions. (*1, 2, 3*) In the conditions of interpolymer association, dramatic changes in the viscosity -or in the elastic character- of the solutions can occur in response to a change in the polymer conformation. Currently, research is being conducted on the modulation of the properties of such polymers by an external stimulus such as a variation of pH, temperature, salt concentration...(*1,4, 5*). The presence of a few mol percent of responsive groups (such as weak acids for pH-responsiveness) or sensitive chains (e.g. chains exhibiting a lower critical solution temperature LCST) along the backbone seems sufficient to modify the self-associating propensity or the conformation of those polymers. For practical developments of highly viscous (>1 Pa.s) or gelified responsive fluids, however, a supplementation with salt or other additives is fraught with technical difficulties. Except in the case of temperature-sensitive systems, such stimuli seem unlikely to propagate rapidly through thick materials.

The properties of polymers have also been triggered by light due to the presence of chromophores in the macromolecules (*6, 7*). Provided that the light could pass through a thick sample, the photo-responsiveness could provide an excellent alternative to rapidly switch the viscosity of a solution. In recent years, several photo-stimulations of physical or chemical properties have been examined although not yet with self-associating chains. Volume changes (*8, 9*), solubility (*10*), chain extension (*11*), enzymatic activity (*12*) have been reversibly switched between two states upon irradiation at two different wavelengths. In such systems, the initial change induced by light (for instance the *trans* to *cis* transconversion of double bonds) is located at molecular level and it needs, therefore, to be amplified up to a supramolecular or macroscopic scale. In the known polymer-based systems, this amplification was achieved by the critical transition between an extended and a collapsed state of the modified macromolecules, resulting either in the precipitation from a solution or in a volume change of a chemical gel. To our knowledge, the sol-gel transition of reversible polymer networks has not been exploited, although cross-links that are sensitive to the absorption of a photon might be very efficient in triggering a reversible gelation. Among advantages of such gels, the extremely low concentration of cross-links would reduce considerably the concentration of chromophores (ideally one-or two per cross-link) down to a value compatible with a low absorbance of thick samples. Decreasing the chromophore content appears crucial owing to the high absorbance of known examples (> 5 OD units per millimeter (*13, 14*) that may limit their use to cases of well stirred solutions or in thin films.

Here we describe reversible photomodulation of the viscosity in various systems containing hydrophobically modified polymers soluble in water. The associative properties of different modified poly acrylic acid (in their sodium form), comprised of varying fractions of azobenzene chromophore and octadecyl hydrophobic groups, were compared. The use of micelles that can connect such macromolecules by hydrophobic association was also investigated as another route to form cross-links in these systems. Aiming at an amplification of the response to irradiation, the possible specific recognition of the chromophores isomers by a protein such as bovine serum albumin was also experimented.

Material and Methods

Materials

Bovine serum albumin, BSA, (fraction V, 95% of purity) was purchased from Sigma Chemical Co. (St. Louis, MO). Sodium dodecyl sulfate, SDS, as >99% pure, was supplied by Poly Labo (France). Proteins and surfactant were used without further purification. Poly(acrylic acid) (PAA) was obtained from Polysciences Inc. (Warrington, PA) and its average molecular weight given by the supplier was 150,000. Gel Permeation Chromatography measurements performed in $LiNO_3$ solution gave a number average Mw of 130,000 and a polydispersity index of 4 for this sodium polyacrylate precursor. The hydrophobically modified polymers were synthesized according to a reaction described below, by grafting octadecylamine and/or the azobenzene group at random along the backbone of the precursor. All the polymers have the same polymerization degree and they were used in the fully neutralized sodium salt form.

Synthesis of polymers

Sets of modified poly(acrylic acid) were synthesized from a single precursor of number average molar mass 130,000 g/mol using a method (in N-methylpyrrolidone) previously described for the grafting of n-alkylamines (*15*). The acid groups along the backbone were grafted at random with octadecylamine in the presence of one equivalent of the coupling agent dicyclohexylcarbodiimide (DCCI) (*15*). 4-phenylazoaniline was subsequently introduced in a 5-fold excess as compared to the desired grafting rate, in the presence of both DCCI and N-hydroxy benzotriazole (HOBt) at a 1:1 molar ratio with the acrylic groups. The reaction bath was maintained at 50°C for at least 2h before being cooled in an ice bath and filtered. The soluble polymer was neutralized and precipitated by a 2-fold excess of sodium methanoate in methanol. It was finally purified by two precipitations in ethyl alcohol from its water solution. The resulting grafting rates were determined by ^{1}H- and ^{13}C-NMR spectroscopy, and by measurement of the absorbance of a solution of polymer in water (determination of the total azobenzene content). The chemical structure of these associative polymers is shown in Figure 1. Their compositions are given in Table 1 with uncertainties that correspond to the difference between the values obtained from the different techniques. The code adopted was as follows: 150-*y*C18-*z azo* is a polymer sample containing *y* mol % of octadecyl hydrophobic units and *z* mol % of azobenzene groups.

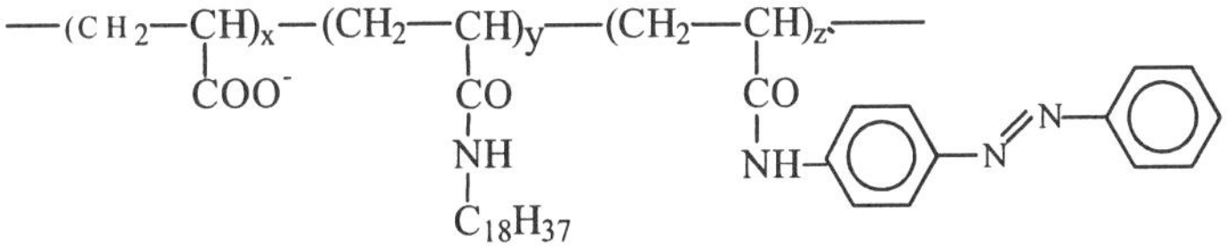

Figure 1. Structure of the azo-modified polymers.

Table 1: Examples of Copolymers and Terpolymers Synthesized

Code	mol% of octadecyl	mol% of azobenzene
150-1.1azo	0	1.1±0.1
150-4.5azo	0	4.5±0.3
150-1C18-0.5azo	1.1±0.1	0.5±0.1
150-1C18-1.2azo	0.9±0.1	1.2±0.1
150-1C18-4.5azo	1.3±0.2	4.5±0.2
150-3C18-0.5azo	3.0±0.2	0.5±0.1

Sample preparation

Stock solutions of polymer in water were routinely prepared under magnetic stirring at least 24 h before use and kept in the dark to ensure that all the azobenzene residues were in the trans form. Protein was dissolved in water 2 h before the preparation of the polymer/protein mixture to avoid bacterial growth. Mixtures were prepared at least 48 h before the measurement and kept in the dark at room temperature.

Instrumentation

Irradiations were carried out with a 350-W mercury arc lamp (Eurosep Instruments, France) combined with a ORIEL interference filter. Figure 2 shows the experimental setup for sample irradiation depending on the technique used to carry out the measurements.

UV Absorption spectra were performed on a Hewlett Packard 8453 UV-Visible Spectrophotometer (Waldbronn, Germany) in the wavelength range 250-700 nm. The system was equipped with a cell stirring module. The sample, immersed in a quartz cell with a path length of 1 cm, was irradiated vertically at 365 nm for at least 10 min (see A on Figure 2) while the spectra were being recorded horizontally at different times.

Rheology. For samples of low viscosity (<0.1 Pa.s), the viscosity measurements were performed with a controlled rate Contraves Low-Shear 30 rheometer in a Couette cell at 25 °C. The shear rate was decreased step by step from 40-100 s^{-1} down to 1 s^{-1} covering a range that encompassed the Newtonian plateau. The samples exposed to light were vertically irradiated for 20 minutes in a cold water bath to prevent evaporation before being loaded. The complete set of measurements, including the thermal equilibration, did not exceed 20 min. Samples exhibiting a higher viscosity were characterized in a home-made transparent Couette-cell fitted to a Haake RS-150 rheometer (Rheo, France). The characteristics of this particular coaxial cylinder sensor system were the following: the radii of the cup and the rotor were 16 and 15 mm, respectively, with a rotor height of 15 mm. Thus, the gap of this cell was 1 mm, and the external wall thickness 0.5 mm (which minimize the adsorption of light). The bottom of the inner cylinder had a convex shape that entrapped a bubble of air. This geometry minimized the shear of the sample located below the bubble which was not irradiated homogeneously (see B on Figure 2). Thus, the measured data corresponded to the viscosity of the sample in the annular gap, neglecting other contributions. The coaxial cylinder viscometer was calibrated by using standard siloxane oils (Brookfield Engineering Laboratories, Stoughton, MA) of viscosities varying from 4.5 to 4790 mPa.s. The calibration was accurate to within 2 % of the oil stated values. The reliable range of the shear rate to get the correct viscosity was also evaluated (10-500 s^{-1} for viscosity samples around 1 Pa.s). The measurements were performed at a constant shear rate (10 s^{-1}) high enough to ensure the rapid homogenization of the sample. In these type of experiments, the samples were horizontally irradiated (see Figure 2) and the influence of the irradiation on their viscosity was followed *in-situ* (the transmission of light toward the cell wall was > 90 % above 365 nm).

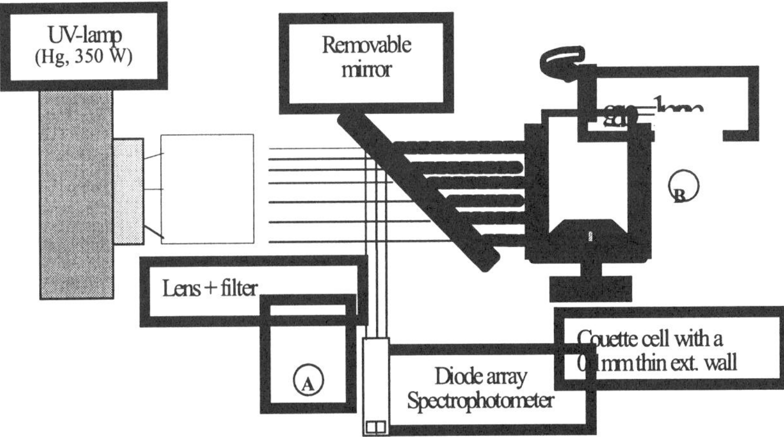

Figure 2. Experimental setup for sample irradiation.

Results

Aqueous Polymer Solutions

Photochromic Properties

The UV-visible absorption spectra of the trans isomer of the azobenzene molecule in solution is characterized by the presence of two bands centered around 320 and 440 nm. These are associated with the electronic transitions ($\pi \rightarrow \pi^*$ or $n \rightarrow \pi^*$) of the azobenzene chromophores.(*16, 17*) Irradiation at 365 these conditions, a polymer solution underwent alternatively increases and decreases of absorbance at 347 nm, as shown in Figure 4. The two processes were fitted by a monoexponential function, either decrease and increase depending on the irradiation light in accordance with first-order kinetics, and in a similar way as reported by Morishima *et al.* (*18*) From the absorbance at 347 nm it was possible to estimate the fraction of trans isomers depending on irradiation wavelength and time. Assuming a cis-to-trans extinction coefficients ratio equals 0.055, as determined by Morishima *et al.* (*18*) it was found that after a few minutes of exposure at 365 nm, the stationary state corresponded to 20 % of the chromophore in their trans form. Upon irradiation at 436 nm, a final fraction of 77 % of trans isomer was recovered.

Rheological Properties

Owing to their similarity with known thickeners (*3*), the hydrophobically modified poly(acrylic acid)s were expected to increase significantly the viscosity of a water solution. Typical viscosity results of hydrophobically modified and precursor poly(acrylic acid) in water as a function of their concentration are given in Figure 5. In all cases, the viscosity increased with the polymer concentration. However, the slope of the curves was much higher in the case of polymers containing the C18 groups. Both the precursor and the copolymers containing only the azobenzene pendant groups, but not the C18 chains, exhibited classic polyelectrolyte behavior: at low polymer concentration, a sharp increase in the viscosity was observed due to the electrostatic repulsions between charged groups along the backbone. By increasing the polymer concentration, the progressive self-screening of the electrostatic interactions led to a slower increase of the viscosity. At a fixed concentration the polymer modified only with the azobenzene group showed a decrease of viscosity with the modification rate. This might reflect the propensity of the azo groups for self-association (inducing a collapse of the chain). Polymers containing C18 alkyl chains along the backbone showed a markedly different behavior. The curves observed were typical of associating polymers: beyond a threshold concentration (which depends on the degree and the type of the modification), the viscosity of the solution increased dramatically, due to interpolymer hydrophobic associations, and gelation may occur at high polymer concentrations. It was found that the introduction of the azobenzene groups did not modify dramatically this self-association threshold, although it shifted the viscosity toward lower values as compared to the 150-1C18. Owing to the high sensitivity of the viscosity to the modification with C18, small fluctuations in the octadecyl alkyl content of the terpolymers studied might explain those shifts. Altogether these measurements nm of an azobenzene solution has been shown to induce the *trans-cis* isomerization of the N=N double bond. This transconversion is known to affect the UV spectrum by shifting the bands towards lower wavelengths and decreasing the peak at 320 nm. Similar variations also attributed to a photo-induced isomerisation in the adsorption spectra have been reported for water based polymeric systems containing azobenzene moieties.(*18, 19*) All the azo-modified polymers given in Table 1 exhibited this property. The only difference was the shift of the maximum of adsorption from 320 to 347 nm, due to the grafting through an amide bond. As shown in Figure 3, the absorbance at 347 nm of a dark adapted polymer solution progressively decreased with irradiation time, reaching a stationary state in about 2 min. It followed first-order kinetics.

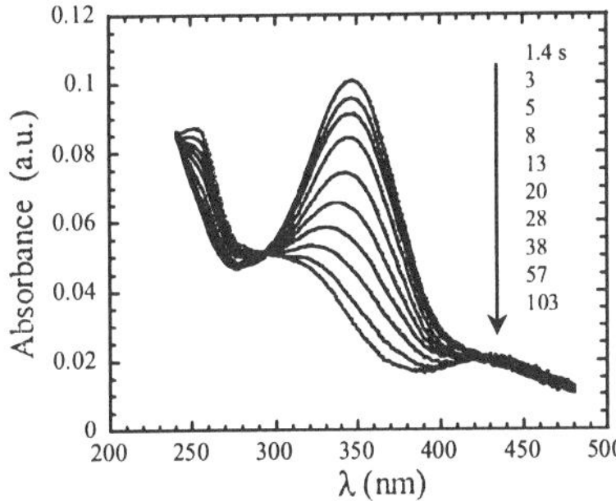

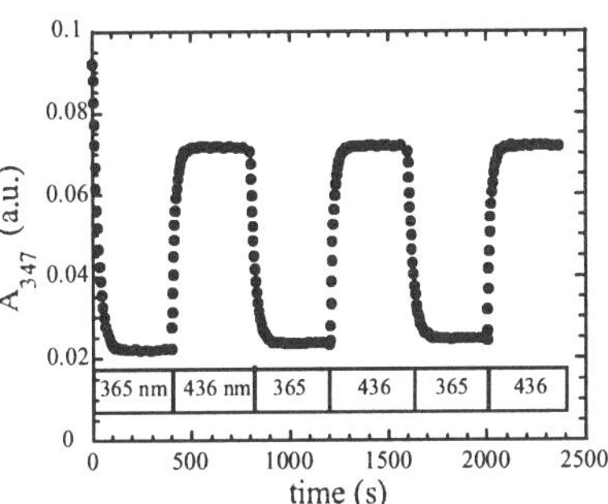

<table>
<tr><td>

Figure 3. *Changes in the UV-visible spectra of the azobenzene residues in a 0.01wt% 150-1C18-0.5azo solution in water upon irradiation at a wavelength of 365 nm. The irradiation times are indicated in the figure.*

</td><td>

Figure 4. *UV absorption at 347 nm of a 0.01 wt% 150-1C18-0.5azo solution as a function of the total irradiation time at 365 and 436 nm.*

</td></tr>
</table>

Two tight isosbestic points (at 296 and 426 nm) were observed, consistent with an equilibrium between two species, namely the *cis* and *trans* isomers of the azobenzene group. This also indicated the absence of polymer photodegradation during the irradiation. Upon a subsequent irradiation at 436 nm of this sample, the absorbance at 347 nm progressively increased, indicating that, under these conditions, the equilibrium between the cis and the trans isomers was shifted in favor of the trans form. Upon exposure to the 436 nm light, the absorbance reached a stationary value within about 1 min but it did not increase up to the initial value of the dark-adapted sample. Cycles were obtained by switching regularly the wavelength of irradiation between 365 nm and 436 nm. Under established that self-associations took place only with terpolymers and beyond a concentration of 3 % for polymers containing about 1 mol % of C18 groups.

The plot of the viscosity *versus* shear rate of a terpolymer solution at a given concentration (Figure 6) revealed its newtonian behavior. This indicated the possible absence of strong interpolymer association or for a very short life-time of possible interpolymer connections. Upon irradiation at 365 nm (for 20 min) only the polymer solutions of high concentrations exhibited a photoresponse. The viscosity was found to decrease at best by 25 % under these conditions. In addition, the viscosity of solutions of copolymers comprising only the azo-chromophore but without C18 pendant groups, were found unmodified by light over the same range of concentration (not shown). Using only polymers in water, it seemed difficult to improve the amplitude of the photoresponsiveness without reaching high polymer concentrations that would increase the absorbance of the sample. The supplementation of the samples with additives that strengthen the thickening properties was therefore studied as a means to modify the viscosity at both low polymer and chromophore content.

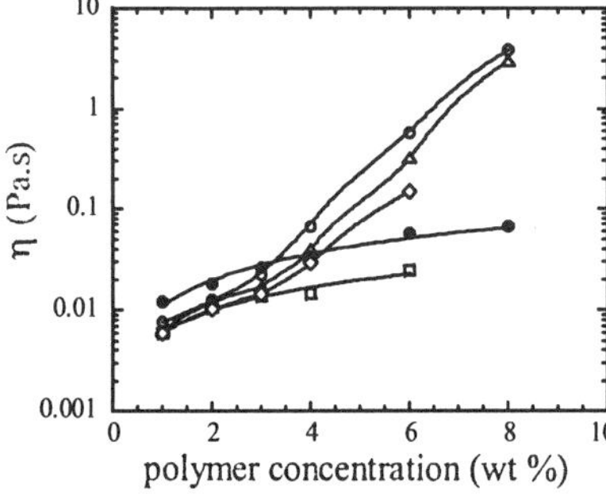

Figure 5. Variation of the viscosity as a function of the polymer concentration for (●)PAA (O) 150-1C18 (□) 150 -4.5azo (Δ) 150-1C18-4azo and (◊)150 -1C18-0.5azo. (Shear rate: 0.06 s⁻¹)

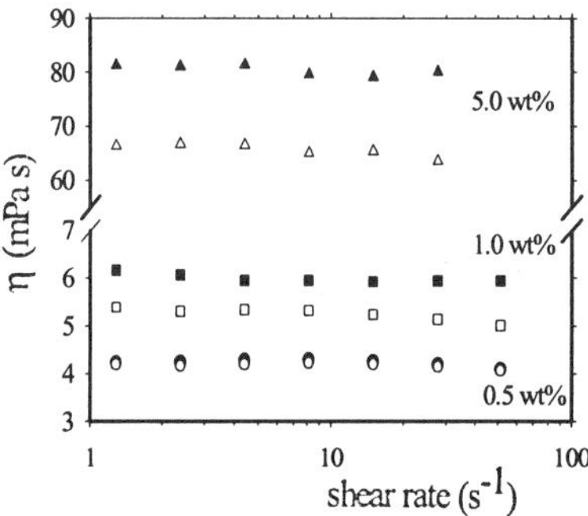

Figure 6. Variation of the viscosity of 150-1C18-0.5azo at different concentrations in water as a function of the shear rate. Filled symbols stand for dark-adapted samples and open symbols for those irradiated at 365 nm.

Polymer/Protein and Polymer/Surfactant Solutions.

Evidence for Complexation

Hydrophobically modified poly(acrylic acid)s are known to associate and form gels with either proteins (*20*) or surfactants. (*21*) Similarly the polymers containing azobenzene moieties were also able to form complexes with the Bovine Serum Albumin (BSA) in dilute solution. (*22*) The binding isotherms on dark adapted polymers depicted in Figure 7 showed that the terpolymers can bind several proteins per chain. In the absence of C18 along the backbone, the azo-modified copolymers were also found to form complexes, although involving a much smaller number of proteins. Azobenzene might nevertheless play an important role in the association with proteins because a small fraction of dye as 0.5 mol % grafted in terpolymers with 3 % of alkyl groups was found to enhance the number of bound BSA per acrylate unit dramatically. C18 and azobenzene pendant groups were both found to bind the BSA with a strong synergism.

Thickening Effect

Due to this association, the increase in viscosity or the gelation was expected to be obtained by a supplementation of a polymer solution with either BSA or sodium dodecylsulfate (SDS) at a concentration above its cmc (critical micelle concentration). A viscosity enhancement was obtained (Figure 8) when a terpolymer solution was supplemented with the protein BSA. For instance, the viscosity of a mixture containing 0.5 wt % of 150-1C18-0.5azo + 0.08 wt % of BSA was close to that of the pure polymer at 5 wt % (see Figure 6). Such a behavior, however, has not been observed in the case of copolymers without C18 pendant groups. Although the BSA is able to associate azo-modified copolymers in dilute solution (see Figure 7), the viscosity of the 150-4.5azo in the presence of increasing amounts of protein remained almost constant (Figure 8). Correlatively, the viscosity of the copolymer/BSA mixtures was not sensitive to light whereas terpolymer/BSA mixtures exhibited photo sensitivity at a polymer concentration of 0.5 wt %.

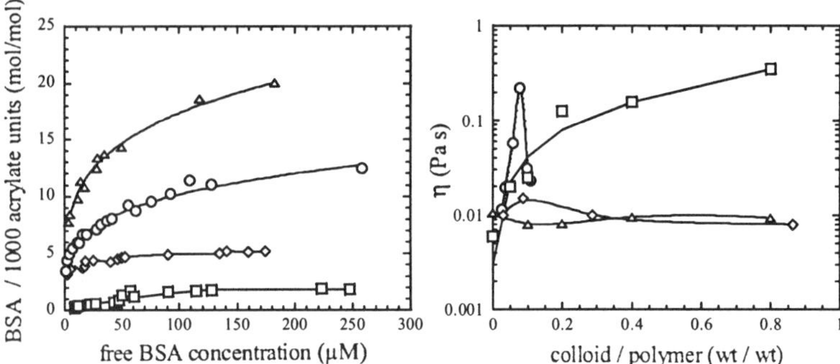

Figure 7. Isotherms of the association between BSA and different modified polymers in 30 mM borate buffer pH 9.2. Symbols stand for the polymers: (O) 150-3C18, (□) 150-1.2azo, (Δ) 150-3C18-0.5azo, and (◊)150-1C18-0.5azo.

Figure 8. Variation of the viscosity as a function of the colloid/polymer ratio. Symbols stand for the systems (O) SDS +2 wt% 150-1C18-0.5azo, (◊) SDS +1wt% 150-1C18, (□) BSA +1 wt% 150-1C18-0.5azo, and (Δ) BSA +1 wt% 150-4.5azo. (Shear rate:10 s⁻¹)

Figure 8 also shows that the presence of SDS in a solution of azo-modified polymers enhanced the viscosity. In this case, the rheological behavior was different to that observed for polymer systems containing BSA but similar to that reported for C18 modified poly(acrylic acid)s. (*23*) Upon addition of surfactant in an azo-modified polymer solution, the viscosity of the mixture increased sharply up to a threshold SDS concentration, beyond which the viscosity decreased. It has been proposed for similar systems that the formation of cross-links between polymer chains, involving mixed micelles, is responsible for such a peculiar rheological behavior. (*21*)

Photoisomerization of the Mixtures

Upon irradiation at 365 nm of a solution of the azo-modified polymers in the presence of either SDS or BSA, the azobenzene chromophores underwent a similar *trans* to *cis* photoisomerization process. Correspondingly shape, intensity and position of the UV adsorption bands were not modified (Figure 9). Indeed the UV spectra recorded at different irradiation times were characterized by the presence of two identical isosbestic points at 296 and 426 nm, thus confirming the presence of *cis* and *trans* isomers of the azobenzene group, and similar photoisomerization properties than those observed with the polymer alone (Figure 3).

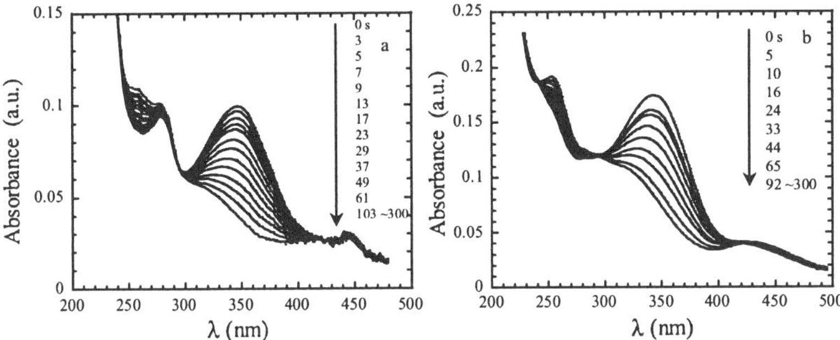

Figure 9. *Influence of an irradiation at 365 nm on the UV spectra of 150-1C18-0.5azo samples in aqueous solution in the presence of (a) 0.06 wt% BSA (polymer at 0.012%) and (b) 0.21 wt% SDS (1cmc) (polymer at 0.025%) upon irradiation at a wavelength of 365 nm. Irradiation times are in the figure.*

Photoisomerization kinetics was investigated in aqueous polymer solutions containing BSA or SDS having an absorbance at the irradiation wavelength lower than 0.2. Experiments were performed by monitoring the absorbance at 347 nm during successive irradiations at either 365 or 436 nm (Figure 10). The presence of the colloidal additives did not affect the behavior as compared to the solutions of polymer alone. Upon exposure to a 365 nm light, the variation of the absorbance at 347 nm *versus* the irradiation time matched with a monoexponential decay for all sample, in accordance with first- order kinetics. When irradiated at 436 nm, the absorbance of the samples steadily increased up to a value slightly lower than that of the dark-adapted sample. From the change in the molar extinction coefficient, it was estimated that the stationnary states (plateaus in Figure 10) corresponded to a fraction of trans-azobenzene equal to 30-35% (exposure to 365nm light) or 75-80% (436nm light).

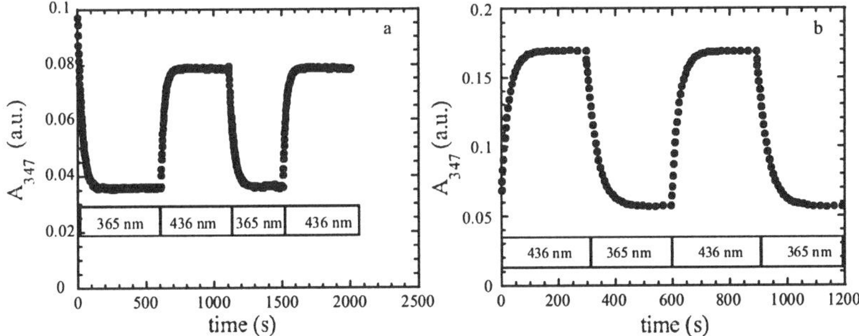

Figure 10. *UV absorption at 347 nm of 150-1C18-0.5azo in the presence of (a) 0.06 wt% BSA (polymer at 0.015 wt%) and (b) 0.021 wt% SDS (polymer at 0.025 wt%) as a function of the total irradiation time at 365 and 436 nm.*

Photoresponsive Viscosity

For viscosity measurements, the polymer/protein or the polymer/SDS mixtures, initially kept in the dark, were loaded in the transparent Couette cell. They were subjected at a constant shear rate of 10 s^{-1} and exposed to sequential irradiations at 365 nm or 436 nm (Figure 11). The initial viscosity measured during the first 100 seconds on the two panels of Figure 11 corresponded to the equilibrium value reached in the dark. When the light was switched on at 365 nm the viscosity dropped, following first-order kinetics. A subsequent irradiation at 436 nm led to a cis-to-trans back isomerization, along with a corresponding rapid increase in the viscosity (within 2 minutes). In addition, after two or three successive 365/436 irradiation cycles, the value of the viscosity at the plateau remained virtually constant.

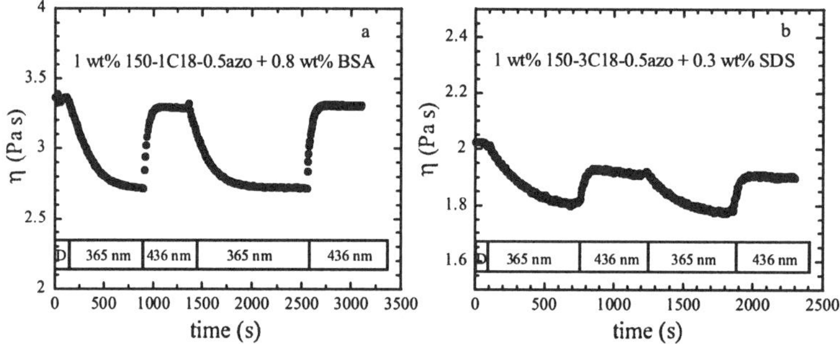

Figure 11. *Viscosity of mixtures containing an azo-modified polymer and (a) BSA or (b) SDS as a function of the total irradiation time at 365 and 436 nm. D stands for the initial period during which the sample is kept in the dark. Shear rate: 10 s^{-1}.*

The variation in viscosity corresponded to a relative value of about 25 % in the best conditions (Figure 11a). Such amplitude, as well as the viscosity itself, were however highly sensitive to the shear rate (increase from 10 to 35 % upon decrease of the shear rate from 100 s^{-1} to 1 s^{-1}). It can be argued that a higher response should be obtained in BSA/polymer mixtures at lower shear rate. The long duration of the experiments at low shear rates has hampered such measurements due to the heating and drying of the samples. At shear rates higher than 1 s^{-1}, the time required for reaching the stationary viscosity value was found to be longer than the time measured by spectrophotometry in dilute solutions. For instance, the absorbance decay in a solution of 150-1C18-0.5azo at 0.01 wt % took place in 2-3 min, whereas the same system at a polymer concentration of 1 wt % in the Couette cell needed more than 10 min to achieve a complete viscosity decrease. The different setup carried out for the irradiation depending on the technique used could be responsible for such a difference. Three origins can be proposed to explain it: i/ a difference in the intensity of light due to the adsorption and reflection at the cell walls and on the mirror. ii/ an influence of the concentration of chromophores (specially due to the adsorption of the light which would limit the penetration of photons in the sample). iii/ an influence of the concentration of chains, 1 wt % being above some threshold concentration (such as C*) that induces a marked change in the dynamics of a conformational reorganization. A more rapid photoresponsive ness at 436 nm (extinction coefficient is much lower than that at 365 nm) and the sensitivity of the decay time of the viscosity to the polymer concentration in the range 0.5-2 wt % (not shown) are points in favor of a major role of the origin ii/. Further works are in progress to achieve a more accurate determination of the parameters that control the kinetics of the photoresponsiveness.

Discussion

Reversibility

For applications such as the regulation of flow or the repetitive release of entrapped materials, the reversibility of the phenomenon is required. The azobenzenes are known to undergo several transconversions without significant degradation. (*18, 19*) In the presence of additives that may exhibit some reactivity, it was nevertheless important to check the absence of side reactions. The tight isosbestic points, observed in all the systems including those with proteins, indicated that one irradiation for at least 20 min did not degrade the chromophores. In addition, cycling between the states reached upon successive irradiations (at 365 and 436 nm) did not induce any significant distortion of the UV-visible spectrum. Reversibility of the transconversion of the N=N bond confirmed this. Variations of the viscosity during irradiations gave additional information on the reversibility of the whole process. Owing to the high sensitivity of the viscosity to the presence of micelles or proteins, any degradation of these components would have resulted in an irreversible change of the rheological properties and a drift of the viscosity upon a long exposure to UV light. On the contrary, the final plateaus of the viscosity did not show any significant drift upon an irradiation for at least 20 min, whereas the initial variation, due to the cis-trans isomerisation, took place within 2 min. These results confirm the absence of any significant influence of a degradation on the coupling between the irradiation and the rheological properties.

Tentative mechanism.

To our knowledge, all the previous systems dealing with photoresponsive polymers in solution have been devised on the basis of the change upon irradiation of the solvent-monomer *versus* monomer-monomer interaction. The photo-controlled collapse/extension of the chains, possibly accompanied by a dissolution/precipitation scheme, has been proposed as the general origin of the coupling between a macroscopic effect and the photo-isomerization. A different mechanism was suspected to be at work here. In the absence of interpolymer association, the rheological response to light vanished both with the 150-1C18-0.5azo below 0.5 wt % and with all the non-associating copolymers up to 8 wt % (those copolymers that were grafted only with azobenzene, from 0.5 % to 7 mol %, and did not contain C18 groups). In other terms, it was not possible to detect any significant change in the chain extension upon irradiation both in dilute solution of a terpolymer or in the absence of C18 hydrophobic groups along the backbone. The photoresponsiveness emerged at the same concentration of 150-1C18-0.5azo, by a supplementation with colloids that form cross-links with the polymers, resulting also in a dramatic thickening. Polymer self-association or cross-linking by micelles or by protein were all found to be efficient in coupling the rheological properties with the isomerization of the chromophores. The present observations were consistent with such a coupling taking place through a displacement of polymer-polymer association (i.e. the formation of cross-links, the stability of which depends on the azobenzene's conformation) without significant change in the conformation of a single chain. It seemed reasonable that the structure of such cross-links was comprised of just a few bound azobenzene groups (tens of links between 2 polymer chains are unlikely to be broken by switching between 20 and 77 % of the trans isomers). Consistent with the participation of a fixed number of azobenzene species per cross-link, the polymers of varying fraction of grafted azobenzene (in the range 0.5 – 3 mol %) did not differ significantly in terms of the relative amplitude of the photo-response in the presence of BSA. Relative variations in viscosity of about 20-30 % were also found to be essentially independent of the concentration of the macromolecules. A structure of cross-links depending only on the protein "connector" and largely independent of the structure of the polymer could explain these observations. Finally, a tentative mechanism is proposed (Figure 12).

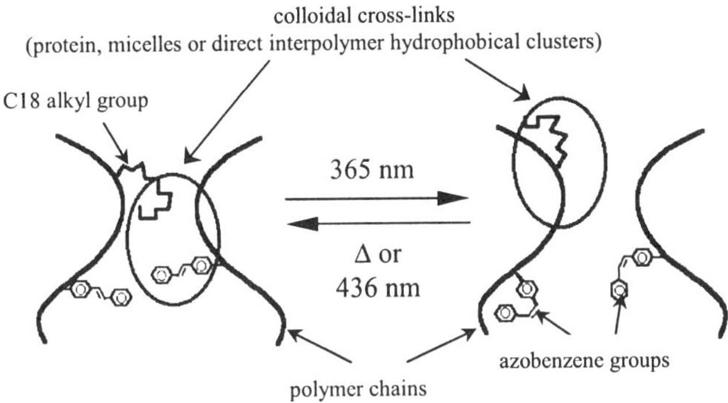

Figure 12. Schematic illustration of the tentative mechanism proposed.

Possible rapid response of thick samples

For practical developments, a chromophore concentration as low as possible would limit the difficulties arising from a high absorbance of the samples, such as heating or inhomogeneities of the flux of photons as a function of depth. The rate of photo-isomerisation being proportional to the intensity of light, the photo-response would be markedly slowed down inside a sample having a high absorbance. In conventional systems, based on the conformational transition of the polymer chains, the ratio of chromophore per monomer unit cannot be decreased without loss of the response. In practice, it has been reported that a fraction as low as 1 mol % can be sufficient (*13*), although most of the systems contained more than 3 mol % of chromophores. (*14*) On the contrary, an extremely low number of cross-links per chain (1 or 2) is sufficient to form gels. Provided that one or two chromophores per cross-link could be efficient in conferring a photo-responsiveness, the minimum concentration of the chromophores in such gels ought to be much lower than in the conventional systems. Among the samples studied in this work, the 1 mm thick systems containing the polymer 150-1C18-0.5azo at a concentration of 0.5 % were shown to respond rapidly to an irradiation at 365 nm, within less than 1 min, due to a maximal absorbance, at 347 nm, lower than 0.6 (per millimeter). For a more accurate estimate of the minimum absorbance that might be reached, it was argued that in concentrated polymer solutions -beyond the critical overlap of the chains- the sol-gel transition is reached at a ratio of 1 cross-link per macromolecule (assuming a functionality of 4). Therefore, only 2 azobenzene species per chains might be sufficient which give, at the overlap concentration C^* of the polymer, a "theoretical" minimum chromophore concentration equal to $2\ C^* / M$ (M being the molar mass of the polymer). With the present polymers, the gelation can take place at a polymer concentration lower than 0.2 %, *(24)* resulting in an estimate of the minimum azobenzene concentration (at 2 hydrophobic groups per chain *i.e.* a grafting rate of 0.14 %) equal to 3.1×10^{-5} mol/L. In this example, the absorbance at 347 nm by a 1 cm thick sample is equal to 0.7. Using our device and at that value of absorbance, the isomerisation through the whole sample can be completed within a fraction of a minute.

Conclusions

A series of hydrophobically modified poly(acrylic acids) grafted by C18 and/or azobenzene groups were synthesized to compare amphiphilic and photosensitive macromolecules having the same length and varying content of pendant groups. Owing to the associating properties of these macromolecules, polymer solutions at low concentration can reach viscosity values of several Pa.s. Upon addition of sodium dodecylsulfate or bovine serum albumin at constant polymer concentration, similar viscosity enhancements were also observed. These expected properties are likely due to the formation of reversible cross-links as already reported in the case of other associating polymers. All those versatile and independent routes for the adjustment of the viscosity (the polymer concentration, hydrophobicity, the presence of micelles or proteins) corresponded to various microscopic states of association between the macromolecules that were compared for their sensitivity to an irradiation by light.

In all mixtures, the azobenzene groups underwent rapid and reversible transconversion (trans to cis isomerization) upon irradiation by a near-UV light. The viscosity of samples was found to be sensitive to the irradiation when the inter-polymer association took place. On the contrary, this coupling between the photo-isomerization of the grafted azobenzene species and viscosity was not observed in the absence of cross-links between the chains. In other terms, an irradiation did not change markedly the extension of the polymer chains. The data collected were consistent with the formation of inter-polymer reversible associations, the stability of which depended on the conformation of the azobenzene groups. Such a mechanism appeared intrinsically different from the origin of photosensitivity reported to date in polymer based systems. Among advantages over the prior systems involving an amplification by the coil-globule or coil-helix transitions of the polymer conformation, significantly lower concentration of the chromophore and rapid responses are expected. The measurement of the dynamic moduli of these samples is in progress aiming at the characterization of cross-links and an assessment of the proposed model of coupling. Future progress is contemplated using more efficient recognition of the chromophore by a specific protein cross-linkers.

Acknowledgments

This project was mainly supported by the CNRS, European program Human Capital and Mobility, TMR Marie Curie grant n° ERBFMBICT983300. We thank B. Cesar for the synthesis of the polymers and G. Ducouret for her technical assistance in the fine performance of the rheology apparatus.

References

1. Magny, B.; Iliopoulos, I.; Audebert, R. *Macromolecular Complexes in Chemistry and Biology*, Dubin, P.; Bock, J.; Davis, M.; Schulz, D.N.; Thies, C. Eds., Springer Verlag, **1994**.
2. Goddard, E.D.; Ananthapadmanabhan, K. P. *Interactions of Surfactants with Polymers and Proteins*, CRC Press: Boca Raton, FL, **1992**.
3. Wang,T.K.; Iliopoulos I.; Audebert R. in *Water Soluble Polymers. Synthesis, Solution Properties and Applications.* Shalaby, S.W.; McCormick, C.L.; Butler, G.B. Eds., ACS Symposium Series, Washington, vol 467, **1991**.
4. Gehrke, S.H.; Agrawal, G.; Yang, M.C. in *Polyelectrolyte Gels*, Harland, R.S.; Prud'homme, R.K. Eds., ACS Symposium Series, Washington **1992**.
5. Hourdet, D.; L'Alloret, F.; Audebert, R. *Polymer* **1994**, *35*, 2625.
6. Eisenback, C. D. *Polymer* **1980**, *21*, 1175.
7. Irie, M.; Hirano, Y.; Hashimoto, S.; Hayashi, K. *Macromolecules* **1981**, *14*, 262.
8. Mamada, A.; Tanaka, T.; Kunwatchakun, D.; Irie, M. *Macromolecules* **1990**, *23*, 1517.
9. Irie, M. in *Advances in Polymer Science, vol. 110,* K. Dusek Ed., Springer Verlag, **1993**.
10. Irie, M.; Tanaka, H. *Macromolecules* **1983**, *16*, 210.
11. Irie, M.; Schnabel, W. *Macromolecules* **1985**, *18*, 394.
12. Willner, I.; Rubin, S. *Angew. Chem. Int. Ed. Engl.* **1996**, *35*, 367.
13. Effing, J. J.; Kwak, J. C. T. *Angew. Chem. Int. Ed. Engl.* **1995**, *34*, 88.
14. Irie, M.; Kunwatchakun, D. *Macromolecules* **1986**, *19*, 2476.

15. Wang, T.K.; Iliopoulos I.; Audebert R. *Polymer Bulletin* **1988**, *20*, 577.

16. Zimmerman, G., Chow, L.Y., Paik, U.J. *J. Am. Chem. Soc.* **1958**, 3528.

17. Bullock, D.J.W.; Cumper, C.W.N.; Vogel, A.I. *J. Chem. Soc.* (London), **1965**, *XLIII*, 5316.

18. Morishima, Y.; Tsuji, M.; Kamachi, M.; Hatada, K. *Macromolecules* **1992**, *25*, 4406.

19. Altomare, A.; Solaro, R.; Angiolini, L.; Caretti, D.; Carlini, C. *Polymer* **1995**, *36*, 3819.

20. Petit, F.; Audebert, R.; Iliopoulos, I. *Colloid Polym. Sci.,* **1995**, *273*, 777.

21. Magny, B.; Iliopoulos, I.; Zana, R.; Audebert, R. *Langmuir* **1994**, *10*, 3180.

22. Porcar, I.; Cottet, H.; Gareil, P.; Tribet, C. *Macromolecules* **1999**, *32*, 3922.

23. Magny, B.; Iliopoulos, I.; Audebert, R.; Piculell, L.; Lindman, B. *Progr. Colloid Polym. Sci.* **1992**, *89*, 118.

24. Borrega, R.; Tribet, C.; Audebert, R. *Macromolecules* **1999**, *32*, 7798.

Chapter 6

Synthesis and Aqueous Solution Behavior of Novel pH Responsive, Zwitterionic Cyclocopolymers

David B. Thomas, R. Scott Armentrout, Charles L. McCormick*

The University of Southern Mississippi, Department of Polymer Science, Hattiesburg, MS 39406

The free radical photopolymerizations of N,N-diallyl-N-methylamine with 3-(N,N-diallyl-N-methylammonio) propane sulfonate, and N,N-diallyl-N,N-dimethylammonium chloride with 4-(N,N-diallyl-N-methylammonio) butanoate have been studied. Reactivity ratios indicate random incorporation of the comonomers. Molecular weights range from 4.13 to 8.42 x 10^4 g mol^{-1} for the sulfobetaine-containing polymers, and from 8.9 to 21.8 x 10^4 g mol^{-1} for the carboxybetaine-containing polymers. Macroscopic phase separation was observed for the sulfobetaine containing polymers with changes in pH, while the carboxybetaine containing polymer solutions remained homogeneous throughout the pH range investigated.

Introduction

Ion-containing water-soluble polymers are the most diverse class of polymers, ranging from biopolymers such as nucleic acids and proteins that mediate life processes to commercial polymers with applications in water remediation, drag reduction, and formulation of pharmaceutics, cosmetics, and coatings. Charged polymers can be arbitrarily divided into two classes: polyelectrolytes and polyampholytes. The former have ionizable functional groups, *either* anionic or cationic, along or pendent to the macromolecular backbone; charges are balanced by small gegenions or counterions. Zwitterionic polymers have *both* cationic and anionic charges along or pendant to the backbone. Copolymers containing cationic and anionic functionality on different mer units are termed polyampholytes while those having both charges on a single mer unit are called polybetaines.

Charge-charge repulsions along the polyion backbone and osmotic effects resulting from counterion mobility are responsible for chain extension and the large hydrodynamic volume of polyelectrolytes in water at low ionic strength. The degree of extension depends greatly on copolymer composition, flexibility, and the effective charge density (*1, 2, 3*).

The solution behavior of polyampholytes is also governed by the charge balance on the polymer chain. A sufficient excess of either charge can cause the polymer to exhibit typical polyelectrolyte behavior (*4, 5, 6, 7, 8, 9*). Polymers with charge balance, on the other hand, typically exhibit antipolyelectrolyte behavior. This behavior is characterized by a collapsed conformation in deionized water and an expansion in the presence of small electrolytes (*5, 9, 10*). Studies of the pH responsiveness of polymers (*4, 6, 8*) containing the carboxylate group or protonated amines confirm this result, as well as reveal potentially useful chemistry for the design of responsive polymer systems in aqueous solution.

Charge density is also an important factor affecting the solution properties of polyampholytes near their isoelectric points. Higher charge density polymers are largely insoluble in pure water and require addition of a critical concentration of salt for solubilization (*5, 8, 9, 11, 12, 13, 14, 15*). Low charge density polyampholytes can be solubilized in water provided comonomers are sufficiently hydrophilic (*16*).

A class of polyampholyte that has received special attention over the last several years is the polybetaine. Zwitterionic monomers have the advantage of providing a well-controlled balance of charge within the polymer. For example, homopolymers synthesized from zwitterionic monomers yield polyampholytes with zero net charge. Both sulfobetaines (*17, 18, 19, 20, 21, 22, 23*) and carboxybetaines (*24, 25, 26, 27, 28*) have been reported in the literature. The carboxybetaine polymers offer the possibility of pH responsiveness through protonation of the carboxylate group, converting the zwitterion to a cation (*27, 29*).

Cyclopolymerization of a number of diallyl quaternary ammonium salts was first reported by Butler in 1949 (*30*). The solubility of the resulting polymers and lack of residual unsaturation led to the hypothesis of alternating intermolecular and intramolecular propagation steps, resulting in a linear polymer with cyclic groups in the backbone (*31*). Butler *et al.* originally proposed a six membered ring based on the prevailing view that the most thermodynamically stable product would form. It was later shown by ^{13}C NMR, however, that the cyclic structure was, in fact, a five membered ring which is believed to be formed by a kinetically controlled mechanism (*32, 33*). Since the discovery of cyclopolymerization, there has been a large volume of work in the area including synthesis of many new types of monomers (*30, 34*)

Recent work by McCormick *et al.* in the area of cyclopolymers includes the synthesis of novel hydrophobically modified polyelectrolytes, (*35*) water soluble polymers containing zwitterionic groups and hydrophobic backbone segments from the cyclopolymerization of *N,N*-diallyl-*N*-methylamine (DAMA), (*36,37, 38*) and the investigation of polymerization in vesicles of diallyl monomers with twin hydrophobic tails (*39*).

In this report we describe two series of pH responsive, zwitterionic cyclocopolymers. The first is a copolymer series consisting of sulfobetaine monomers copolymerized with a tertiary amine monomer; the latter provides the pH response in the form of a hydrophilic to hydrophobic transition upon deprotonation. The second series of copolymers consists of the quaternary amine DADMAC copolymerized with a carboxybetaine monomer. In this case, the pH response involves a transition from polyampholyte to polyelectrolyte upon protonation.

Experimental

Monomer Synthesis

N,N-diallyl-*N,N*-dimethylammonium chloride (**1**) was purchased from Aldrich Chemical Company and used as received. *N,N*-diallyl-*N*-methylamine (**2**), (*40*) 3-(*N,N*-diallyl-*N*-methylammonio) propane sulfonate (**3**), (*41*) and 4-(*N,N*-diallyl-*N*-methylammonio) butanoate (**4**) (*42*) were prepared as previously reported.

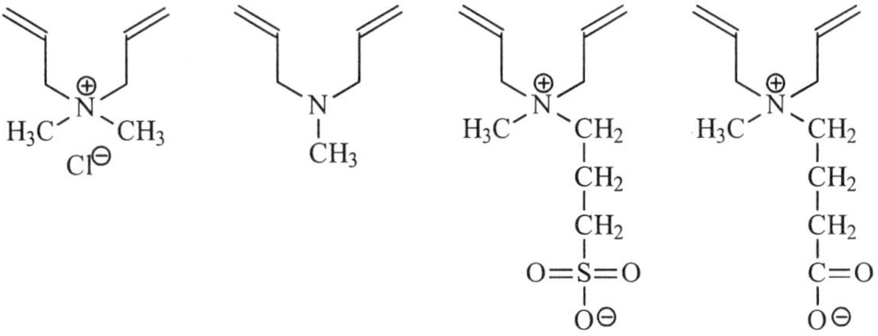

Scheme 1. *Monomers utilized in cyclocopolymerization*

Polymer Synthesis

Polymers were prepared in $0.5M$ NaCl aqueous solution at 35°C using the photoinitiator, 2-hydroxy-1-[4-(hydroxy-ethoxy)phenyl]-2-methyl-1-propane (Irgacure 2959) (Ciba). Total Monomer concentration was held constant at 2.5 M. The pH of polymerization solutions containing **2** was adjusted to pH 4 so that the monomer was in the ionized form. A low conversion sample, <30%, was removed during the course of the polymerization for reactivity ratio studies. The polymerizations were terminated at <50% conversion as a precaution against copolymer drift. Polymers were purified by dialysis against deionized water followed by lypholization.

Polymer Characterization

Solution studies were performed with a Contraves LS-30 rheometer. Inverse gated decoupled 13C NMR was used to determine copolymer compositions. Molecular weight studies were performed on a Brookhaven Instruments 128-channel BI-2030 AT digital correlator using a spectra Physics He-Ne laser operating at 632.8 nm. Refractive index measurements were carried out using a Chromatix KMX-16 differential refractometer.

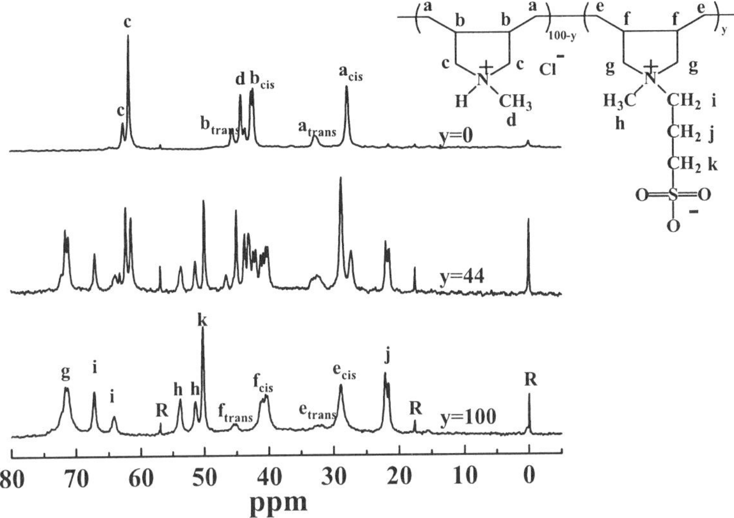

Figure 1. Representative inverse gated decoupled ^{13}C NMR spectra of the 2-co-3 cyclocopolymers in D_2O (1M NaCl, pH=4). R = 3-(trimethylsilyl)-1-propanesulfonic acid, sodium salt as a reference.

Results and Discussion

Polymer Synthesis and Compositional Analysis

The use of the photoinitiator Irgacure 2959 for the polymerization of the diallyl type monomers allows the rapid production of a high concentration of radicals and thus more rapid conversion than in thermal polymerization of diallyl salts. Polymerization reactions follow the general Scheme 2. No phase separation or precipitation has been observed for any of the copolymer systems studied thus far. Homogeneous reaction conditions throughout the polymerization can, therefore, be inferred.

Representative inverse gated decoupled ^{13}C NMR spectra for poly(**2**-co-**3**) and poly(**1**-co-**4**) are shown in **Figures 1** and **2,** respectively. The five-membered ring structure and the *cis* conformation were found to predominate, as in the DADMAC homopolymer, in both copolymer series (*43*). (**Figure 1.**)

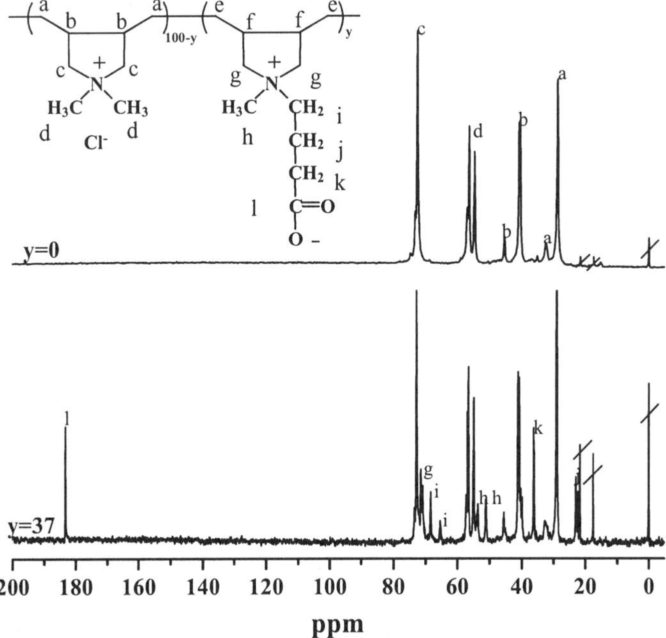

R$_1$=H, CH$_3$
R$_2$=C$_3$H$_6$SO$_3^-$, C$_3$H$_6$CO$_2^-$

[Monomers]$_o$ = 2.5 *M*
x = 0, 10, 25, 40, 60, 80, 100 mol%

Scheme 2. *Synthetic pathway for preparation of DAMA-co-DAMAPS (pH = 4) and DADMAC-co-DAMAB cyclocopolymers.*

Figure 2. *Representative inverse gated decoupled* ^{13}*C NMR spectra of the 1-co-4 cyclocopolymers in D$_2$O (1M NaCl).*

Copolymer compositions were determined by inverse gated decoupled ^{13}C NMR according to **equation 1** for poly(**2**-co-**3**).

$$\text{Mol\% } 3 = I_i \left(\frac{1}{I_a + I_e} + \frac{1}{I_c + I_g} \right) \qquad (1)$$

In this equation the quantities I_i, I_a, I_e, I_c, and I_g represent the integrated intensities corresponding to carbons i, a, e, c, and g respectively (**Figure 1**). From the experimental compositions, reactivity ratios of $r_1 = 0.67 \pm 0.05$ and $r_2 = 1.13 \pm 0.05$ for $\mathbf{2}(M_1)/\mathbf{3}(M_2)$ were determined by a nonlinear least squares method (*44*). The experimentally measured values of copolymer composition as a function of feed composition are shown in **Figure 3**; the dashed line represents ideal incorporation. These data are also given in **Table 1**.

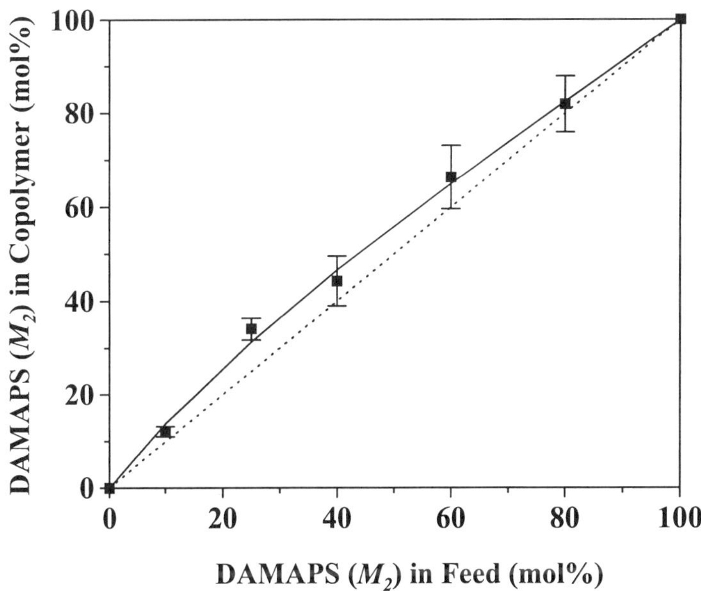

Figure 3. *Mol% sulfobetaine incorporation in 2-co-3 copolymers as a function of mol% sulfobetaine in the feed as determined by inverse gated decoupled ^{13}C NMR.*

108

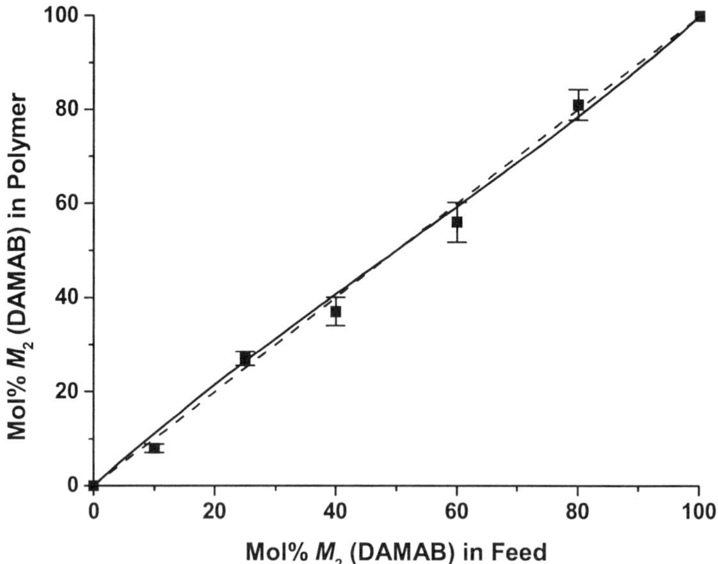

Figure 4. Mol% carboxybetaine incorporation in 1-co-4 copolymers as a function of mol% carboxybetaine in the feed as determined by inverse gated decoupled ^{13}C NMR.

Copolymer compositions for poly(1-co-4) were likewise determined from inverse gated decoupled ^{13}C NMR experiments. Equation 2 was used to determine the incorporation of **4**.

$$Mol\%4 = \frac{4}{3}\left(\frac{I_l + I_i + I_k}{I_a + I_b} \right) \quad (2)$$

The quantities I_l, I_i, I_k, I_a, and I_b represent the integrated intensities of carbon atoms l, i, k, a, and b respectively (**Figure 2**). The compositions, determined using **equation 2** and shown in **Table 2**, were used to determine the reactivity ratios, r_1=0.86 ± 0.03 r_2=0.99 ± 0.03 (M_1 = **1**, M_2 = **4**), using the same method as for the copolymers of **2** and **3**. **Figure 4** shows the experimentally determined values of the copolymer composition as a function of feed composition with the dashed line once again representing the ideal incorporation.

Table 1. Polymer composition and light scattering data for poly(2-co-3).

Polymer	3:2 in Feed	3:2 in Polymer	M_w x 10^{-4} (g mol^{-1})	A_2 x 10^4 (ml mol g^{-2})
DAMA	0	0^a	4.13	2.19
DAMS12	10	12	4.00	4.62
DAMS34	25	34	5.36	3.75
DAMS44	40	44	4.32	2.85
DAMS66	60	66	5.81	2.42
DAMS82	80	82	6.83	2.29
DAMAPS	100	100^a	8.42	2.14

a. Theoretical value.

Table 2. Polymer composition and light scattering data for poly(1-co-4).

Sample	1:4 in Feed	1:4 in Polymer	M_w x 10^{-4} (g mol^{-1})	A_2 x 10^4 (ml mol g^{-2})
DADMAC	0	0^a	3.0	8.78
DADC8	10	8.0 ±0.9	10.7	3.84
DADC28	25	28 ±2	10.6	2.46
DADC37	40	37 ±3	21.8	1.62
DADC56	60	56 ±4	8.9	2.67
DADC80	80	80 ±3	9.5	5.33
DAMAB	100	100^a	11.0	5.36

a.   Theoretical value.

Given the very similar structure of the reactive groups on all the diallyl monomers, it is not surprising that both copolymer series have reactivity ratios close to ideal, $r_1=r_2=1$.

Multi-angle laser light scattering

The weight-average molecular weights and second virial coefficients (A_2) for poly(2-co-3) and poly(1-co-4) are given in **Tables 1** and **2**, respectively. These values were determined using Berry plots at 25°C in 1M NaCl for both copolymer series. The solution pH was adjusted to 4.0 for the poly(2-co-3)

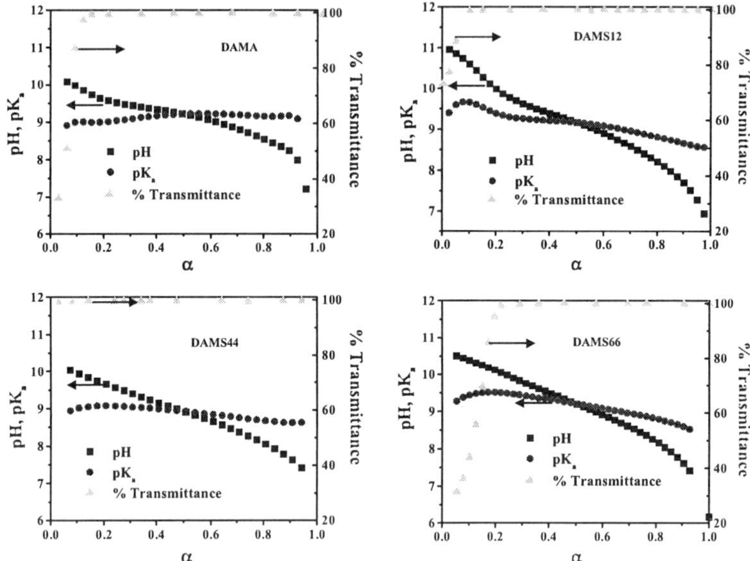

Figure 5. *Potentiometric and turbidimetric titrations of DAMA and three copolymers of DAMA and DAMAPS (α represents the degree of ionization of the DAMA mer unit).*

series. Molecular weights for the various sulfobetaine-containing polymers varied from 4.13 to 8.42 x 10^4 g mol^{-1}, while molecular weights for the carboxybetaine containing polymers varied from 8.9 to 21.8 x 10^4 g mol^{-1}. These values are typical of cyclopolymers from diallyl ammonium salts, which likely suffer from facile chain transfer to the labile allylic hydrogen atoms. Second virial coefficients for the sulfobetaine copolymers at pH=4 varied between 2.14 and 4.62 x 10^{-4} ml•mol g^{-2}, while the carboxybetaine copolymers had second virial coefficients between 1.62 and 5.36 ml•mol g^{-2} in 1 M NaCl. The generally low values for the second virial coefficient indicate the polymer coils were in a state close to theta conditions. Light scattering data for poly(**2**-co-**3**) and poly(**1**-co-**4**) polymers are included **Tables 1** and **2,** respectively.

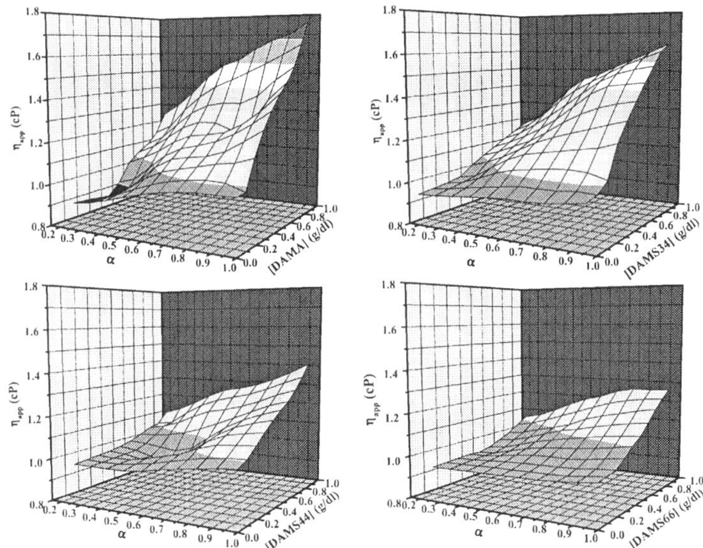

Figure 6. Apparent viscosity for the homopolymer of DAMA and three copolymers of DAMA and DAMAPS as functions of polymer concentration and the degree of ionization of the DAMA mer unit (α) (T = 25°C, γ = 5.96s^{-1}).

Solution Behavior

Phase behavior of the poly(**2**:**3**) series was studied with potentiometric and turbidimetric titration as well as with low shear viscometry. Copolymers of **2** and **3** with 66 mol% **3** or less were found to be soluble in deionized water at high degrees of ionization of the **2** mer unit. This behavior is similar to copolymers of DADMAC and **3** (*45*). As the degree of ionization decreases, the polymer solutions of DAMA, DAMS12, and DAMS66 become turbid (**Figure 5**). The homopolymer, DAMA, precipitates from solution at less than 10 mol% ionization of the mer units. Apparently the hydrophobic character of the tertiary amine dominates the phase behavior at these levels of ionization. Macroscopic phase separation, rather than precipitation, is observed for the DAMS12 copolymer. In this case the presence of the **3** mer unit is apparently stabilizing the hydrophobic portions of the polymer in solution. Copolymers DAMS34 and DAMS44 were found to be soluble throughout the ionization range. DAMS66 exhibits macroscopic phase separation at less than 20 mol% ionization. This behavior is attributed to dipole-dipole attractions between zwitterionic units rather than hydrophobic effects. Low shear viscosity experiments reveal a decrease in the apparent viscosity of the polymers as the degree of ionization decreases (**Figure 6**). Polymer chain expansion due to charge-charge repulsion at high degrees of ionization is responsible for the elevated viscosity in this state. As the degree of ionization decreases, hydrophobic effects cause the polymer to adopt a collapsed conformation, reducing the apparent viscosity of the system. Incorporation of the sulfobetaine monomer reduces the apparent viscosity at high degrees of ionization due to dipole-dipole attractions. At low degrees of ionization, incorporation of the sulfobetaine monomer results in an increase in the apparent viscosity up to 44 mol%. Apparently at these lower incorporations, the zwitterion enhances the hydrophilicity of the polymer coil, resulting in a more expanded conformation in solution.

Polymers containing the carboxybetaine monomer, **4**, can, upon protonation of the carboxylate anion, have the zwitterionic, attractive mer units converted to cationic, repulsive mer units. All polymers containing the carboxybetaine mer unit have a pK_a of 3.6, indicating half the carboxylate groups would be protonated at this pH. No obvious macroscopic phase changes are observed for this system; even the homopolymer of **4** is completely soluble in deionized water in the zwitterionic form.

Conclusions

In this chapter, we have illustrated the effects of pH responsive mer units on the solution behavior of cyclopolymers based on diallyl ammonium salts. In the case of those polymers containing a tertiary amine as the responsive moiety and a sulfobetaine moiety as the zwitterionic segment, the cationic-hydrophobic transition that occurs upon deprotonation results in a decrease in the coil volume which, in some cases, effects macroscopic phase separation. In the case of the carboxybetaine containing polymers, no phase separation is observed for any of the copolymers or homopolymers at any degree of protonation. The transition from zwitterionic units to cationic units was observed with potentiometric titration.

Acknowledgements

Support for this research by the Office of Naval Research and the U.S. Department of Energy is gratefully acknowledged.

References

1. Molyneux, P. *Water Soluble Synthetic Polymers: Properties and Behavior* Vol. II 1984 CRC Press, Inc. Boca Ratan FL.
2. Flory, P. J. *Principles of Polymer Chemistry*; Cornell University Press; Ithaca NY 1953.
3. Katchalsky, A. Kunzle, O. Kuhn, W. *J. Polm. Sci.* **1950**, 5, 283.
4. McCormick, C. L. Salazar, L. C. *Macromolecules* **1992**, 25, 1896.
5. Copart, J. Candau, F. *Macromolecules* **1993**, 26, 1333.
6. McCormick, C. L. Salazar, L. C. *Journal of Applied Polymer Science* **1993**, 48, 1115.
7. Peiffer, D. G. Lundberg, R. D. *Polymer* **1985**, 26, 1058.
8. McCormick, C. L. Johnson, C. B. *Macromolecules* **1988**, 21, 694.
9. Nisato, G.; Munch, J. P.; Candau, S. J. *Langmuir,* **1999**, 15, 4236.
10. Higgs, P. G.; Joanny ,J. F. *J. Chem. Phys.* **1991**, 94, 1543.
11. Kathman, E. Davis, D. McCormick, C. L. *Macromolecules* **1994**, 27, 3156.
12. Knoesel, R. Ehrmann, M. Galin, J. C. *Polymer* **1993**, 34, 1925.
13. Salamone, J. C. Volksen, W, Olson, A. P. Israel, S. C. *Polymer* **1978**, 19, 1157.
14. Soto, V. M. M. Galin, J. C. Polymer **1984**, 25, 254.
15. Skouri, M. Munch, J. P. Candau, S. J. Neyret, s. Candau, F. *Macromolecules* **1994**, 27, 69.
16. Peiffer, D. G. Lundberg, R. D. *Polymer* **1985**, 26, 1058.
17. Soto, V.M. M. Galin, J. C. *Polymer* **1984**, 25, 121.
18. Zheng, Y. Galin, M. Galin, J. *Polymer* **1988**, 29, 724.
19. Soto, V. M. M. Galin, J. C. Polymer **1984**, 25, 254.
20. Schulz, D. N.; Peiffer, D. G.; Agarwal, P. K.; Larabee, J.; Kaladas, J. J.; Soni, L.; Handwerker, B.; Garner, R. T.; *Polymer* **1986**, 27, 1734.
21. Salamone, J. C. Volksen, W, Israel, S. Olson, A. P. Raia, D. C. *Polymer* **1977**, 18, 1058.
22. Hart, R. Timmerman, D. *Journal of Polymer Science* **1958**, 28, 637.
23. Wielema, T. A. Engberts, J. B. *European Polymer Journal* **1987**, 23, 947.
24. Ladenheim, H. and Morawetz, H. *J. Polym. Sci.* **1957**, 26, 251.
25. Kathmann, E. E. and McCormick, C. L. *Polym. Prepr. Am. Chem. Soc. Div. Polym. Chem.* **1994**, 35, 641.
26. Rosenheck, K. and Katchalsky, A. *Journal of Polymer Science* **1958**, 32, 511.

27. Kathmann, E. E. White, L. A. and McCormick, C. L. *Polymer* 1997, **38**, 871.

28. Babilis, D. Dais, P. Margaritis, L. H. and Paleos C. M. *J. Polym. Sci.: Polym. Chem. Ed.* **1985**, 23, 1089.

29. Kathmann, E. E. White, L. A. and McCormick, C. L. *Polymer* 1997, **38**, 879.

30. Butler, G. B. *Cyclopolymerization and Cyclocopolymerization* 1992 Marcel Dekker, New York.

31. Butler, G. B. Angelo, R. J. *J. Am. Chem. Soc.* **1957**, 79, 3128.

32. Johns, S. R. Willing, R. I. *J. Macromol. Sci.-Chem.* **1976**, A10(5), 875.

33. Lancaster, J. E. Baccei, L. and Panzer, H. P. *Polymer Lett. Ed.* **1976**, 14, 549.

34. Butler, G. B. *Acc. Chem. Res.* **1982**, 15, 370.

35. Chang, Y. and McCormick, C. L. *Polymer* **1994**, 35, 3503.

36. McCormick, C. L.; Armentrout, R. S. *Polymer Preprints* **1998**, 39, 617.

37. Kaladas, J. J.; Schulz, D. N.; U. S. Patent 4,708,998 (November 24, 1987)

38. Kaladas, J. Kastrup, R. Schulz, D. N.; *Polymer Preprints* **1998**, 39, 619.

39. Richardson, M. F.; McCormick, C. L. *Polymer Preprints* **1998**, 39, 312.

40. Chang, Y.; McCormick, C.L. *Polymer* **1992**, 33, 4617.

41. Armentrout, R. S.; McCormick, C. L. *Polymer Preprints* **1998**, 39, 617.

42. Thomas, D. B.; Armentrout, R. S.; McCormick, C. L. *Polymer Preprints* **1999**, 40, 275.

43. Lancaster, J. E.; Baccei, L.; Panzer, H. P. *Polymer Letters* **1976**, 14, 549.

44. Tidwell, P. W.; Mortimer, G. A. *J. Polym. Sci.: Part A* **1965**, 2, 369.

45. Armentrout, R. S. and McCormick, C. L. Submitted to *Macromolecules* **1999**.

Chapter 7

Synthesis of Novel Shell Cross-Linked Micelles with Hydrophilic Cores

V. Bütün[1,2], and S. P. Armes[1,*]

[1]School of Chemistry, Physics and Environmental Science, University of Sussex, Falmer, Brighton, BN1 9QJ, UK.
[2]Current Address: Osmangazi University, Faculty of Arts and Science, Department of Chemistry, TR-26020, Eskişehir, Turkey.

We review our recent progress in the synthesis of novel shell cross-linked knedel (SCK) micelles. The reversible self-assembly of hydrophilic-hydrophilic block copolymers in aqueous media is exploited in order to develop three new classes of SCK micelles. In each case the micelle cores can be tuned to become hydrophilic or hydrophobic depending on the solution pH, electrolyte concentration or temperature. In our initial studies (1), a tertiary amine methacrylate diblock copolymer was used to prepare SCK micelles of 20-40 nm diameter, with shell cross-linking being achieved using a bifunctional quaternizing agent. Secondly, an ampholytic diblock copolymer was used to prepare zwitterionic SCK micelles with two distinct particle morphologies: **Type I** micelles, which comprise cationic coronas and anionic cores and **Type II** micelles, which comprise anionic coronas and cationic cores (2). Finally, a novel tailor-made triblock copolymer was synthesized and utilized in SCK micelle syntheses at high solids (> 10 %) (3). This 'proof-of-concept' experiment suggests that commercial applications of these fascinating nanostructures may be a realistic possibility.

Introduction

In a recent series of pioneering papers, Wooley and co-workers reported the synthesis of shell cross-linked 'knedel' (SCK) micelles (*4-12*). According to these authors, these novel supramolecular structures appear to be a hybrid between dendrimers, hollow spheres, latex particles, and block copolymer micelles. A broad range of applications, in areas as diverse as solubilization, catalysis, fillers, coatings and delivery, were suggested for these fascinating new materials (*4*). Wooley's group reported several synthetic routes to SCK micelles. In their original communication styrene-4-vinyl pyridine (St-4VP) diblock copolymers were first partially quaternized with 4-chloromethylstyrene and then cross-linked *via* uv irradiation at 254 nm in the presence of 4,4'-azobis(4-cyanovaleric acid) initiator (*4*). In later papers, both esterification and carbodiimide coupling chemistry were exploited for shell cross-linking of the solvated carboxylic acid residues of St-acrylic acid (St-AA) (*6*) and isoprene-acrylic acid (IP-AA) (*9*), diblock copolymers. The factors affecting the dimensions and properties of the SCK micelles such as the degree of shell cross-linking, comonomer ratio in the precursor block copolymer, and the degree of quaternization of the 4VP residues were also examined (*10*). The encapsulation of small hydrophobic molecules by these SCK micelles has also been examined (*11*).

Ding and Liu (*12*) synthesized SCK micelles using styrene-2-cinnamoylethyl methacrylate (St-CEMA) diblock copolymers, with the St block forming the micelle core and the CEMA block forming the micelle corona in a THF/acetonitrile solvent mixture. The CEMA residues were cross-linked [or dimerized] by uv irradiation. The effect of the degree of CEMA conversion on the micelle diameter and the degree of intermicellar cross-linking was investigated. It was demonstrated that the micellar structure became 'locked in' at a CEMA conversion of ~ 10%. However, these SCK micelles were not stable at higher CEMA conversions due to the reduced solubility of the polyCEMA chains in the corona after extensive inter- and intra-chain dimerization. The hydrodynamic diameter increased from 31 nm up to 326 nm, indicating the formation of multi-micellar aggregates. The same research group has also reported the synthesis of 'hollow nanospheres' based on IP-CEMA-t-butyl acrylate (IP-CEMA- tBuA) triblock copolymers. Here the IP residues initially formed the micelle core but were subsequently quantitatively degraded using ozonolysis after shell cross-linking (*13*). Wooley's group have also prepared hollow SCK micelles, 'nanocages', using a similar synthetic strategy (*14*).

Although the synthesis of SCK micelles is quite straightforward (indeed, rather attractive compared to multi-step dendrimer syntheses), two problems are nevertheless apparent. Firstly, the shell cross-linking reaction is generally carried out at relatively high dilution in order to avoid inter-micellar cross-linking. Secondly, all examples of SCK micelles described to date either contain permanently hydrophobic micelle cores or are 'hollow spheres' (i.e. the hydrophobic micelle core has been removed by ozonolysis) (*13,14*). However, if these nanocapsules were to be used as delivery vehicles, an explicit release mechanism or 'trigger' would be desirable.

The Sussex Polymer Group has gained considerable experience in the synthesis of water-soluble diblock copolymers over the last few years. We have published a series of papers describing the behaviour of hydrophilic-hydrophilic block copolymers based on vinyl ethers (*15-16*), methacrylates (*17-22*) and styrenic monomers (*23*). Herein we report the synthesis of three new classes of SCK micelles based on tertiary amine methacrylates. In each case the micelle cores can be reversibly hydrated or dehydrated depending on the solution temperature and/or electrolyte concentration. The ability to control the degree of hydration of the micelle cores suggests possible uptake/release applications.

Experimental

Block copolymer Syntheses.

The preparation of the 2-(dimethylamino)ethyl methacrylate-2-(*N*-morpholino)ethyl methacrylate (DMA-MEMA) (*18-20,24*), DMA-2-tetrahydropyranyl methacrylate (DMA-THPMA) (*17,25*) and DMA-methacrylic acid (DMA-MAA) (*25*) diblock copolymers have been described in detail in earlier publications. The synthesis of the poly(ethylene oxide)-DMA-MEMA (PEO-DMA-MEMA) triblock copolymer was carried out either using oxyanionic polymerization (*26,27*) or atom transfer radical polymerization (ATRP) (*28*).

Characterization Techniques and Instrumentation.

Dynamic light scattering (DLS) studies were carried out using a Malvern 4700 instrument equipped with a 80 mW argon ion laser and analyzed using CONTIN software. All measurements were made between 25°C and 60°C on 0.3 to 1.0 wt. % copolymer solutions at a fixed angle of 90°. Proton NMR spectra were recorded using either a 250 MHz or 300 MHz NMR spectrometer in either D_2O or $CDCl_3$. In certain experiments, the solution pH was adjusted using DCl or NaOD as required. The extent of deprotection of the THPMA residues was assessed using a 95:5 D_2O: d_8THF mixture and variable temperature NMR

118

studies were conducted between 25 and 60°C in D_2O at pH 10. Gel permeation chromatography (GPC) data were obtained under the following conditions: refractive index detector, THF eluent, Polymer Labs Mixed 'E' columns, poly(methyl methacrylate) (PMMA) standards, flow rate 1.0 ml min^{-1}. Fourier transform infra-red (FTIR) spectra (64 scans, 4 cm^{-1} resolution) of the dried SCK micelles were recorded using a Nicolet Magna 550 spectrometer equipped with a diamond ATR Golden Gate accessory. Electrophoresis measurements were carried out using a Malvern ZetaMasterS instrument. Zeta potentials were calculated from mobilities using the Henry equation and determined as a function of solution pH (0.01 M NaCl; 25°C). The pH was adjusted by adding HCl or NaOH. Transmission electron microscopy (TEM) studies were carried out (without staining) on dilute suspensions of the SCK micelles dried onto TEM grids using a Hitachi 7100 instrument. Static light scattering (SLS) measurements were carried out using a Sofica instrument, equipped with a He-Ne laser (vertically polarized light; $\lambda = 633$ nm) in the angular range of 30-150°. The refractive index increments, *dn/dc*, of the unimers (in THF) and the SCK micelles in aqueous media were determined using a Brice-Phoenix differential refractometer ($\lambda = 633$ nm).

Results and Discussion

1. SCK Micelles Based on DMA-MEMA Diblock Copolymers

The overall M_n and M_w/M_n of the DMA-MEMA block copolymer (see Figure 1) was 36,000 and 1.10 respectively (*18-20, 24*), as measured by GPC. The DMA content of the block copolymer was 35 mol %, as determined by proton NMR spectroscopy in $CDCl_3$ (the peak integral of the six dimethylamino protons of the DMA residues at δ 2.3 was compared to that due to the four methylene protons of the MEMA residues at δ 3.7). Partial *selective* quaternization of the DMA residues was achieved using methyl iodide in THF (25°C for 1 h). Subsequently, proton NMR spectrocopy indicated a degree of quaternization of 30% for the DMA block. Control experiments with DMA and MEMA homopolymers confirmed that the MEMA block was not quaternized under these reaction conditions (*29*). This quaternization step, although not strictly essential, provided better discrimination between the hydrophilic DMA and MEMA block sequences in aqueous media, since the partially quaternized DMA block no longer exhibits inverse temperature-solubility behaviour. Thus the derivatized block copolymer is expected to undergo reversible micellization in aqueous media on raising the solution temperature above the cloud point of the MEMA block, i.e. the MEMA block forms the hydrophobic micelle core

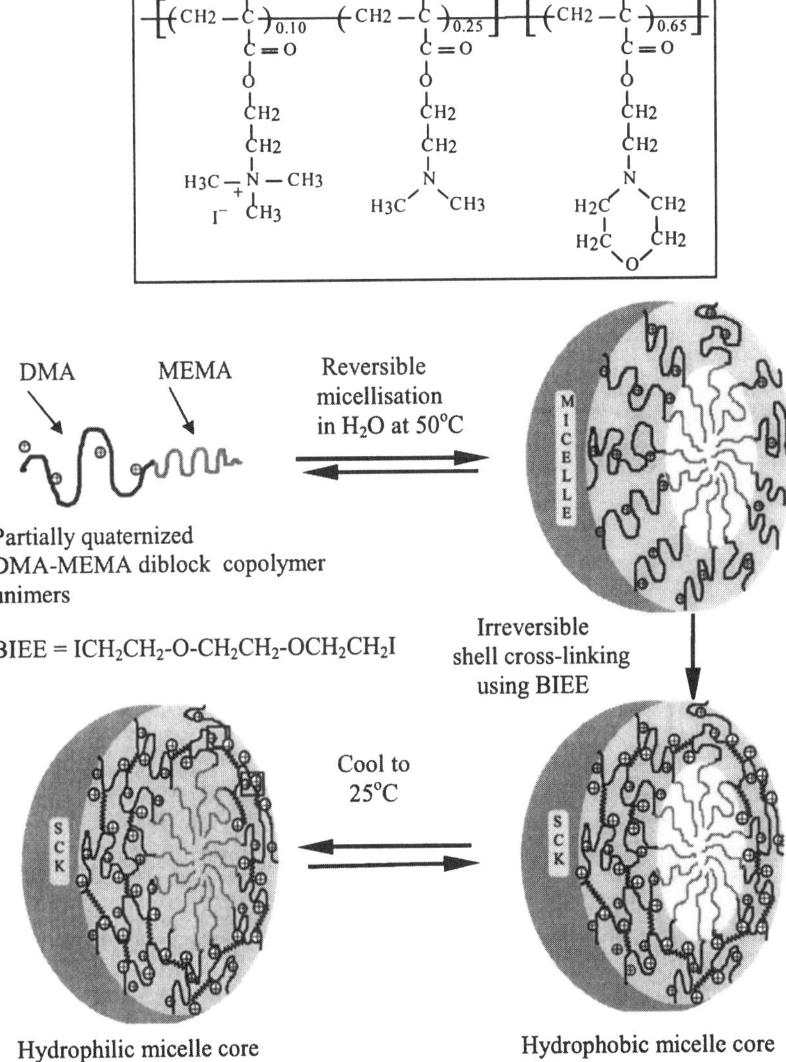

Figure 1. Chemical structure of the partially quaternized DMA-MEMA block copolymer and a reaction scheme for the synthesis of SCK micelles using the partially quaternized DMA-MEMA block copolymer.

and the partially quaternized DMA block forms the solvated corona. DLS studies at 60°C indicated an intensity-average micelle diameter of *ca.* 36 nm. Shell cross-linking of the unquaternized DMA residues in the corona was achieved by adding 1,2-bis-(2-iodoethoxy)ethane (BIEE) (40 mol% BIEE based on unquaternized DMA residues) to the micellar solution at 60°C. SCK micelles are then readily prepared by selective cross-linking of the unquaternized DMA residues in the micellar corona (see Figure 1).

A 0.50 wt. % aqueous solution of the partially quaternized DMA-MEMA block copolymer in 0.10 M Na_2SO_4 at pH 10 was heated to 60°C. Under these conditions the MEMA block becomes dehydrated and forms the non-solvated micelle core, with the partially quaternized DMA block forming the solvated corona. This is confirmed by variable temperature proton NMR studies: proton signals due to the MEMA block are significantly reduced in intensity relative to the DMA signals at 60°C (compare the spectra shown in Figure 2a and 2b). At 60°C the intensity-average micelle diameter during the shell cross-linking reaction was 28 nm. However, after cooling this reaction solution to 25°C, dynamic light scattering studies indicated an intensity-average diameter of *ca.* 30 nm, which confirms the formation of SCK micelles. If shell cross-linking had been unsuccessful no micelles would exist at room temperature since, under these conditions, the MEMA block is hydrophilic and there would be no reason for the block copolymer chains to remain aggregated. The slight increase in micelle diameter on cooling from 60°C to 25°C indicates a small degree of swelling due to the ingress of water into the now-hydrophilic MEMA cores. A proton NMR spectrum of these SCK micelles recorded in D_2O at 60°C and 0.25 M Na_2SO_4 indicated that shell cross-linking was incomplete since there is a residual signal due to the unquaternized DMA residues at δ 2.2-2.3 (see Figure 2c). Comparison of this peak integral with that due to the quaternized DMA residues at δ 3.1-3.3 indicated that around 30% of the DMA residues remained unquaternized after shell cross-linking. Taking into account the original degree of quaternization of 30 % obtained using methyl iodide, we estimate an upper limit of 14 % for the degree of shell cross-linking of the block copolymer. However, the actual degree of shell cross-linking is most likely somewhat lower, since a significant proportion of the -CH_2I groups of the bifunctional cross-linking agent probably remains unreacted.

Although the effect of temperature on the SCK micelle diameter described above is relatively small, it was found to be both reproducible and reversible. Significantly larger changes in micelle diameter were observed on varying the solution pH. Thus, 30 nm micelles at 25°C and pH 10 swelled up to 38 nm at pH 2 on addition of HCl. This is presumably due to extensive protonation of the MEMA core. Such pH-induced micelle swelling is reversible: returning to pH

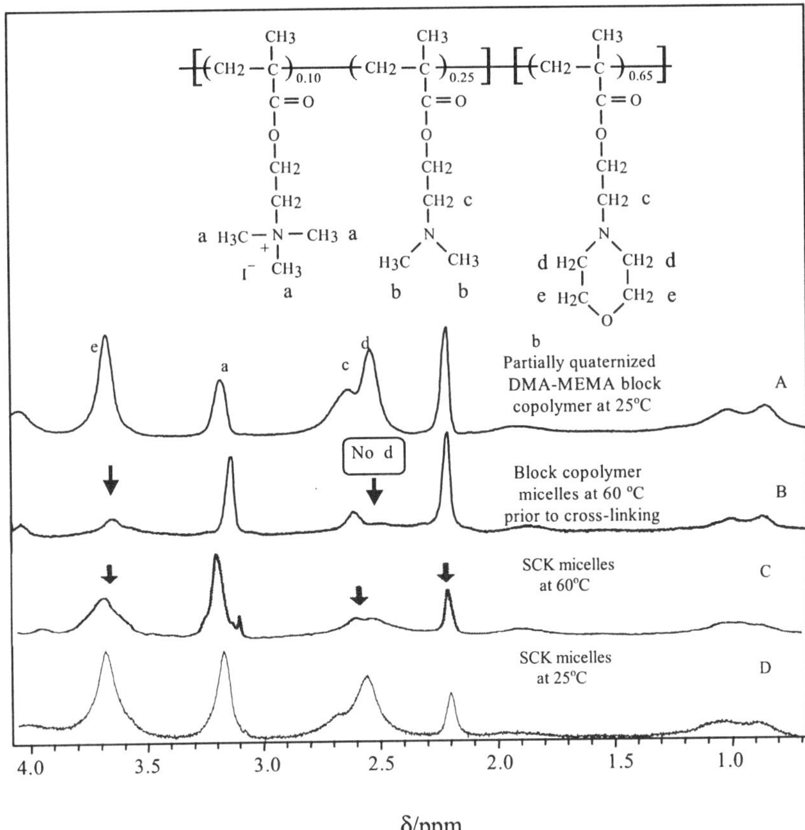

Figure 2. Variable temperature proton NMR studies of a 1.0 % D$_2$O solution of a partially quaternized DMA-MEMA block copolymer [35 mol % DMA, M$_n$ = 36,000, 30 % quaternized using methyl iodide] at pH 10: (a) unimers at 25°C (b) micelles at 60°C in the presence of 0.10 M Na$_2$SO$_4$; (c) shell cross-linked micelles at 60°C in the presence of 0.25 M Na$_2$SO$_4$; (d) SCK micelles on cooling 25°C in the presence of 0.25 M Na$_2$SO$_4$.

10 by addition of NaOD produces micelles of 31 nm. Finally, heating an acidic (pH 2) micellar solution from 25°C to 60°C apparently leads to micelle de-swelling, since the intensity-average micelle diameter decreases from 38 nm to 32 nm. This is a surprising observation since the non-cross-linked, partially quaternized DMA-MEMA block copolymer remains completely soluble as

unimers under the same conditions, which suggests that temperature-induced dehydration of the MEMA block does not occur at pH 2.

An electron micrograph of a dilute suspension of SCK micelles dried onto a TEM grid is shown in Figure 3. This confirms the reasonably narrow size distribution and spherical morphology of the SCK micelles and suggests a number-average micelle diameter of *ca.* 20 nm. Allowing for dehydration and polydispersity effects, this value is in reasonable agreement with the intensity-average diameter of 30 nm obtained from the DLS measurements.

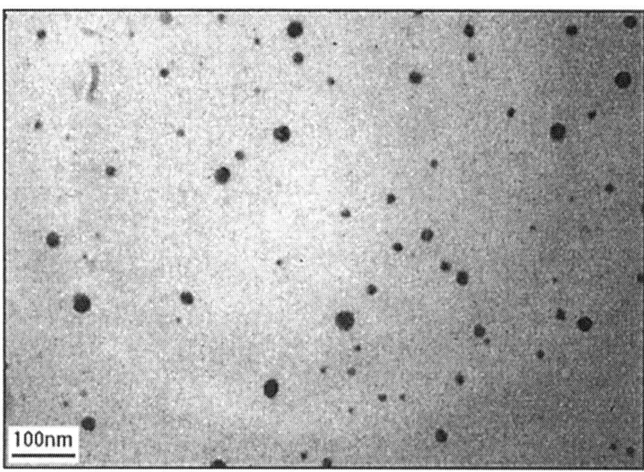

Figure 3. Transmission electron micrograph of SCK micelles prepared using a partially quaternized DMA-MEMA block copolymer [35 mol % DMA, M_n = 36,000, 30 % quaternized using methyl iodide].

Variable temperature NMR spectra recorded for a 0.50 wt% solution of the partially quaternized DMA-MEMA block copolymer in D_2O (pH 10, 0.1 M Na_2SO_4) are shown in Figure 2. At 25°C NMR signals due to both the DMA block (δ 3.1-3.3) and MEMA block (δ 3.6-3.9 and δ 2.5-2.7) are evident, as expected (see Figure 2a). On heating this solution to 60°C, the MEMA signal is reduced in intensity relative to the DMA signal (see Figure 2b). This indicates that the MEMA block becomes less solvated under these conditions, which is consistent with dehydration of the micelle core. A similar NMR spectrum was obtained after shell cross-linking although, since the overall degree of quaternization is now higher, the signal at δ 3.1-3.3 due to the three methyl groups of the quaternized DMA residues increases relative to that at δ 2.2-2.3 due to the two methyl groups of the unquaternized DMA residues (see Figure

2c). Since this MEMA signal does not disappear, it seems that the cores of the SCK micelles do not become fully dehydrated under these conditions (see Figure 2c). On cooling the solution to 25 °C, the MEMA signals at δ 3.6-3.9 and δ 2.5-2.7 regain their original intensities: this confirms hydration of the micelle cores (see Figure 2d). Temperature cycling confirmed that core (de)hydration was reversible.

The effect of addition of electrolyte at 25°C is also demonstrated for these SCK micelles in Figure 4. Under these conditions the two NMR signals due to

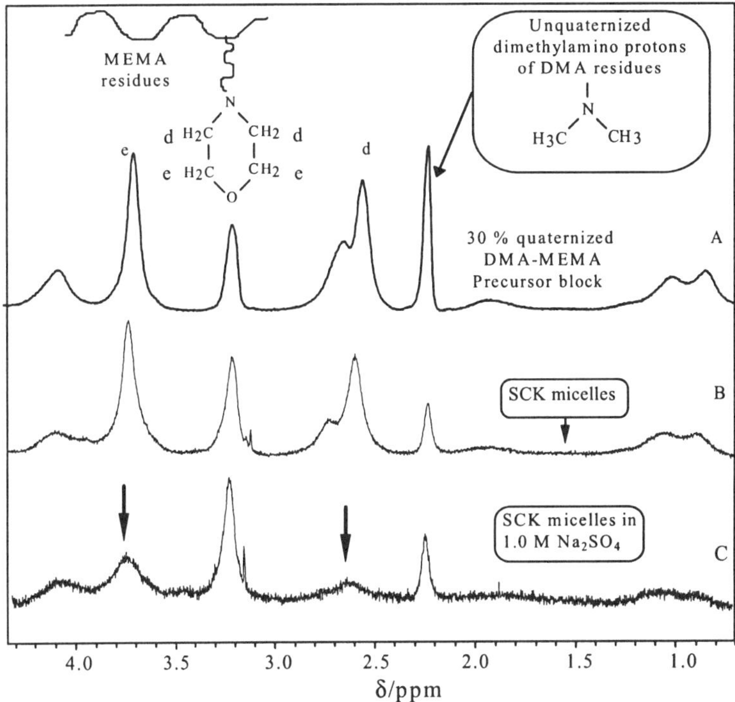

Figure 4. Proton NMR studies of the effect of salt on partially quaternized SCK micelles at 25°C: (a) DMA-MEMA precursor block copolymer [35 mol% DMA, M_n = 36,000, 30 % quaternized using methyl iodide] in D_2O at pH 10; (b) SCK micelles in D_2O at 25°C; (c) SCK micelles in the presence of 1.0 M Na_2SO_4. Note the marked suppression of the MEMA signals in the micellar core as this block becomes dehydrated in the presence of salt.

the MEMA block at δ 2.6-2.7 and δ 3.7-3.9 are again suppressed, indicating substantial dehydration of the SCK micelle cores at 25 °C due to the high salt concentration (compare Figures 4b and 4c; Figure 4a is a spectrum of the precursor diblock copolymer and is included as a reference). Again, this effect is reversible: removal of the salt *via* dialysis restores the MEMA signals in the NMR spectrum.

Very recently we have taken advantage of the MEMA block's low tolerance to added electrolyte to prepare SCK micelles without partial quaternization of the DMA-MEMA precursor block. Sufficient discrimination between the DMA and MEMA blocks can be achieved simply by careful control of the solution pH and the electrolyte concentration. Thus, a 0.40 % aqueous solution of an *unquaternized* DMA-MEMA block copolymer (48 mol % DMA; overall block copolymer M_n, 21,500) in 0.30 M Na_2SO_4 at pH 7.5 was heated to 55°C. Variable temperature NMR studies (not shown) confirmed that, under these conditions, the MEMA block forms the non-solvated micellar core, with the partially protonated DMA block forming the solvated corona. DLS studies at 50 °C indicated an intensity-average micelle diameter of *ca.* 23 nm. Successful shell cross-linking of the DMA residues was achieved by adding BIEE to this micellar block copolymer solution at 55°C. Thus, judicious selection of the aqueous solution conditions allows a more efficient synthesis of SCK micelles since the quaternization step may be omitted if desired.

The partially quaternized SCK micelles can be redispersed in water after drying. DLS measurements indicated only a modest increase in intensity-average micelle diameter, from 23 nm for the as-made SCK micelles up to *ca.* 30-40 nm for the redispersed micelles. This apparent increase in size most likely indicates some degree of aggregation of the micelles.

2. Zwitterionic Shell Cross-linked Micelles

In this section we report the first synthesis of zwitterionic SCK micelles. These new nanoparticles also have hydrophilic micelle cores. Depending on the reaction conditions used, two classes of zwitterionic SCK micelles can be obtained: **Type I** micelles, which have anionic cores and cationic coronas and **Type II** micelles, which have cationic cores and anionic coronas (see Figure 5). Both micelle types exhibit isoelectric points (IEP's) and thus can be precipitated from aqueous solution by judicious adjustment of the solution pH.

The two DMA-THPMA diblock copolymer precursors were synthesized as described previously (*17-25*). One had a DMA content of 51 mol % and an overall M_n of 34,000 and the other had a DMA content of 43 mol % and an overall M_n of 42,400. These linear copolymers are readily converted into zwitterionic

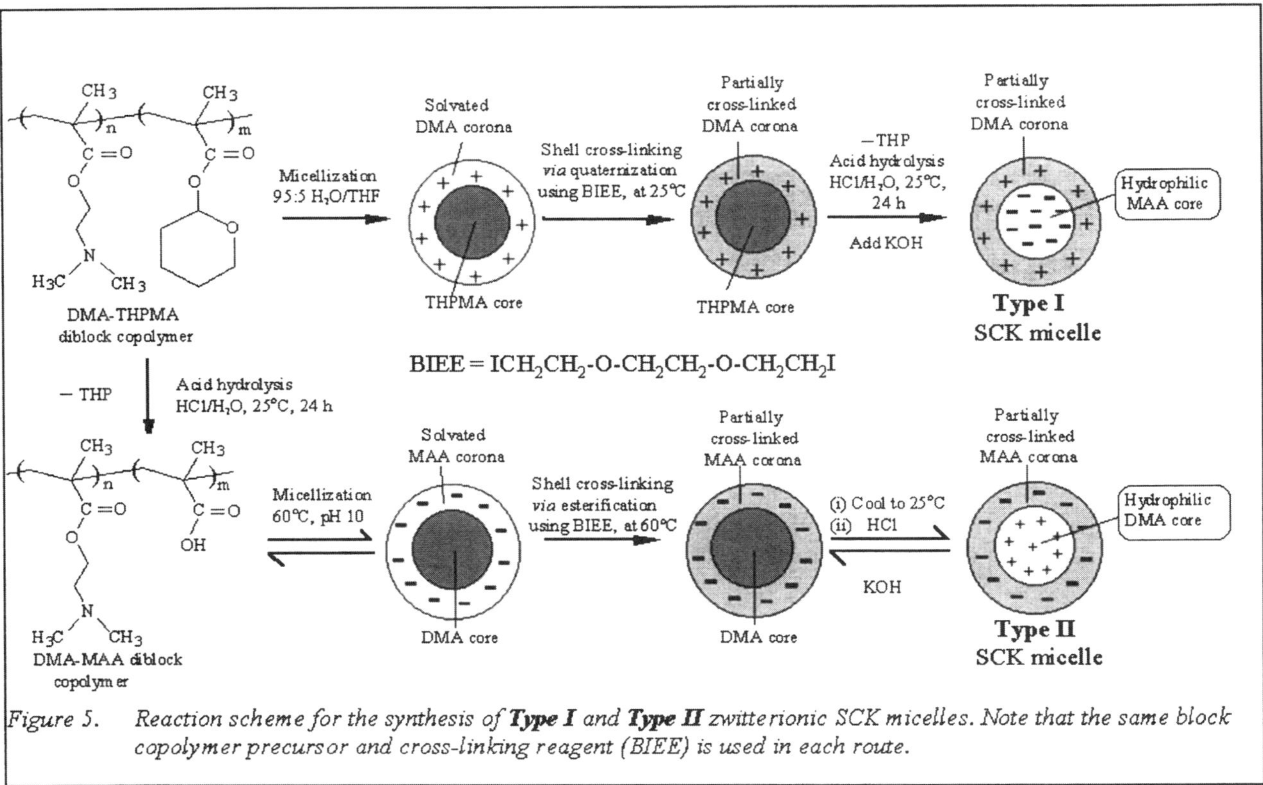

Figure 5. Reaction scheme for the synthesis of **Type I** and **Type II** zwitterionic SCK micelles. Note that the same block copolymer precursor and cross-linking reagent (BIEE) is used in each route.

126

DMA-MAA blocks by acid hydrolysis (*25*). Depending on the reaction *sequence*, two types of zwitterionic SCK micelles could be prepared from the DMA-THPMA block copolymer precursors using the BIEE cross-linker (see Figure 5). For the synthesis of the Type I SCK micelles (*30*), which have anionic MAA cores and cationic DMA coronas, conventional micelles of the DMA-THPMA precursor were initially formed in aqueous solution using 5 vol.% THF. This co-solvent, though not essential, ensured that the hydrophilic-hydrophobic precursor block copolymer forms well-defined micelles with THPMA cores (*31*). Shell cross-linking of the DMA residues in the micelle corona was achieved at 25°C and pH 10 by quaternization (see Figure 6) using BIEE. Finally, the THP groups were removed by acid hydrolysis (*17-25*) to obtain the zwitterionic SCK micelles. Initially, the methacrylic acid residues were obtained in their un-ionized form, but addition of base leads to carboxylate formation. **Type II** SCK micelles (cationic cores and anionic coronas) were synthesized (*32*) by first removing the THP groups from the DMA-THPMA precursor by acid hydrolysis to give a DMA-MAA block copolymer. This zwitterionic block copolymer forms micelles with DMA cores and MAA coronas in aqueous solution at elevated temperature (*17-25*). Shell cross-linking of these micelles was achieved at 60°C using BIEE. It is noteworthy that this shell cross-linking chemistry involves *esterification* of the MAA residues (see Figure 7), rather than quaternization (see Figure 6) (*33*). The former cross-linking chemistry leads to the micellar shell becoming less hydrophilic in nature, whereas the latter leads to increased hydrophilicity, and hence improved colloidal stability.

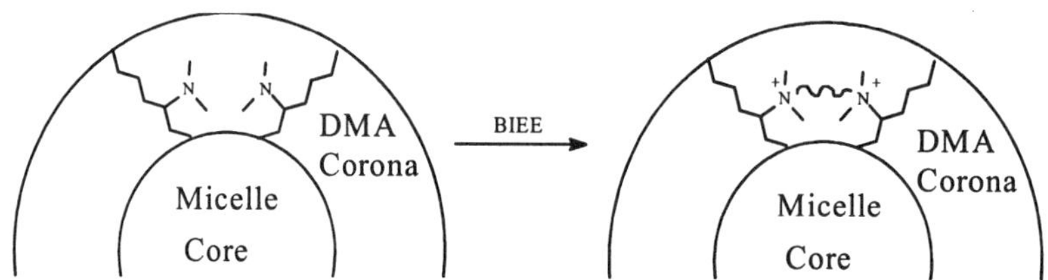

*Figure 6. Shell cross-linking of DMA residues via quaternization leads to increased hydrophilicity of the micelle corona in **Type I** micelles.*

Control experiments with MAA and DMA homopolymers confirmed that the MAA residues are much more reactive towards BIEE than the DMA residues (MAA homopolymer reacted near-quantitatively within 5-10 minutes, whereas only a low degree of quaternization was obtained for DMA homopolymer after several days). Relatively high degrees of shell cross-linking (typically 40-50 %) were sought in order to ensure that stable zwitterionic SCK micelles were obtained. However, given the pronounced difference in reactivities observed, sub-stoichiometric quantities of BIEE (based on the concentration of MAA residues present) should nevertheless ensure that any cross-linking between the DMA residues in the cores of the **Type II** micelles is negligible. Certainly, there was no NMR evidence to suggest that the DMA chains in the micellar core had become cross-linked by reacting with the BIEE.

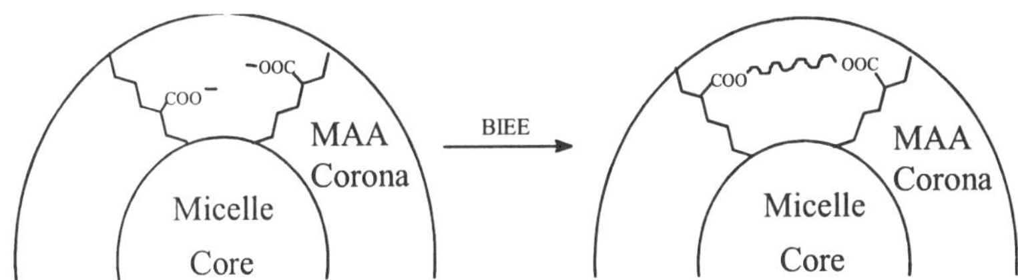

*Figure 7. Shell cross-linking of ionized MAA residues via esterification leads to decreased hydrophilicity of the micelle corona in **Type II** micelles.*

Proton NMR spectroscopy was used to determine the compositions of the precursor copolymers and also to verify that the THPMA block initially formed the non-solvated micelle cores during the synthesis of the **Type I** SCK micelles. In addition, complete deprotection of the THPMA residues was confirmed, since a commensurate increase in the methacrylate backbone signal due to the now-solvated MAA residues was observed after acid hydrolysis. Finally, variable temperature NMR studies during the synthesis of the **Type II** SCK micelles verified that the DMA blocks were substantially de-solvated at 60°C prior to shell cross-linking. On cooling to 25°C, the DMA signals reappeared as the SCK micelle cores became hydrated, as expected (*13, 21*).

FTIR spectroscopy studies of the dried **Type I** SCK micelles proved useful for assessing the conversion of the THPMA residues into MAA residues. After acid hydrolysis of the DMA-THPMA precursor SCK micelles, the zwitterionic micelles were isolated from *alkaline* solution. Two bands, which were absent in the IR spectrum of the DMA-THPMA SCK micelles prior to hydrolysis, were observed at 1555 and 1450 cm^{-1} (see Figure 8).

These new features are characteristic of carboxylate anion and indicate that deprotonation of the acidic micelle cores can be easily achieved. It is also noteworthy that the mass loss due to the removal of the THP groups from the SCK micelle cores is substantial (more than 50 %). Thus quantitative deprotection of the THPMA residues must lead to significant core porosity.

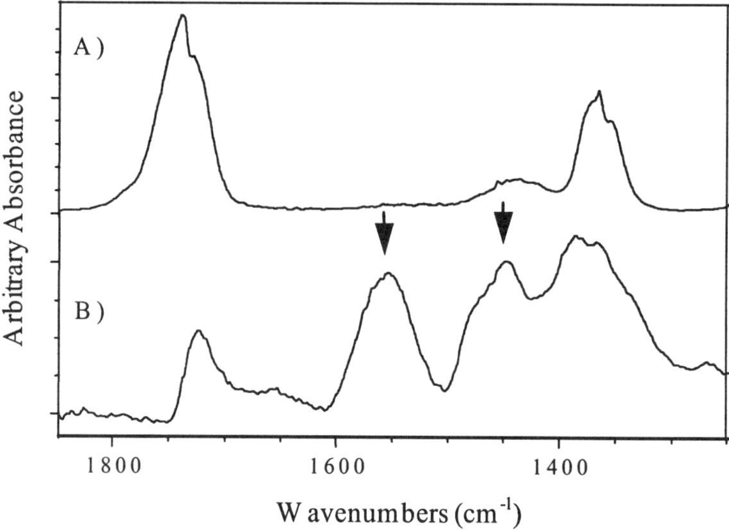

Figure 8. *FTIR spectra for: a)* ***Type I*** *SCK micelles prepared using the 51:49 DMA-THPMA copolymer precursor; b) the zwitterionic SCK micelles obtained after hydrolysis of the THP groups to produce methacrylic acid residues. In the latter case, the micelles were isolated from alkaline solution, leading to the formation of carboxylate anions. This gives rise to the two new bands at 1555 and 1450 cm⁻¹, respectively.*

Dilute suspensions of the two zwitterionic SCK micelles (**Types I** and **II**) were dried onto TEM grids and examined directly. Representative electron micrographs of both **Type I** and **Type II** SCK micelles are shown in Figure 9. Image analysis indicated a mean number-average diameter of approximately 18 nm for the **Type I** SCK micelles. Allowing for dehydration and polydispersity effects, this value is in reasonable agreement with the intensity-average diameter of 33 nm obtained from DLS measurements. TEM studies of the **Type II** SCK micelles (see Figure 9b) suggested somewhat larger micelles of around 25 nm diameter, which is reasonably consistent with a diameter of 40 nm obtained from DLS. The larger average size of the **Type II** micelles compared to the **Type I** micelles is probably due to the relatively hydrated nature of the DMA cores prior to shell cross-linking in the former system.

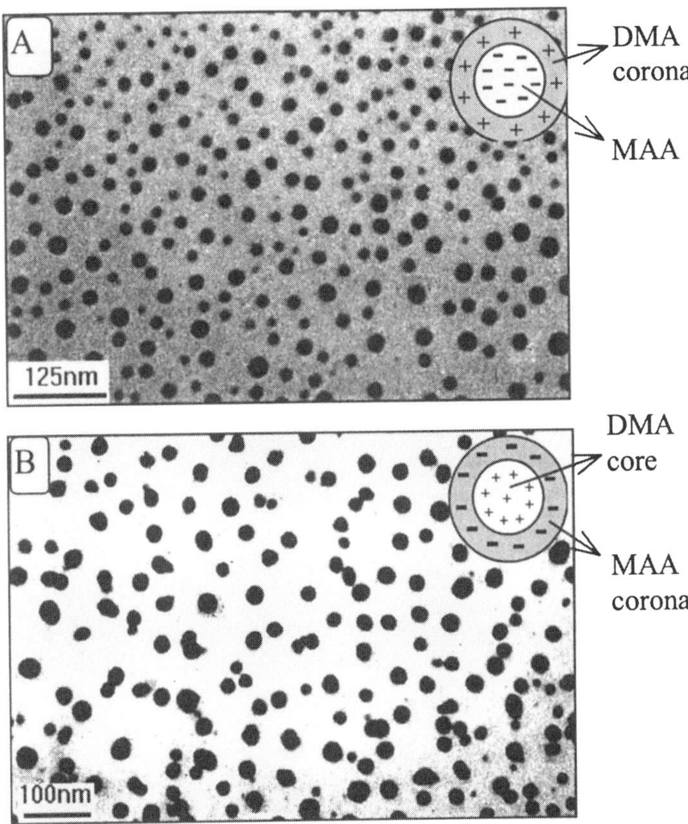

Figure 9. Transmission electron micrographs of zwitterionic SCK micelles: a) Type I SCK micelles prepared with a 51:49 DMA-THPMA block copolymer (average micelle diameter is 18 nm); b) Type II SCK micelles prepared using a 43:57 DMA-MAA block copolymer (average micelle diameter is 25 nm).

One fascinating aspect of these zwitterionic SCK micelles is that they exhibit IEP's in aqueous solution (see Figure 10). Thus, at a certain critical pH (the IEP) the micelles become electrically neutral and are precipitated quantitatively from aqueous solution. In this sense they behave like synthetic proteins. Addition of acid or base leads to complete redissolution of the micelles. This precipitation-dissolution behaviour is well known for conventional proteins and their synthetic analogues *(34)*. However, as far as we are aware, it has not been reported for micelles, SCK micelles or nanoparticles. The precise IEP depends mainly on the relative block composition (*i.e.* the DMA/MAA molar ratio). A secondary factor is the degree of shell cross-linking: quaternization of the DMA residues with BIEE leads to a permanent increase in the cationic charge density of the **Type I** micelles, whereas esterification of the MAA residues with BIEE leads to an irreversible reduction in the anionic charge density of the **Type II** micelles. Aqueous electrophoresis is a useful method for determining the IEP of an SCK micelle. Three zeta potential *vs.* pH curves are shown in Figure 10. The first curve (filled circles) represents a **Type I** SCK micelle synthesized using a 51:49 DMA-THPMA copolymer prior to removal of the THP groups from the micelle cores. As expected, no IEP is observed and the surface charge remains cationic across the entire pH range since the only ionizable groups are the DMA residues in the

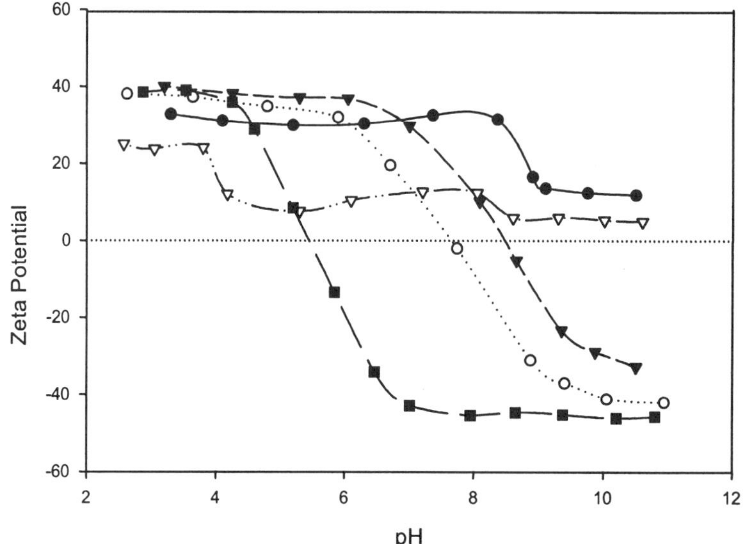

*Figure 10. Zeta potential vs. pH curves obtained for: (a) **Type I** SCK micelles prepared using a 51:49 DMA-THPMA copolymer prior to acid hydrolysis (●); (b) **Type I** zwitterionic SCK micelles prepared using the same copolymer after acid hydrolysis (o); (c) **Type II** zwitterionic SCK micelles prepared using a 43:57 DMA-MAA copolymer (▼).*

micelle coronas. The reduction in cationic surface charge observed at around pH 8.5 corresponds approximately to the pK_a reported for DMA homopolymer by Hoogeveen *et al.* (*35*). After deprotection of the THPMA residues in the micelle core, the resulting zwitterionic micelles (open circles) have an IEP at approximately pH 7.60, as expected. Given the relatively high degree of shell cross-linking of the DMA residues, this IEP is quite close to the value of 6.74 reported earlier for the identical *linear* DMA-MAA block copolymer (*17-25*). Similarly, **Type II** SCK micelles synthesized using a 43:57 DMA-MAA block copolymer have an IEP at pH 5.50 (filled triangles). A lower IEP is expected in this case due to the increased proportion of MAA residues in the copolymer.

Clearly this unusual aqueous solution behaviour offers considerable scope for the isolation, purification and harvesting of zwitterionic SCK micelles in various applications. Since these novel nanoparticles contain both acidic (MAA) and basic (DMA) binding sites, they are expected to be suitable delivery vehicles for a wide range of ionic 'actives' (e.g. drugs or pesticides). They may also prove to be useful as novel DNA vectors for gene therapy (*36*).

3. Synthesis of Shell Cross-linked Micelles at High Solids in Aqueous Media

A major drawback in the synthesis of conventional SCK micelles is that shell cross-linking must be carried out at high dilution (typically 0.1 to 0.5 % solids) in order to avoid extensive inter-micellar cross-linking. Unless this problem is addressed, it is unlikely that the synthesis of SCK micelles on an industrial scale will be commercially viable, even for speciality applications. Herein we report the first synthesis of well-defined SCK micelles in concentrated solution (10 % solids). This is achieved in aqueous media using a novel PEO-DMA-MEMA triblock copolymer (see Figure 11) (*3*).

Aqueous GPC studies (using PEO standards) indicated that the original PEO block had a low polydispersity ($M_w/M_n < 1.10$) and a number-average molecular weight of approximately 2,000, which corresponds to a degree of polymerization of 45. Treating this PEO block as an 'end-group', inspection of the proton NMR ($CDCl_3$) spectrum of the triblock copolymer indicated average degrees of polymerization for the DMA and MEMA blocks of 57 and 62, respectively. Thus the overall number-average molecular weight of this triblock copolymer [designated PEO-DMA-MEMA(1)] was calculated to be 23,600. GPC studies of the triblock copolymer indicated a polydispersity (M_w/M_n) of 1.33. This triblock copolymer can be molecularly dissolved in water at 20°C but the chains undergo self-assembly to form a three-layer 'onion' micelle in the presence of Na_2SO_4. This electrolyte selectively 'salts out' the MEMA block, with the hydrated DMA residues forming the inner shell and the ethylene oxide residues forming the outer shell (see Figure 11). This structure is supported by proton NMR studies; the signals associated with the MEMA residues disappear (see Figure 12a and 12b), whereas those due to the ethylene oxide and DMA residues remain visible.

The hydrodynamic diameter of the diluted 'onion' micelles is around 68 nm by DLS.

Inner shell cross-linking was achieved in 1.0 M Na_2SO_4 at 20°C for 3-7 days using BIEE [6-20 mol % based on DMA residues]. This reagent selectively quaternizes the more reactive DMA residues in the presence of the sterically hindered MEMA residues (24). The PEO block in the outer layer of the SCK micelle is sufficiently long to prevent the DMA residues of one SCK micelle coming into contact with those in a neighbouring SCK micelle. As micelles approach each other in solution, inter-mixing of the PEO chains in the respective outer shells is both entropically (due to restricted conformations) and enthalpically (PEO-water interactions are more favourable than PEO-PEO interactions since water is a good solvent for PEO) unfavourably. Thus, even at high copolymer concentrations, cross-linking of the DMA residues occurs exclusively <u>within</u> micelles and aggregation via inter-micelle cross-linking is prevented by the well known 'steric stabilization' mechanism (37). In a control experiment, shell cross-linking of an analogous DMA-MEMA diblock copolymer at 10 % solids led to macroscopic gelation due to extensive inter-micellar cross-linking. This failure serves to emphasize the importance of the outer shell of PEO chains for the preparation of well-defined shell cross-linked micelles at high solids. It is also noteworthy that these PEO chains may confer 'stealth' properties on these SCK micelles, which could be beneficial in drug delivery applications (38).

Shell cross-linking was verified as follows. After exposure to BIEE for several days at 20°C, the 10 % micellar solution was diluted ten-fold and adjusted to pH 2 using HCl. If no shell cross-linking had occurred, dissolution of the micelles into the individual triblock copolymer chains would be expected, since the MEMA block becomes hydrophilic under these conditions (this was confirmed by control experiments in the absence of BIEE). However, DLS studies indicated the continued presence of the micelles, which confirmed that shell cross-linking had been successful. In one series of experiments conducted with a similar triblock copolymer of higher molecular weight [PEO-DMA-MEMA(2)] (39), the BIEE concentration was systematically increased from 3 to 30 mol % (based on DMA residues) and shell cross-linking was carried out in D_2O to facilitate NMR studies (see Figure 11). Aliquots of the SCK micelles were periodically extracted, diluted with D_2O and analyzed by DLS. It was found that robust, if somewhat polydisperse, SCK micelles of approximately 90 nm were obtained even at 6 mol % BIEE. This value is significantly lower than the minimum degree of cross-linking of 10 % reported by Ding and Liu for the uv-induced cross-linking of CEMA residues (12).

In practice, the true degree of shell cross-linking will almost certainly be less than 6 mol % because (1) it is improbable that both iodoalkyl groups of each BIEE molecule react with DMA residues and (2) some degree of intra-chain quaternization most likely occurs. Unfortunately overlapping peak integrals prevented estimation of the degree of quaternization of the DMA residues by proton NMR spectroscopy. However, NMR proved very useful for monitoring

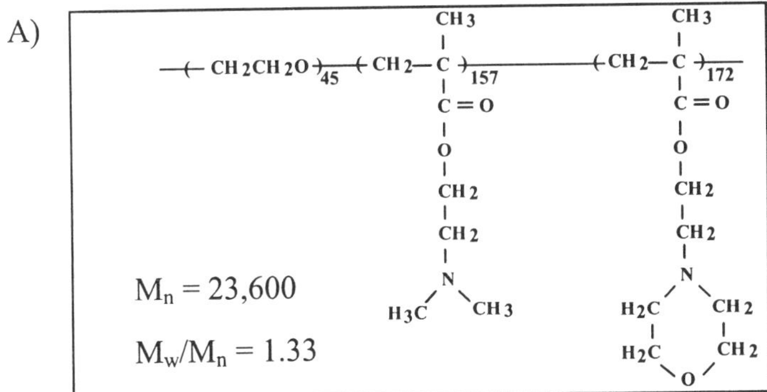

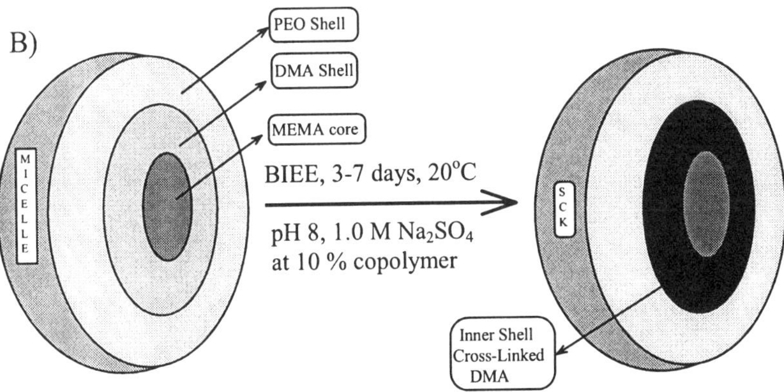

Figure 11. a) Chemical structure of the PEO-DMA-MEMA triblock copolymer. b) Schematic formation of SCK micelles at high solids using the PEO-DMA-MEMA triblock copolymer. Cross-linking of the inner shell is indicated by the darker shading.

134

the extent of hydration of the MEMA residues in the cores of the SCK micelles on varying the electrolyte concentration and/or solution temperature. For example, in the presence of 1.0 M Na_2SO_4 at 20°C the NMR signals at δ 2.5 and δ 3.6 due to the MEMA residues disappeared. These signals reappeared after dialysis and/or dilution (see Figure 12c and 12d).

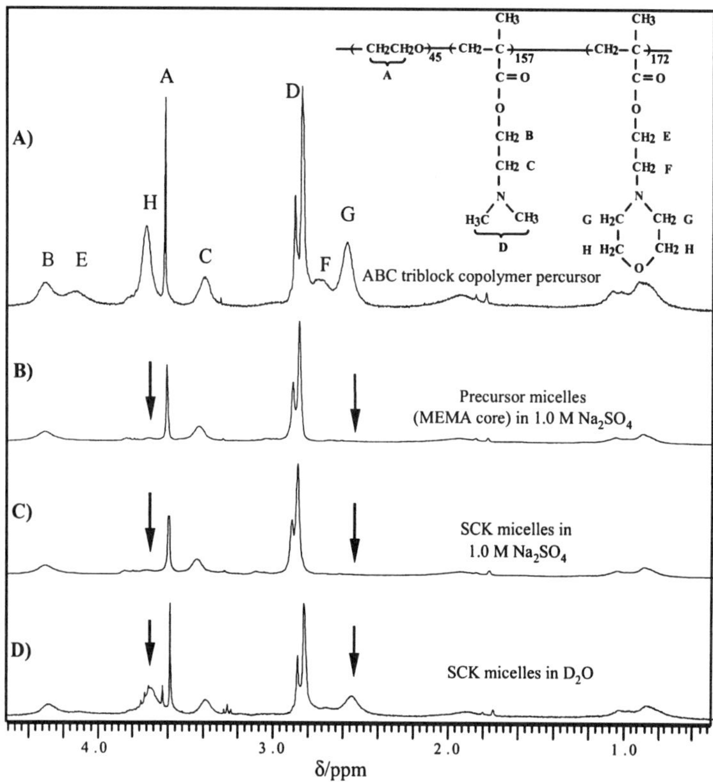

Figure 12. Proton NMR spectra in D_2O at 20 °C for: a) triblock copolymer PEO-DMA-MEMA(2); b) the same copolymer in the presence of 1.0 M Na_2SO_4 (note the absence of any signals due to the MEMA residues, indicating that this block forms the core of the onion micelles, as expected); c) SCK micelles prepared from this copolymer at 10 % solids in the presence of 1.0 M Na_2SO_4; d) the same SCK micelles in the absence of Na_2SO_4 (note the reappearance of the MEMA signals at δ 2.5 and δ 3.6, indicating rehydration of the core).

A transmission electron micrograph of a dilute suspension of SCK micelles originally prepared at 10 % solids using PEO-DMA-MEMA(1) is shown in Figure 13. The particles have a spherical morphology, a high degree of dispersion (i.e. essentially no inter-micelle cross-linking occurs) and a mean number-average particle diameter of 40 nm. This size is somewhat smaller than the hydrodynamic diameter of 68 nm obtained from DLS studies. Some discrepancy would be expected simply due to solvation effects but the relatively large difference observed probably also reflect the polydisperse nature of these particles.

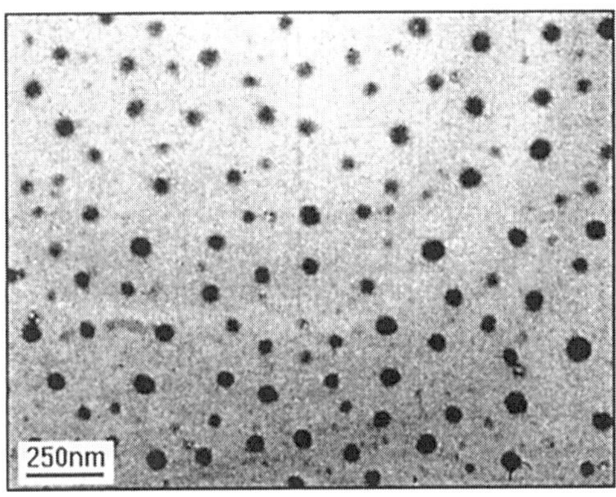

Figure 13. Transmission electron micrograph of the SCK micelles obtained after inner-shell cross-linking of the DMA residues at 10% solids. The particles range in size from 20 nm to 60 nm. Note that there is no evidence for inter-micelle cross-linking.

SLS studies of diluted SCK micelles prepared using PEO-DMA-MEMA(1) gave a high quality Zimm plot which yielded a radius of gyration of 21 ± 2 nm and a micelle mass of approximately 11.5 x 10^6 g mol^{-1}. Given that the weight-average molecular weight of the original triblock copolymer is estimated to be 31,400, this indicates a micelle aggregation number of 340.

Other ABC triblock copolymers gave similar results. In particular, a copolymer, PEO-DMA- t-butylaminoethyl methacrylate (PEO-DMA-BAEMA) (*40*) comprising EO, DMA and BAEMA residues dissolved molecularly in water at 4°C but formed 'BAEMA-core' micelles on warming to 20°C. Thus, one advantage of this type of copolymer is that micellization can be induced without addition of electrolyte. Shell cross-linking at 20°C was achieved with BIEE at 10 % solids with minimal inter-micelle cross-linking, as judged by DLS. We anticipate that at some critical copolymer concentration between 20 and 50 % there should be a phase transition from spherical to non-spherical micelles (i.e. worm-like or rod-like particles). Shell cross-linking under these conditions is expected to lead to new anisotropic supramolecular structures with interesting rheological properties (*41*).

Conclusions

In summary, we have demonstrated that several new classes of SCK micelles with hydrophilic cores can be readily prepared from various DMA-based block copolymers. The aqueous solution behaviour of the zwitterionic SCK micelles looks particularly interesting, since reversible precipitation at the IEP may offer opportunities for the purification and loading of these nanocapsules. Our preliminary 'proof of concept' experiments have also verified that sterically-stabilized SCK micelles with tunable core hydrophilicity can be readily prepared at high solids in aqueous media using ABC triblock copolymers. This is an important advance which suggests that SCK micelles, perhaps hitherto regarded as an academic curiosity, may be amenable to commercial exploitation.

Prospect

In future work we will focus on the synthesis of SCK micelles at high solids using ABC triblock copolymers. In particular, we will examine whether the latest advances in synthetic polymer chemistry will allow more environmentally-friendly and cost-effective routes to SCK's. Our long-term goal is the one-pot synthesis of SCK micelles at high solids directly in aqueous media starting from the requisite monomers. In principle, such syntheses should be feasible using ATRP. If we are successful in this regard it would represent a significant step forward in realizing the commercial potential offered by SCK micelles. Additional characterization techniques which will be utilized include both analytical ultracentrifugation and x-ray photoelectron spectroscopy. The former should provide detailed information on particle size distributions and inter-micelle interactions, whereas the latter will be used to verify the core-shell nature of the SCK micelles.

Acknowledgements. VB thanks Osmangazi University, Turkey for funding a PhD studentship. Important contributions from past and present members (Prof. N. C. Billingham, Dr. X.–S. Wang, Dr. A. B. Lowe, Dr. M. V. Paz de Banez and Ms. K. L. Robinson) of the Sussex Polymer Group are gratefully acknowledged. The expertise of Dr. Z. Tuzar in the static light scattering studies is much appreciated. The Royal Society is thanked for a travel grant to visit Dr.

Z. Tuzar at the Institute of Macromolecular Chemistry in Prague. Laporte Specialites (Hythe, UK) is thanked for its generous donation of the PEO.

References

1. Bütün, V.; Billingham, N. C.; Armes, S. P. *J. Am. Chem. Soc.*, **1998**, 120, 12135-12136.
2. Bütün, V.; Lowe, A. B.; Billingham, N. C.; Armes, S. P. *J. Am. Chem. Soc.*, **1999**, 121, 4288-4289.
3. Bütün, V.; Wang, X.-S.; de Paz Banez, M. P.; Robinson, K. L.; Billingham, N. C.; Armes, S. P.; Tuzar, Z. *Macromolecules*, in press, **1999**.
4. Thurmond, K. B.; Kowalewski, T.; Wooley, K. L. *J. Am. Chem. Soc.*, **1996**, 118, 7239-7240.
5. Wooley, K. L. *Chem. A - Europ. J.*, **1997**, 3(9), 1397-1399.
6. Huang, H. Y.; Kowalewski, T.; Remsen, E. E.; Gertzmann, R.; Wooley, K. L. *J. Am. Chem. Soc.*, **1997**, 119, 11653-11659.
7. Thurmond, K. B.; Kowalewski, T,; Wooley, K. L. *J. Am. Chem. Soc.*, **1997,** 119, 6656-6665.
8. Huang, H. Y.; Remsen, E. E.; Wooley, K. L. *Chem. Commun.*, **1998**, 13, 1415-1416.
9. Huang, H. Y.; Wooley, K. L. *ACS Polym. Prepr.*, **1998,** 39(1), 239.
10. Thurmond, K. B.; Kowalewski, T.; Wooley, K. L. *ACS Polym. Prepr.*, **1997**, 38(2), 592-593.
11. Thurmond, K. B.; Wooley, K. L. *ACS Polym. Prepr.*, **1998**, 39(1), 303.
12. Ding, J.; Liu, G. *Macromolecules*, **1998**, 31, 6554-6558.
13. Stewart, S.; Liu, G. *Chem. Mater.*, **1999**, 11, 1048-1054.
14. Huang, H.; Remsen, E. E.; Kowalewski, T.; Wooley, K. L. *J. Am. Chem. Soc.*, **1999**, 121, 3805-3806.
15. Forder, C.; Patrickios, C. S.; Armes, S. P.; Billingham, N. C. *Macromolecules*, **1996**, 29, 8160-8169.
16. Forder, C.; Patrickios, C. S.; Billingham, N. C.; Armes, S. P. *Chem. Commun.*, **1996**, 883-884.
17. Lowe, A. B.; Armes, S. P.; Billingham, N. C. *Macromolecules*, **1998**, 31, 5991-5998.
18. Bütün, V.; Billingham, N. C.; Armes, S. P. *Chem. Commun.*, **1997**, 671-672.
19. Bütün, V.; Bennett, C. E.; Vamvakaki, M.; Lowe, A. B.; Billingham, N. C.; Armes, S. P. *J. Mater. Chem.*, **1997**, 7(9), 1693-1795.
20. Bütün, V.; Billingham, N. C.; Armes, S. P. *J. Am. Chem. Soc.,* **1998**, 120, 11818-11819.
21. Lee, A. S.; Gast, A. P.; Bütün. V.; Armes, S. P. *Macromolecules*, **1999**, 32, 4302-4310.

22. Bütün, V.; Vamvakaki, M.; Billingham, N. C.; Armes, S. P. *Polymer*, in press, **2000**.

23. Gabaston, L. I.; Furlong, S. A.; Jackson, R. A.; Armes, S. P. *Polymer*, **1999**, 40, 4505-4514.

24. Bütün, V. *PhD thesis*, University of Sussex, Brighton, UK, **1999**.

25. Lowe, A. B.; Armes, S. P.; Billingham, N. C. *Chem. Commun.*, **1997**, 1035-1036.

26. Nagasaki, Y.; Sato, Y.; Kato, M. *Macromol. Rapid Commun.*, **1997**, 18, 827-835.

27. Vamvakaki, M.; Billingham, N. C.; Armes, S. P. *Macromolecules*, **1999**, 32, 2088-2090.

28. Zhang, X.; Matyjaszewski, K. *Macromolecules*, **1999**, 32, 1763-1766.

29. Bütün, V.; Billingham, N. C.; Armes, S. P. to be submitted to *Macromolecules*.

30. Synthesis of **Type I** micelles (DMA corona, MAA core). A 51:49 DMA-THPMA diblock copolymer (0.30 g) was dissolved in 5.0 ml THF and de-ionized water (145 ml) was then added to the stirred solution. The resulting micellar solution (22 nm intensity-average diameter by dynamic light scattering) was maintained at 25°C prior to shell cross-linking using BIEE (40 µl; 48 mol % based on DMA residues; 25°C for 4 days). The THPMA residues in the micellar cores were deprotected using HCl (3 ml of a 3.0 M solution) at 25°C for three days. After neutralization of the excess HCl by KOH, the final zwitterionic **Type I** SCK micelles were either dialyzed against de-ionized water or precipitated by adjusting the solution pH to the isoelectric point.

31. Baines, F. L.; Armes, S. P.; Billingham, N. C.; Tuzar, Z. *Macromolecules*, **1996**, 29, 8151-8159.

32. Synthesis of **Type II** zwitterionic SCK micelles (MAA corona, DMA core). A zwitterionic 43:57 DMA-MAA diblock copolymer (0.20 g) was dissolved in de-ionized water (80 ml) and the solution pH was adjusted to pH 10 using KOH. On raising the solution temperature to 60°C, this block copolymer formed **Type II** micelles of 56 nm diameter, with the DMA residues forming the core and the MAA residues in the corona. In some cases electrolyte was added to these micellar solutions to obtain narrower size distributions. The MAA residues were then shell cross-linked by addition of BIEE (40 µl; 42 mol % based on MAA residues; 60°C for 2 h). After cooling to 25°C, the resulting zwitterionic micelles were isolated by precipitation from aqueous solution at their isoelectric point.

33. (a) Iizawa, T.; Matsuda, F. *Polym. J.*, **1998**, 30, 155-157, (b) Matsuda, F.; Matsuno, N.; Iizawa, T. *Kobunshi Ronbunshu*, **1998**, 55, 440-447.

34. Patrickios, C. S.; Hertler, W. R.; Abbott, N. L.; Hatton, T. A. *Macromolecules*, **1994**, 27, 930-937.

35. Hoogeveen, N. G.; Stuart, M. A. C.; Fleer, G. J.; Frank, W.; Arnold, M. *Macromol. Chem.*, **1996**, 197, 2553-2564.

36. Wolfert, M. A.; Schact, E. H.; Toncheva, V.; Ulbrich, K.; Nazarova, O.; Seymour, L. W. *Human Gene Therapy*, **1996**, 7, 2123-2133.

37. Napper, D. H. *Polymeric Stabilization of Colloidal Dispersions*, Acadmic Press, London, **1983**.

38. (a) Coombes, A. G. A.; Tasker, S.; Lindblad, M.; Holmgren, J.; Hoste, K.; Toncheva, V.; Schacht, E.; Davies, M. C.; Illum, L.; Davis, S. S. *Biomaterials*, **1997**, 18, 1153-1161; (b) Stolnik, S.; Felumb, N. C.; Heald, C. R.; Garnett, M. C.; Illum, L.; Davis, S. *Colloids Surf. A - Physicochem. Eng. Aspects*, **1997**, 122, 151-159; (c) Tobio, M.; Gref, R.; Sanchez, A.; Langer, R.; Alonso, M. J. *Pharmaceutical Research*, **1998**, 15, 270-275.

39. The average degrees of polymerization for this triblock copolymer were 45 for PEO, 157 for DMA and 172 for MEMA.

40. The average degrees of polymerization for this triblock copolymer were 45 for PEO, 49 for DMA and 71 for BAEMA.

41. Won, Y. Y.; Davis, H. T.; Bates, F. S. *Science*, **1999**, 283, 960-963.

Chapter 8

Controlled Polymerization of Acrylamides

Marina Baum, Mical E. Pallack, Jude T. Rademacher, and William J. Brittain

Department of Polymer Science, The University of Akron Akron, OH 44325-3909

Polyacrylamides are an important class of water-soluble polymers, which are used in a variety of industries such as medicine, oil, textile, and waste management. Interest in the development of new polymerization methods for well-defined molecular architectures of polyacrylamides is growing due to the versatility of these polymers. Anionic polymerization has long been known to provide the control needed for synthesis of such molecules. However, stringent reaction conditions and high purity of reagents make this synthetic method unattractive for industrial use. Over recent years, several more robust "controlled" radical polymerization processes have been developed. Among them are atom transfer radical polymerization (ATRP), reversible addition fragmentation chain transfer process (RAFT), and nitroxide-mediated stable free radical polymerization (SFRP). All of these methods utilize reversible activation-deactivation methods to control the concentration of growing radicals in order to decrease termination. This chapter provides an overview of research done in the area of "controlled" polymerization of acrylamides.

Research in living free radical polymerization (*1, 2*) has become an active area of polymer synthesis. Work has been published on nitroxide-mediated processes, (*3, 4, 5*) atom transfer radical polymerization (ATRP), (*1*) and reversible addition fragmentation chain transfer (RAFT). (*6*) Most work on living radical polymerization has focused on styrene and (meth)acrylates. Only recently, has the living radical polymerization of acrylamide-based monomers become a topic of interest. (*7*) Polyacrylamides are an important class of water-soluble polymers with a wide range of applications in a variety of industries. (*8*) Due to the non-toxicity and biocompatibility of these polymers, polyacrylamides are used in medical applications for drug-delivery, implants, and DNA sequencing gels. As water-soluble polymers, polyacrylamides are used as viscosifiers in industries such as oil, irrigation pumping, and fire fighting. They are also widely utilized as flocculation agents in solid recovery, waste removal, and water clarification. Due to their adhesive and film-forming properties, polyacrylamides are important in textile and paper industries. (*9*) Methods for controlled polymerization would be very valuable for many of these applications. Conventional polymerizations suffer from uncontrolled chain transfer and termination, which lead to poor control of molecular weight distribution, end-group functionalities, and chain architecture. Controlled processes allow the synthesis of specific molecular weights and well-defined molecular architectures.

Anionic Polymerization

Anionic polymerization has long been known to proceed in the absence of chain transfer and termination and result in narrow molecular weight distribution polymers. Xie and Hogen-Esch (*10*) have described the anionic polymerization of *N,N*-dimethylacrylamide (DMA) and *N*-acryloyl-*N'*-methylpiperazine. These workers produced polyacrylamides with controlled molecular weights and narrow molecular weight distributions when polymerizations were conducted at low temperatures (-78 °C) in THF initiated in the presence of triphenylmethyl cesium (*Scheme 1*).

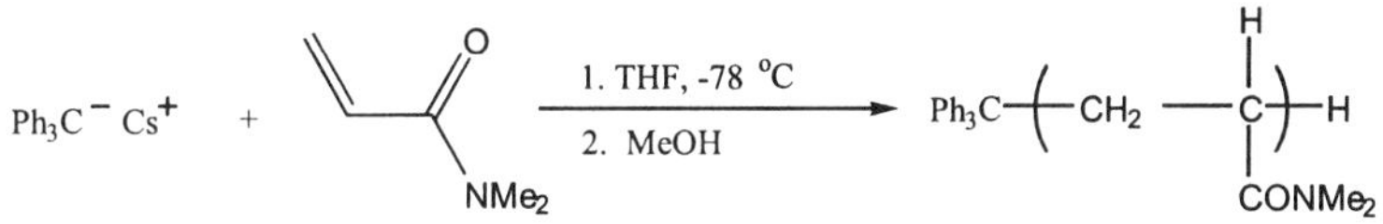

Scheme 1. Anionic polymerization of DMA.

Recently, Nakahama and co-workers (*11*) anionically prepared stereospecific poly(*N, N*-dialkylacrylamides) with narrow polydispersity in the presence of Et_2Zn. These reactions were conducted in THF at 0 °C for 1 h.

The disadvantages of anionic polymerization include extensive purification of reagents and stringent reaction conditions. More robust methods for controlled polymerizations are desired.

Radical Polymerization

Radical polymerizations are known to be very robust systems and are widely used in industry. However, conventional free radical polymerization methods do not offer the degree of control that we seek. Over the past few decades, controlled radical polymerization methods have been developed and include SFRP, ATRP, and RAFT. SFRP and ATRP reactions are governed by the same general scheme (*Scheme 2*). In this reaction the persistent species (does not initiate polymerization) reversibly couples with the transient species (propagating species) to form the "dormant" species. To achieve control, the equilibrium lies strongly to the left reducing the number of active species and thus minimizing the number of termination reactions. Consequently this process also lowers the rate of propagation. In RAFT, while an active polymer chain end is growing, the reversible capping agent is a non-radical dithioester species.

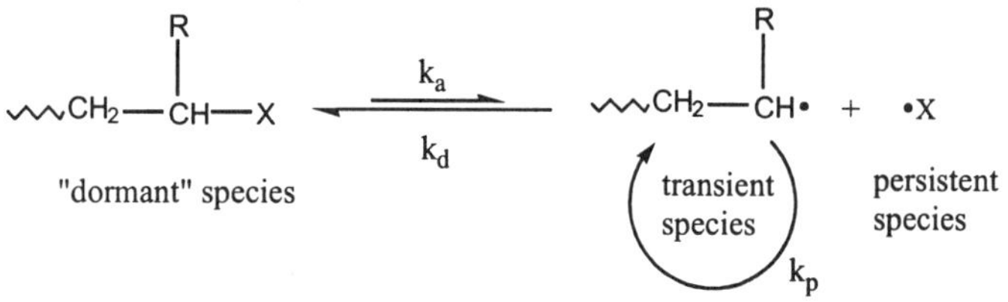

Scheme 2. General scheme for SFRP and ATRP.

The living free radical polymerization of acrylamides has not been widely studied.

Nitroxide-Mediated SFRP

A classic example of controlled radical polymerization is SFRP, where a growing radical reversible recombines with a scavenging radical forming a dormant species (*Scheme 3*). In this reaction, the carbon-oxygen bond of the "dormant" species undergoes thermal fragmentation to give a stable, persistent nitroxide radical and a reactive polymeric radical. Benoit, Chaplinski, Braslau, and Hawker (*3*) reported nitroxide-mediated polymerization of DMA by use of an alkoxyamine initiator. These reactions were carried out in bulk using 2, 2, 5-trimethyl-3-(1-phenylethoxy)-4-phenyl-3-azahexane initiator with the addition of 0.05 eq of the corresponding nitroxide radical at 120 °C. Polymers with narrow molecular weight distributions and molecular weights ranging from 4,000-50,000 g/mol were synthesized.

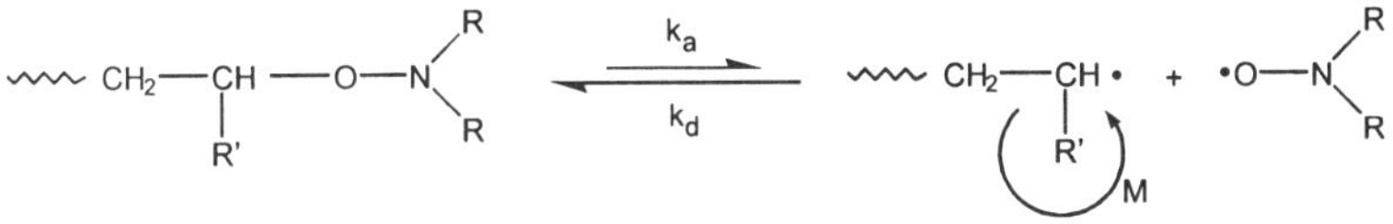

Scheme 3. *Nitroxie-mediated SFRP scheme.*

Reversible Addition Fragmentation Chain Transfer Technique

Another method for controlled free radical polymerization is RAFT. Controlled character is obtained by a reversible transfer of a dithio moiety between active and "dormant" species (*Scheme 4*). Le, Moad, Rizzardo, and Thang (*12*) reported the RAFT polymerization of DMA using 2-phenylprop-2-yl dithiobenzoate as chain transfer agent. These reactions were conducted in benzene at 60 °C using AIBN as the initiator to afford near-monodisperse polymers with specific molecular weights. The researchers reported low conversions (13-31%) after 1 h.

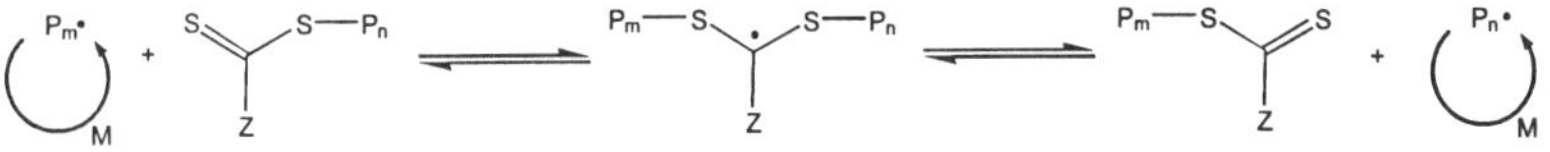

Scheme 4. *General RAFT scheme.*

Atom Transfer Radical Polymerization

ATRP is a controlled free radical polymerization based on a reversible redox activation-deactivation method to reduce the concentration of growing radicals (*Scheme 5*). Here, a metal halide complexed with a ligand reversibly donates a halogen to a growing chain-end forming a "dormant" species and thus minimizing the number of reactive species. Matyjaszewski and co-workers (*1*) have extensively investigated ATRP of styrenes and acrylates. They have demonstrated that this is a viable method for the synthesis of near-monodisperse polymers and copolymers with predetermined molecular weights.

$$R\text{—}X \;+\; Mt^n / Ligand \;\underset{k_d}{\overset{k_a}{\rightleftharpoons}}\; R\bullet \;+\; X\text{-}Mt^{n+1} / Ligand$$

$$\xrightarrow{k_t} R\text{-}R \qquad +M \quad k_p$$

Scheme 5. *General ATRP scheme.*

These successful living polymerizations of styrene- and acrylate-based monomers led to the investigation of ATRP for acrylamido monomers. Recent work by Teodorescu and Matyjaszewski (*13*) and Brittain and co-workers (*14*) have shown that copper-based ATRP of DMA is not controlled. Both research groups investigated various solvents, ligands, initiators and copper halides and observed broad polydispersities and lack of agreement between experimental and theoretical M_n. The uncontrolled nature of ATRP of DMA was confirmed by the polymer growth resumption experiment in which only a small fraction of the chain ends increased in molecular weight. (*14*) It was found that a variety of solvents could be used to obtain high DMA conversions including DMF, water and toluene. Polymerizations conducted in water proceeded most rapidly. In fact, complete monomer conversions could be obtained in minutes. This is consistent with literature reports on the conventional radical polymerization of (meth)acrylamides where water exerts an accelerating effect on polymerization. (*15*) Of course, these fast reaction times are undesirable for controlled/"living" radical polymerization. An important mechanistic feature of controlled radical polymerizations is the ability to maintain a low radical concentration in order to minimize spontaneous, bimolecular termination. Rapid polymerizations indicate a high concentration of radicals.

Interestingly, Wirth and co-workers (*16*) have claimed a living polymerization of acrylamide using ATRP with surface-immobilized initiators. The evidence for living polymerization was based on a correlation between film thickness and monomer concentration; also narrow molecular weight distributions were cited by Wirth and co-workers as evidence that their polymerization was living. A truer test of living behavior in Wirth's system would have been polymer growth resumption. It is possible that acrylamide behaves differently than DMA or that surface-initiated polymerizations are mechanistically different than solution polymerizations.

Teodorescu and Matyjaszewski (*13*) offered several reasons why there is lack of control in the copper-mediated ATRP of (meth)acrylamides. These reasons include: 1) deactivation of catalyst by polymer complexation, 2) strong bond between bromine and the terminal monomer unit in the polymer, and 3) nucleophilic displacement of terminal bromine by the penultimate amide group. They concluded that the first two reasons could not be solely responsible for lack of control in the polymerizations. Brittain and co-workers (*14*) cited the organic literature which supports the contention that the uncontrolled polymerization is due to slow deactivation of the chain ends. Several examples have been reported of radical generation from alkyl bromides where the reaction is promoted by complexation of Lewis acids to amide bonds. (*17*) A particularly relevant example is that of Sibi and Ji (*18*) where the alkyl bromide (*Scheme 6*) is more readily transformed to the corresponding radical when a Lewis acid is present (e.g., $MgBr_2$ or a lanthanide). A similar example of this effect was observed by Porter and co-workers. (*19*) Both of these reports demonstrate that the presence of a Lewis acid can have a profound effect on C-Br bond lability in chemical structures bearing an imide group alpha to the bromine. Using the reactivity-selectivity principle, it was concluded that the formation of a C-Br bond from an amide-substituted radical would be less favorable in the presence of a Lewis acid. ATRP involves metals that can serve as Lewis acids. It is the presence of the metal and its complexation to the amide functionality that slows deactivation in ATRP of (meth)acrylamides and makes the process an uncontrolled polymerization.

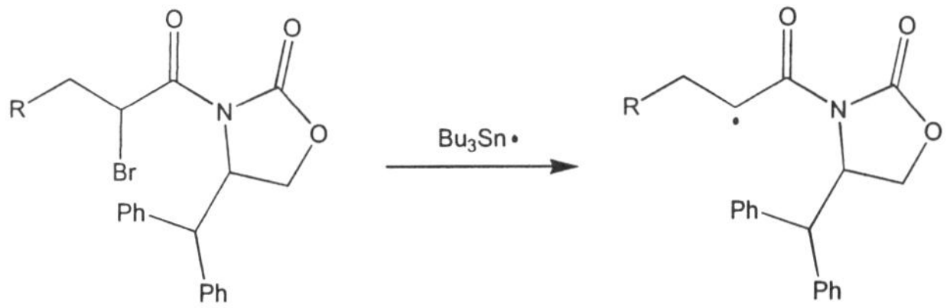

Scheme 6. Radical formation in the presence of a Lewis acid.

Sawamoto and co-workers (*20*) investigated ATRP of DMA using $RuCl_2(PPh_3)_3$ in conjunction with an alkyl halide as an initiator in the presence of $Al(Oi\text{-}Pr)_3$ in toluene at 60-80 °C. They observed faster polymerization rates

146

than those of methacrylates and obtained polymers with controlled molecular weights and polydispersities of ~1.6. Polymerization proceeded *via* the activation of the terminal carbon-halide bond by the ruthenium complex. Living character was determined by polymer growth resumption experiments in which most of the polymer chains resumed growth upon addition of the second charge of monomer. Success of these polymerizations suggests that the metal catalyst has a profound effect on the nature of these reactions. It seems that Ru metal does not have the same complexing characteristic as Cu with the amide group.

In summary, the most promising methods for controlled radical polymerization of acrylamides are SFRP and RAFT. More in-depth research into these areas is necessary for better understanding of the controlled nature of these reactions. Copper-based ATRP is not a viable system for controlling acrylamide polymerization. However, new results in ruthenium-based ATRP show promise.

References

1. Matyjaszewski, K. *Controlled Radical Polymerization;* ACS Symposium Series 685; American Chemical Society: Washington, DC 1998.
2. Hawker, C. J. *Acc. Chem. Res.* **1997**, *30*, 373.
3. Benoit, D.; Chaplinski, V.; Braslau, R.; Hawker, C. J. *J. Am. Chem. Soc.* **1999**, *121*, 3904.
4. Georges, M. K.; Veregin, R. P. N.; Kazmaier, P. M.; Hamer, G. K. *Trends Polym. Sci.* **1994**, *2*, 66.
5. Odell, P. G.; Veregin, R. P. N.; Michalak, L. M.; Georges, M. K. *Macromolecules* **1997**, *30*, 2232; and references contained therein.
6. Chong, Y. K.; Le, T. P. T.; Moad, G.; Rizzardo, E.; Thang, S. H. *Macromolecules* **1999**, *32*, 2071.
7. Li, D.; Brittain, W. J. *Macromolecules* **1998**, *31*, 3952.
8. Shalaby, S. W.; McCormick, C. L.; Butler, G. B. *Water-Soluble Polymers: Synthesis, Solution Properties and Applications,* ACS Symposium Series 467, American Chemical Society: Washington, DC 1991.
9. *Polymeric Materials Encyclopedia*; Salamone, J. C., Ed.; CRC Press, Inc.: Boca Raton, **1996**, 47.
10. Xie, X.; Hogen-Esch, T. E. *Macromolecules* **1996**, *29*, 1746.
11. Kobayashi, M.; Okuyama, S.; Ishizone, T.; Nakahama, S. *Macromolecules,* **1999**, *32*, 6466.
12. Le, T.; Moad, G.; Rizzardo, E.; Thang, S. H. *Intern. Pat.* WO 98/01478, 1998.
13. Teodorescu, M.; Matyjaszewski, K. *Macromolecules* **1999**, *32*, 4826.

14. Rademacher, J. T.; Baum, M.; Pallack, M. E.; Brittain, W. J.; Simonsick, W. J. *Macromolecules* **2000**, *33*

15. Hamilton, C. J.; Tighe, B. J. In *Comprehensive Polymer Science*; Eastmond, G. C.; Ledwith, A.; Russo, S.; Sigwalt, P., Ed.; Pergamon Press: New York, 1989; Vol. 3, Chapter 20.

16. Wirth, M. J.; Huang, X. *Macromolecules* **1999**, *32*, 1694.

17. Sibi, M.; Porter, N. A. *Acc. Chem. Res.* **1999**, *32*, 163; Mero, C. L.; Porter, N. A. *J. Am. Chem. Soc.* **1999**, *121*, 5155.

18. Sibi, M. P.; Ji, J. *Angew. Chem., Int. Ed. Engl.* **1996**, *35*, 190.

19. Wu, J. H.; Radinov, R.; Porter, N. A. *J. Am. Chem. Soc.* **1995**, *117*, 11029.

20. Senoo, M.; Kotani, Y.; Kamigaito, M.; Sawamoto, M.; *Macromolecules*, **1999**, *32*, 8005.

Chapter 9

Water-Soluble and Water Dispersible Polymers by Living Radical Polymerisation

Stefan A.F. Bon, Kohji Ohno and David M. Haddleton*

University of Warwick, Department of Chemistry, Coventry CV4 7AL, UK.

An overview of recent advances in living radical polymerisation under aqueous conditions and the synthesis of water-soluble and dispersible polymers by living radical polymerisation techniques is given. Living radical polymerisation involving nitroxide stabilised radicals, transition metal mediated and reversible addition fragmentation (RAFT) is covered. This overview is followed by some specific examples of the use of copper(I) mediated living radical polymerisation for the synthesis of some water soluble/dispersible polymers. Firstly the use of a modified SPAN surfactant to polymerise 2-(dimethylamino)ethyl methacrylate (DMAEMA) producing polymers with narrow polydispersity and controlled Mn is described (Mn = 6700, PDI = 1.27). These polymers disperse in acidic aqueous media with a CAC of 0.16 gL^{-1}. Derivatised solketal is described as an initator for sequential atom transfer polymerisation of methyl methacrylate (MMA) and DMAEMA to give Y-shaped water soluble polymers. These two examples serve to illustrate the range of topology and hydrophilic functionality which can easily be incorporated into vinyl polymers through living radical methodology.

INTRODUCTION

Living radical polymerisation has emerged as a very effective method of producing macromolecules with a high level of control over the polymer architecture and topology. Recent advances can be traced to the use of stable nitroxides, first introduced by Moad (*1*), developed by Georges (*2*) and more recently by Hawker (*3, 4*) for nitroxide mediated radical polymerisation (NMRP). NMRP enabled polystyrene to be produced with high molecular mass and narrow polydispersity and has subsequently been used to develop a range of block copolymers and other polymers with unusual architecture. Following on from this, transition metal mediated living radical polymerisation was introduced by both Sawamoto (*5*) and Matyjaszewski (*6*) with Ru(II) and Cu(I) catalysts respectively. These major breakthroughs have led to an enormous amount of work in this area in a relatively short period of time (*7, 8*). This has produced a variety of different transition metal catalysts including Ni(II) (*9*), Rh(I) (*10*), Cu(0) (*11*), Pd(0) (*12*), Fe(II) (*13*), etc. A plethora of different polymer types have been reported ranging from statistical copolymers (*14*), block copolymers (*15*), star copolymers (*16, 17*) and even a new type of polymer, gradient copolymers (*14*). New ligands have been introduced as have different processes designed to help with removal and recycling of catalysts (*18, 19*). Indeed more recently further advances have been reported by Rizzardo and Moad using a range of dithio reagents as reversible addition fragmentation transfer agents (RAFT) which is a metal free system for living radical polymerisation (*20*).

Probably the greatest attraction of a living radical process over an ionic system is the tolerance to impurities and/or functionality present in monomers, solvents and reagents. The removal of the need for protecting group chemistry and potentially expensive purification steps has obvious advantages. Increasingly this has meant that water soluble or dispersible polymers have been targeted, as has the use of water as a solvent or dispersing medium (suspension or emulsion polymerisation).

This present contribution will be in two parts. Firstly there will be short (non-comprehensive) review highlighting both the use of water as a reaction medium and the synthesis of highly hydrophilic/functional polymers designed for applications involving water. Secondly this will be illustrated with some new examples from our own laboratory so as to provide experimental details and an idea of the potential of these exciting processes.

Recent developments in living radical polymerisation in aqueous media

One of the first significant demonstrations of living radical polymesization in water was carried out heterogeneously under emulsion polymerisation conditions by Bon and coworkers (*21*). They used a nitroxide system (TEMPO) in a seeded emulsion polymerisation of styrene under pressure at 398 K. This resulted in an evolution of molecular mass with conversion albeit with a broadening of polydispersity, especially to low mass. The related cobalt mediated catalytic chain transfer polymerisation, which is not a living polymerisation but does give controlled polymerisation to vinyl terminal macromonomers, is most effectively carried out under emulsion conditions or indeed aqueous solution (*22, 23*). The macromonomers produced by CCTP act as addition fragmentation reagents themselves giving rise to living polymerisation (*24*). This has been used to prepare poly(methacrylate)s where the molar mass increases linearly with conversion and also poly(methacrylate) block copolymers under emulsion polymerisation conditions. Polymers produced via this macromonomer addition fragmentation show a broadening of polydispersity due to the low chain transfer coefficients of the macromonomers towards propagating methacrylate radicals, typically of the order of 0.1. More recently addition fragmentation agents based on thio esters and xanthates have been shown to give very effective living polymerisation and the facile production of block copolymers. This reversible addition fragmentation chain transfer (RAFT) process is particularly useful under emulsion polymerisation conditions and represents a significant step forward in the production of block copolymers in heterogeneous processes. For example, many different AB and ABA block copolymers of methacrylates, acrylates and styrene with Mn's over 100 K and PDi less than 1.20 have been synthesised containing both hydrophilic and hydrophobic segments (*20, 25*).

Transition metal mediated living radical polymerisation has probably received the most attention in this area to date, since its inception in 1995. As well as the tremendous advances in organic solution progress has been made with emulsion, suspension and aqueous solution processes. Sawamoto described living radical polymerisation of MMA in mixtures of alcohols and water under suspension polymerisation conditions mediated by $RuCl_2(PPh_3)_3$ to give polymers with controlled molar masses and narrow PDI (*26*). This reinforced the observation that dry solvents are not necessary for effective use of this chemistry. Indeed copper(I) has been shown to be tolerant of trace organic acids (*27*). Gaynor and Matyjaszsewsi have described the application of a copper(I) bipyridyl catalyst for the living radical polymerisation in water-borne systems demonstrating the linear evolution of M_n with conversion for butyl methacrylate and butyl acrylate under emulsion/mini-emulsion polymerisation conditions using a non-ionic surfactant (*28*). This work addressed the problem of residual catalyst in an elegant way by addition of Dowex MSC-1 ion-exchange resin upon completion of polymerisation, which removes the intense green colour very effectively.

Armes has addressed two of the outstanding concerns of living radical polymerisation, namely polymerisation in aqueous solution and polymerisation of acidic monomers. The use of water-soluble macro-initiators, based on methoxy poly(ethylene glycol) (PEG) with the sodium and potassium salts of methacrylic acid leads to living polymerisation with controlled Mn and narrow PDi. The system is also very effective for neutral monomers based on PEG in aqueous solution, showing enhanced rates of polymerisation. The use of water-soluble macro-initiators has also been demonstrated with water-soluble acid salts of styrenic monomers. This work demonstrates that the problem with the free acid monomers is protonation of the ligands/catalysts rendering them inactive and not complexation of the carboxylate groups at the metal. Thus, water is an effective medium for living radical polymerisation in both homogeneous and inhomogeneous systems. Indeed due to the large increase in rate of polymerisation, water may well be the solvent of choice in certain specific cases.

Synthesis of hydrophilic polymers and block copolymers by living radical polymerisation.

Most living radical polymerisation chemistry is inert to the majority of functional groups. Perhaps one of the only main exceptions to this is low pH with transition metal mediated polymerisation systems where there is a tendency for either the ligands or metal complex to be hydrolytically unstable. This tolerance allows many functional monomers to be polymerised which has led to an ever increasing number of reports concerning hydrophilic and amphiphilic block and statistical copolymers. Both hydroxyethyl acrylate (*29*) and methacrylate (*30*) have been polymerised directly to homopolymers in MEK/propan-1-ol solvent mixtures and in the bulk and to block copolymers with hydrophobic monomers such as MMA. Direct polymerisation gives polymers with a relatively broad polydispersity (approx. 1.3) which can be narrowed by using trimethylsilyl protected monomer. Protected hydroxyl functional monomers have been used to prepare a range of amphiphilic block copolymers with butyl acrylate (*31*). Tertiary amine containing monomers such as 2-(dimethylamino)ethyl methacrylate undergo living polymerisation in a range of solvents to yield water-soluble polymers in both the quaternized and non quaternized states. There have been several attempts to prepare poly(acrylamide)s by atom transfer polymerisation (*32, 33*), which unfortunately proved elusive to date due to premature termination.

One of the simplest amphiphilic block copolymers has been reported by Kops, who has used a poly(ethylene glycol) macroinitiator (*34*), in a similar fashion to Armes (*35*), to prepare ABA block copolymers where A is poly(styrene). More sophisticated structures have been synthesised by Fukuda and Ohno with sugar containing monomers to prepare glycopolymer and artificial glycolipids (*36*) by both NMRP (*37*) and transition metal mediated polymerisation (*38*). Indeed adenosine and uridine functional methacrylates and acrylates have recently been polymerised successfully in both the protected and unprotected forms (*39*).

Hence a wide range of functionality can be introduced relatively easily into statistical copolymers, block copolymers and polymers of other more unusual topologies. Living radical polymerisation is currently being used to target "effect polymers" with specific and often high added value, applications. Invariably these new polymers contain hydrophilic groups and are either water-soluble or water-dispersible and have potential use in a range of products including personal products, laundry detergents, new surfactants, contact lenses, etc. It is in this area where living radical polymerisation has the most attractive potential applications. The remainder of this paper will give brief examples to illustrate these points.

Novel polymers from atom transfer polymerisation using glyco-initiators

The initiators used in our group for atom transfer polymerisation are usually of general structure **A**.

These are readily synthesised via the esterification of a hydroxyl functional compound with 2-bromoisopropionic acid (or preferably its acyl bromide or anhydride analogue). Two different procedures, one using the acyl bromide of 2-bromoisopropionic acid and the other using its anhydride derivative, to synthesize initiators of structure **A** have been used in this current work. The first comprises the dropwise addition of 1.1 mole equivalent of 2-bromo-2-methylpropionylbromide to a stirred solution of the hydroxyl compound in tetrahydrofuran in presence of a fourfold excess of triethylamine under anhydrous conditions at 0 °C. The second method is to react an excess amount of 2-bromoisopropionic anhydride with the hydroxyl compound in pyridine as a solvent at room temperature under anhydrous conditions in presence of a small amount of dimethylaminopyridine (DMAP) as catalyst (see Experimental section for details).

Monofunctional carbohydrate initiators.

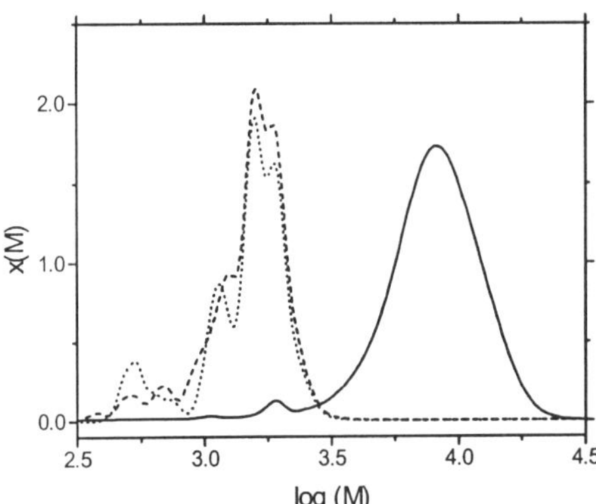

Structure of 1

Compound **1** was synthesised from sorbitan tristearate (span 65), which is one species of a range of commercial available cyclic natural fatty-acid modified sorbitol derivatives. **1** was used as the initiator for the atom transfer polymerisation of 2-(dimethylamino)ethyl methacrylate. The reaction was carried out at 90 °C for a period of 4 h to reach nearly complete conversion, after which the product was purified by passing the solution over basic alumina. The SEC spectra of the original Span 65, its modified initiator analogue and the final PDMAEMA are given in Figure 1. It can be seen that Span 65 is in fact a mixture of sorbitan derivatives (Sigma catalogue states 50% stearic acid; primarily balanced palmitic acid). It seems that all species are modified to monofunctional atom transfer polymerisation initiators, which is seen from the narrow MWD of the resulting amphiphilic PDMAEMA ($<M_n> = 6.7 \times 10^3$; PDI = 1.27).

Figure 1. *Differential log MWD of Span 65 (·····), **1** (-----) and the PDMAEMA amphiphilic polymer (———) acquired with DRI detection.*

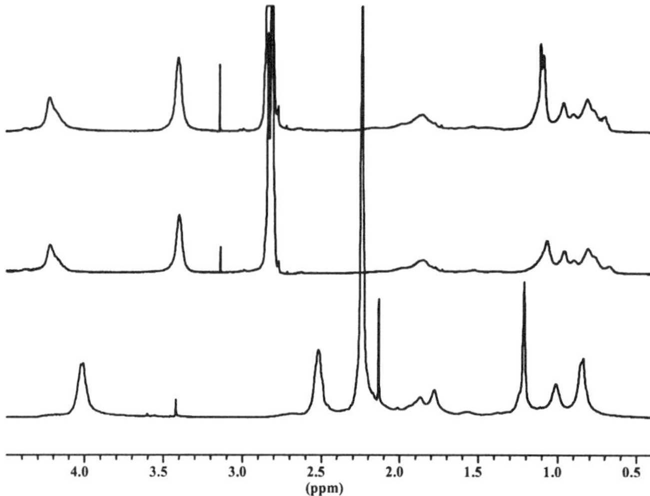

Figure 2. *^{1}H NMR spectra of the amphiphilic PDMAEMA in neat CDCl$_3$ (lower spectrum), acidified D$_2$O (2 drops conc. HCl (aq.); middle spectrum), and acidified D$_2$O in presence of a small amount of CDCl$_3$ (top spectrum).*

The polymer was expected to undergo aggregation in acidified water, due to protonation of the PDMAEMA. This aggregation in water will lead to the formation of hydrophobic domains consisting of the aliphatic chains from the Span 65 part of the polymer molecule. Whether the structure of the aggregates is unimicellar or of higher order, is currently being studied. The formation of aggregates was investigated by ^{1}H NMR analysis. When a fixed pulse angle and relaxation delay are used the intensity of peak can be correlated to the mobility of the specific hydrogen atoms. Aggregation reduces mobility and, thus results in a reduced observed signal.

Figure 2 shows ^{1}H NMR spectra of the amphiphilic PDMAEMA recorded in neat CDCl$_3$ (lower spectrum), acidified D$_2$O (2 drops conc. HCl (aq.); middle spectrum), and acidified D$_2$O in presence of a small amount of CDCl$_3$ (top spectrum). The ^{1}H NMR spectra recorded in neat CDCl$_3$ was used as reference. It can be seen that the intensity of the aliphatic chains decreases upon analysis of the amphiphilic polymer in acidified D$_2$O as a result of aggregation (*ca.* 1.2 ppm in CDCl$_3$ and 1.1 ppm in D$_2$O). Addition of CDCl$_3$ to the aqueous solution swells the aggregates and, therefore, increases the mobility of the hydrophobic regions. This is clearly demonstrated in the ^{1}H NMR spectrum. Moreover, the addition of the chloroform leads to a non-transparent solution, clearly indicating surfactant properties for the Span 65 end-functionalized PDMAEMA.

An alternative method to study the aggregation behavior of amphiphilic polymers and to determine the concentration of polymer in water at which aggregates are being formed, *i.e.*, the critical aggregation concentration (CAC), has been developed and widely explored for a variety of polymers by Winnik (40, 41). Part of this method is based on the partitioning of pyrene between the waterphase and the newly formed hydrophobic phase and the implications of changes in both its emission and absorbance fluorescence spectra.

Figure 3 represents $I_{338.5}/I_{333}$ taken from the excitation spectrum of pyrene *vs.* the concentration of the Span 65 end-functionalized PDMAEMA in water kept at a pH of 3.60. In our opinion a good value for the critical aggregation concentration can be obtained by fitting the experimental data with a sigmoidal function and taking the first maximum of its second derivative. This results in a value for the CAC of 0.16 g L^{-1}.

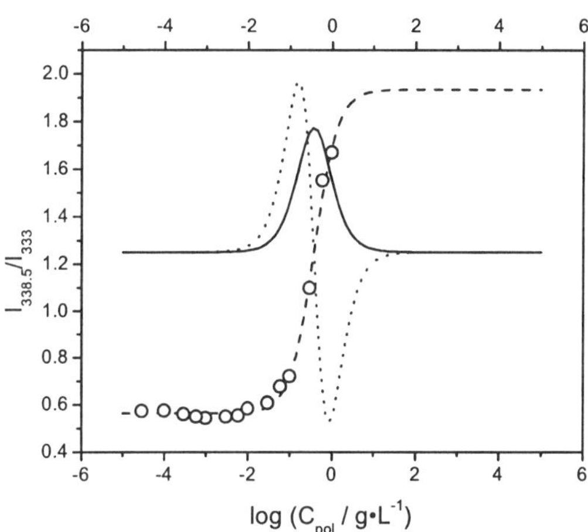

Figure 3. *(O) $I_{338.5}/I_{333}$ from the excitation spectrum of pyrene vs. the concentration of the Span 65 end-functionalized PDMAEMA in water (pH = 3.60) The lines represent the sigmoidal fit and its 1st and 2nd derivative.*

Synthesis of Y-shaped amphiphilic copolymers

Compound **2c** was synthesised in a two step procedure from solketal (**2a**). This consisted of esterification with 2-bromo-2-methylpropionyl bromide to yield **2b,** in nearly quantitative yields, and subsequent removal of the acetonide protective group. The use of 80% formic acid resulted in partial elimination of HBr, and stirring in methanol in presence of catalytic amount iodine failed completely. Heating a solution

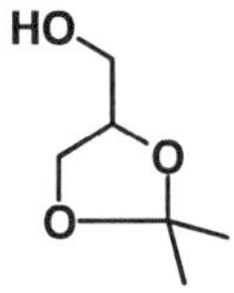

2a 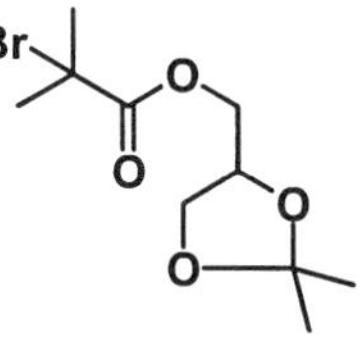   2b 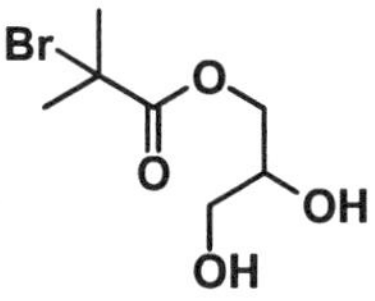   2c

of compound **2b** in a mixture of glacial acetic acid and water (15/40 v/v%) in the presence of a small amount of methoxybenzene at 80 °C for 30 min resulted in complete removal of the acetonide group to yield **2c** quantitatively. Compound **2c** was used as initiator in the atom transfer polymerisation of MMA in toluene at 90 °C as a first step in the preparation of Y-shaped polymers. The targeted molecular weight at complete monomer conversion was 5.0×10^3 g mol^{-1}. The reaction was quenched after 2 h to yield, following column chromatography and precipitation purification, PMMA having an $<M_n>$ of 6.15×10^3 g mol^{-1} and PDI of 1.07. There is a marked rate enhancement, which is ascribed to co-ordination of the hydroxyl groups at the copper centre.

The second stage consisted of reacting the two hydroxyl groups at the α–terminus of the polymer chain with 2-bromo-2-methylpropionyl bromide. The reaction mixture was stirred overnight to yield a polymer chain with further two further initiator sites at its α–terminus and one initiator site at its ω–terminus. This PMMA macroinitiator was used in the subsequent atom transfer polymerisation of DMAEMA (DMAEMA:toluene 9.33:20 v:v%) at 90 °C for 30 min, carried out at an initiator site concentration of 16.6 mmol L^{-1}. This yielded the targeted Y-shaped poly(2-dimethylamino]ethyl methacrylate)-*block*-poly(methyl methacrylate) (**3**; $<M_n>$ = 7.76×10^3; PDI = 1.08).

3

Summary

Living radical polymerisation has emerged as a very attractive route for the synthesis of hydrophilic polymers with controlled structures. The robustness of the chemistry employed often means that protecting groups are not required. Transition metal mediated systems have been used in heterogeneous and aqueous solution. A range of water soluble polymers have already been prepared including relatively simple AB and ABA block copolymers to polymers containing potential bio-recognition sites. Undoubtedly a plethora of new water-soluble polymers will become available over the next few years.

Experimental Section

All manipulations were carried out using standard Schlenk line techniques under nitrogen. Reagents were used without further purification unless otherwise stated. NMR spectroscopy on polymers was performed on a Bruker AC 400 MHz using a 30° pulse angle and long relaxation delays. All other NMR analyses were performed on a Bruker DPX 300 MHz. Size exclusion chromatography (SEC) analyses were carried out either on a system equipped with a guard column and one mixed bed E column (Polymer Labs) or equipped with a guard column and two mixed bed C columns (Polymer Labs). Both systems were operated using tetrahydrofuran as eluent at a flow rate of 1 mL min^{-1} at room temperature. The molecular weight distributions were detected using differential refractive index and were calibrated specifically using narrow MWD PMMA standards. Fluorescence measurements were carried out on a Perkin-Elmer LS50 luminescence spectrometer thermostated at 25 ± 0.1 °C with

158

a band-width for both excitation and emission of 2.5 nm. For the fluorescence emission spectra, λ_{ex} was 339 nm, and for the excitation spectra λ_{em} was 390. Eight scans were accumulated for each final spectrum.

Sorbitan-5-oxy-(2-bromo-2-methylpropionyl)-2,3,6-tristearate, (**1**). This was prepared on a 0.8 M scale using a similar procedure as given below for the synthesis of compound (**2b**). Yellowish solid was obtained in quantitative yields. ^{1}H NMR (CDCl$_3$) δ 0.86 (t, 9H, C$\underline{H}_3$-ailph.), 1.23 (br. s., approx. 90H), 1.59 (br. s., 6H), 1.93 (s, 6H, CBr(C$\underline{H}_3$)$_2$), 2.31 (m, 6H).

2,2-Dimethyl-1,3-dioxolane-4-methoxy-(2-bromo-2-methylpropionyl) (**2b**). This was prepared by esterification of 2,2-dimethyl-1,3-dioxolane-4-methanol with 2-bromo-2-methylpropionylbromide. Triethylamine was dried over KOH pellets prior to use and the glassware dried at 150 °C overnight. 2,2-Dimethyl-1,3-dioxolane-4-methanol (0.04 mol, 5.29 g), triethylamine (0.08 mol, 8.10 g) and anhydrous tetrahydrofuran (50 mL) were charged in a 100 mL round-bottomed flask equipped with a magnetic stirrer and cooled to 0 °C with an ice-bath. 2-Bromo-2-methylpropionylbromide (0.044 mol, 5.44 mL) was added dropwise via syringe. Next the mixture was stirred for 45 min and allowed to reach room temperature. After this the reaction mixture was poured into an excess of cold water and extracted with 50 mL of diethylether. The organic layer was subsequently washed with a saturated aqueous solution of Na$_2$CO$_3$, acidified water (pH ≈ 4.5), and again the saturated aqueous solution of Na$_2$CO$_3$. The organic layer was dried over anhydrous Na$_2$SO$_4$. Finally, the Na$_2$SO$_4$ was removed by filtration and the solvent was removed using a rotary evaporator in order to isolate the title compound in quantitative yield as a slightly yellowish oil. ^{1}H NMR (DMSO-d_6) δ 1.30 and 1.37 (each s, 3H, OC(C$\underline{H}_3$)$_2$O), 1.92 (s, 6H, C(C$\underline{H}_3$)$_2$Br), 3.75 (dd, 1H, C5$\underline{H}_a$H$_b$, J_{ab} = 8.55 Hz, J_Z = 6.10 Hz), 4.05 (dd, 1H, C5H$_a\underline{H}_b$, J_{ab} = 8.55 Hz, J_E = 6.41 Hz), 4.15 (dd, 1H, C$\underline{H}_a$H$_b$, J_{ab} = 11.60 Hz, J_E = 4.89 Hz), 4.22 (dd, 1H, CH$_a\underline{H}_b$, J_{ab} = 11.60 Hz, J_Z = 3.97 Hz), 4.31 (m, 1H C4$\underline{H}$). ^{13}C NMR (DMSO-d_6) δ 25.6 and 26.8 (1C, OC($\underline{C}H_3$)$_2$O), 30.6 (2C, C($\underline{C}H_3$)$_2$Br), 57.3 (1C, $\underline{C}$(CH$_3$)$_2$Br), 65.5 (1C, $\underline{C}$5H$_a$H$_b$), 65.7 (1C, $\underline{C}$H$_a$H$_b$), 73.3 (1C, $\underline{C}$4H), 109.2 (1C, O$\underline{C}$(CH$_3$)$_2$O), 171.0 (1C, C=O).

1,2-dihydroxypropane-3-oxy-(2-bromo-2-methylpropionyl) (**2c**): A mixture of **2** (5.0 g), glacial acetic acid (15 mL), water (40 mL) and a catalytic amount of methoxybenzene was stirred for 30 min at 80 °C. The solution was subsequently cooled to room temperature, prior to the addition of diethylether (50 mL), the aqueous layer was saturated with sodium hydrogen carbonate by slow addition (CAUTION: formation of CO$_2$). The layers were separated and the aqueous layer washed with diethylether (50 mL). The crude product, initially a yellowish oil, was obtained in quantitative yield from the combined organic layers by removal of the ether using a rotary evaporator. Crystallization occurred overnight upon standing at room temperature. The crude yellowish solid was recrystallized from toluene (1 g in 25 mL) to yield pearl crystals in quantitative yield. ^{1}H NMR (DMSO-d_6) δ ^{1}H NMR (DMSO-d_6) δ 1.91 (s, 6H, C(C$\underline{H}_3$)$_2$Br), 3.40 (app. d, 2H, C$\underline{H}$2OH J = 5.27 Hz), 3.70 (m, 1H, C$\underline{H}$OH), 4.04 (dd, 1H, C$\underline{H}_a$H$_b$OC=O, J_{ab} = 11.11 Hz, J_E = 5.93 Hz), 4.15 (dd, 1H, CH$_a\underline{H}_b$OC=O, J_{ab} = 11.11 Hz, J_Z = 4.53 Hz), 4.69 (br. s, 1H CH$_2$O$\underline{H}$), 4.97 (br. S, 1H, CHO$\underline{H}$). ^{13}C NMR (DMSO-d_6) δ 30.7 (2C, C($\underline{C}$H$_3$)$_2$Br), 57.6 (1C, $\underline{C}$(CH$_3$)$_2$Br), 62.8 (1C, $\underline{C}$H$_2$OH), 67.3 (1C, $\underline{C}$H$_2$OC=O), 69.4 (1C, $\underline{C}$HOH), 171.1 (1C, C=O).

Typical procedure for polymerisations: Polymerisations were carried out at 90 °C using *n*-propyl-2-pyridylmethanimine as ligand for copper(I). A typical polymerisation recipe is based on 30 wt% monomer in toluene. The ratio of initiator:CuBr:ligand is 1:1:2 on a molar basis. The specific amounts of initiator, monomer, toluene and ligand were charged into a Schlenk tube and placed under an inert nitrogen atmosphere via three freeze-pump-thaw cycles (pressure < 10^{-2} mbar). Next this mixture was transferred into a Schlenk containing CuBr under a nitrogen atmosphere. The reaction mixture was placed in an oil bath at 90 °C and left for a specific reaction time while stirring. Samples were quenched in liquid nitrogen, passed over a basic alumina column prior to analysis using tetrahydrofuran as eluent to remove the copper and ligands. Final polymer products were further purified by precipitation from a THF solution in *n*-heptane.

References

1. Moad, G.; Solomon, D. H. *Aust. J. Chem.* **1990**, *43*, 215.
2. Georges, M. K.; Veregin, R. P. N.; Hamer, G. K.; Kazmaier, P. M. *Trends Polym. Sci.* **1994**, *2*, 66.
3. Hawker, C. J. *Acc. Chem. Res.* **1997**, *30*, 373.
4. Benoit, D.; Chaplinski, V.; Braslau, R.; Hawker, C. J. *J. Am. Chem. Soc.* **1999**, *121*, 3904.
5. Kato, M.; Kamigaito, M.; Sawamoto, M.; Higashimura, T. *Macromolecules* **1995**, *28*, 1721.
6. Wang, J. S.; Matyjaszewski, K. *J. Am. Chem. Soc.* **1995**, *117*, 5614.
7. Sawamoto, M.; Kamigaito, M. *Trends Polym. Sci.* **1996**, *4*, 371.
8. Matyjaszewski, K. *Curr. Opin. Sol. St. Mat. Sci.* **1996**, *1*, 769.
9. Granel, C.; Teyssie, P.; DuBois, P.; Jerome, P. *Macromolecules* **1996**, *29*, 8576.
10. Moineau, G.; Granel, C.; Dubois, P.; Jerome, R.; Teyssie, P. *Macromolecules* **1998**, *31*, 542.

11. Percec, V.; Barboiu, B.; vanderSluis, M. *Macromolecules* **1998**, *31*, 4053.

12. Lecomte, P.; Drapier, I.; DuBois, P.; Teyssie, P.; Jerome, R. *Macromolecules* **1997**, *30*, 7631.

13. Matyjaszewski, K.; Wei, M. L.; Xia, J. H.; McDermott, N. E. *Macromolecules* **1997**, *30*, 8161.

14. Matyjaszewski, K. *J. Macromol. Sci., Pure Appl. Chem.* **1997**, *A34*, 1785.

15. Kotani, Y.; Kato, M.; Kamigaito, M.; M, S. *Macromolecules* **1996**, *29*, 6979.

16. Haddleton, D. M.; Edmonds, R.; Heming, A. M.; Kelly, E. J.; Kukulj, D. *N. J. Chem.* **1999**, *23*, 477.

17. Angot, S.; Murthy, K. S.; Taton, D.; Gnanou, Y. *Macromolecules* **1998**, *31*, 6748.

18. Haddleton, D. M.; Kukulj, D.; Radigue, A. P. *Chem.l Commun.* **1999**, 99.

19. Haddleton, D. M.; Duncalf, D. J.; Kukulj, D.; Radigue, A. P. *Macromolecules* **1999**, *32*, 4769.

20. Chiefari, J.; Chong, Y. K.; Ercole, F.; Krstina, J.; Jeffery, J.; Le, T. P. T.; Mayadunne, R. T. A.; Meijs, G. F.; Moad, C. L.; Moad, G.; Rizzardo, E.; Thang, S. H. *Macromolecules* **1998**, *31*, 5559.

21. Bon, S. A. F.; Bosveld, M.; Klumperman, B.; German, A. L. *Macromolecules* **1997**, *30*, 324.

22. Kukulj, D.; Davis, T. P.; Suddaby, K. G.; Haddleton, D. M.; Gilbert, R. G. *J. Polym. Sci., Polym. Chem.* **1997**, *35*, 859.

23. Davis, T. P.; Kukulj, D.; Haddleton, D. M.; Maloney, D. R. *Trends Polym. Sci.* **1995**, *3*, 365.

24. Krstina, J.; Moad, G.; Rizzardo, E.; Winzor, C. L. *Macromolecules* **1995**, *28*, 5381.

25. Chong, B. Y. K.; Le, T. P. T.; Moad, G.; Rizzardo, E.; Thang, S. H. *Macromolecules* **1999**, *32*, 2071.

26. Nishikawa, T.; Kamigaito, M.; Sawamoto, M. *Macromolecules* **1999**, *32*, 2204.

27. Haddleton, D. M.; Heming, A. M.; Kukulj, D.; Duncalf, D. J.; Shooter, A. J. *Macromolecules* **1998**, *31*, 2016.

28. Gaynor, S. G.; Qiu, J.; Matyjaszewski, K. *Macromolecules* **1998**, *31*, 5951.

29. Coca, S.; Jasieczek, C. B.; Beers, K. L.; Matyjaszewski, K. *J. Polym. Sci., Polym. Chem.* **1998**, *36*, 1417.

30. Beers, K. L.; Boo, S.; Gaynor, S. G.; Matyjaszewski, K. *Macromolecules* **1999**, *32*, 5772

31. Muhlebach, A.; Gaynor, S. G.; Matyjaszewski, K. *Macromolecules* **1998**, *31*, 6046.

32. Teodorescu, M.; Matyjaszewski, K. *Macromolecules* **1999**, *32*, 4826.

33. Rademacher, J. T.; Baum, R.; Pallack, M. E.; Brittain, W. J.; Simonsick, W. J. *Macromolecules* **2000**, *33*, 284.

34. Jankova, K.; Chen, X. Y.; Kops, J.; Batsberg, W. *Macromolecules* **1998**, *31*, 538.

35. Ashford, E. J.; Naldi, V.; O'Dell, R.; Billingham, N. C.; Armes, S. P. *Chem. Commun.* **1999**, 1285.

36. Ohno, K.; Fukuda, T.; Kitano, H. *Macromol. Chem. Phys.* **1998**, *199*, 2193.

37. Ohno, K.; Tsujii, Y.; Miyamoto, T.; Fukuda, T.; Goto, M.; Kobayashi, K.; Akaike, T. *Macromolecules* **1998**, *31*, 1064.

38. Ohno, K.; Tsujii, Y.; Fukuda, T. *J. Polym Sci., Polym. Chem.* **1998**, *36*, 2473.

39. Marsh, A.; Khan, A.; Haddleton, D. M.; Hannon, M. J. *Macromolecules* - accepted for publication.

40. Winnik, M. A.; Winnik, F. M. *Ad. Chem. Ser.s* **1993**, 485.

41. Winnik, F. M. *Chem. Rev.* **1993**, *93*, 587.

Chapter 10

Grafting of Functionalized Water Soluble Polymers on Gold Surfaces

Stable Stimuli-responsive Thin Hydrogel Films Exhibiting a LCST or a UCST

B. S. Heinz[1], A. Laschewsky[1,*], E. D. Rekaï[1], E. Wischerhoff[2], T. Zacher[2]

[1]Université catholique de Louvain, Dept. of Chemistry, Place L. Pasteur 1, B-1348 Louvain-la-Neuve, Belgium
[2]BioTuL AG, Gollierstrasse 70, D-80339 München, Germany

Stimuli-responsive water soluble acrylamide polymers which exhibit a lower critical solution temperature (LCST), or an upper critical solution temperature (UCST) were prepared by free radical polymerization, or by chemical modification of precursor polymers. The obtained nonionic thermosensitive polymers are employed for the preparation of thin hydrogel films on gold surfaces. Two different strategies were studied. The first one consists of the immobilization of polymers functionalized with disulfide end-groups onto gold (grafting to). The second strategy consists of the immobilization of a disulfide functionalized initiator followed by the polymerization in situ from gold surfaces (grafting from). The grafting reactions in water or in ethanol were followed by surface plasmon resonance (SPR), ellipsometry and contact angle measurements.

INTRODUCTION

Polymers responding to external stimuli, such as temperature, are discussed by virtue of their potential applications in various fields (*1,2*). Many polymers and hydrogels show thermoresponsive behavior in aqueous systems, undergoing a phase separation at a lower critical solution temperature (LCST), or at an upper critical solution temperature (UCST). The LCST is a critical temperature beyond which polymer and solvent separate into two phases; whereas, the UCST is the critical temperature below which the polymer and solvent separate. LCST and UCST are linked, within other parameters, to the balance of hydrophilic and hydrophobic groups in the polymer (*3*). The occurence of a LCST is widespread for nonionic polymers in water (*4*). Poly(*N*-vinyl caprolactam) (*5*), polyethyleneoxide, polyvinylmethylether and poly(*N*-isopropylacrylamide) (P-NIPAM) (*6*) are notorious examples. Typically, the LCST is adapted by varying the chemical structure of the monomer employed (an approach which is limited) or by copolymerization (*7*). The latter approach suffers in general from the inherent chemical heterogeneity of the statistical copolymers obtained, which broadens the thermal transition. This problem may be overcome by the chemical modification of nonionic precursor polymers, resulting in most cases in random copolymers. But this approach has been rarely used so far (*8*).

Within this line of reasoning, we synthesized copolymers of *N*-isopropyl acrylamide and *N*-tris(hydroxymethyl)methylacrylamide (P-NIPAM-co-THMA) by free radical copolymerization (Figure 1). Additionally, we have employed chemical modification, namely the acylation of the water soluble poly[*N*-tris-(hydroxymethyl)methylacrylamide] (P-THMA) and poly[*N*-2-hydroxypropyl-methacrylamide] (P-HPMA) that do not exhibit a LCST under atmospheric pressure themselves. Appropriately acetylated and cinnamoylated copolymers however exhibit a LCST in water which can be tailored in the full temperature range between 0°C and 100°C by controlling the extent of esterification (*9*).

Instead of a LCST, thermoresponsive water soluble polymers can exhibit a UCST. UCSTs are frequently observed for sulfobetaine polymers such as poly[*N*-(3-sulfopropyl)-*N*-methacryloyloxyethyl]-*N*,*N*-dimethyl ammonium betaine] (*10*) or poly[*N*,*N*-dimethyl-*N*-methacrylamidopropyl-*N*-(3-sulfopropyl) ammonium betaine] (P-SPM) **10** (*11*). UCSTs are exhibited by more complex structures, too, such as poly [*N*,*N*-dimethyl-*N*-11-methacryloyl oxy-undecylammonio propanesulfonate] (*12*), or as naphtalene-labeled copolymers of styrene and *N*,*N*-dimethylmaleimido propylammonium propanesulfonate (*13*). We used P-SPM **10** which has a UCST of about 70°C in pure water for our investigations. This polymer contains an amide moiety and is therefore relatively stable towards hydrolysis.

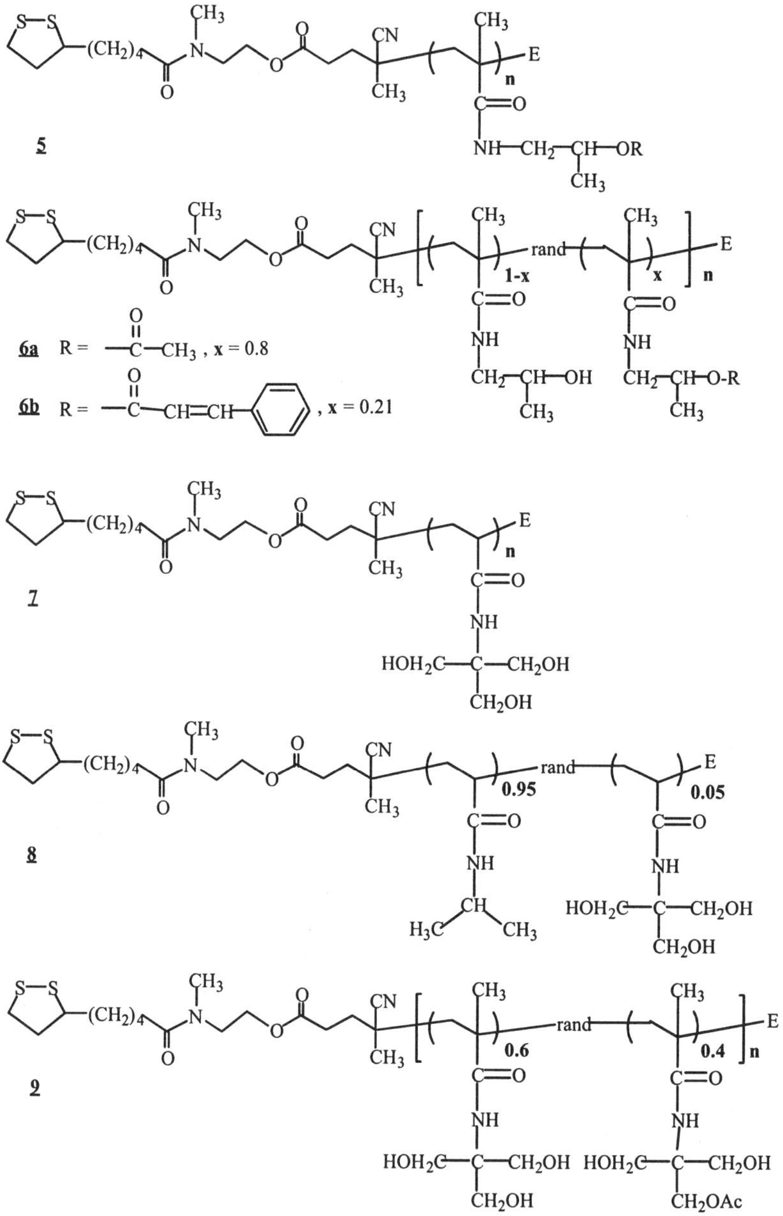

164

Figure 1. *Disulfide functionalized water soluble polymers.*

We wanted to explore methods to prepare stable thin, stimuli-responsive hydrogel films on surfaces by using the polymers shown in Figure 1. Generally, two different pathways can be followed to graft polymer monolayers on surfaces (cf. Figure 3). The first approach is based on the adsorption of functionalized polymers from solution to surfaces (*14*). Numerous systems in which the polymer is covalently attached to a solid surface have been obtained by this "grafting to" technique. However, the growth of films by this technique shows some limitations and disadvantages, e.g. low graft densities, due to steric hindrance of the grafting by already attached polymer chains and grafting against a concentration gradient. The second concept is based on the grafting of an initiator to the surface, followed by a thermal or photochemically activated polymerization in situ. This "grafting from" approach offers higher graft densities of polymers on surfaces (*15*).

We have studied both of these approaches to build monolayer films on gold surfaces using P-NIPAM-co-THMA, acetylated and cinnamoylated P-THMA or P-HPMA, respectively, and P-SPM which are all thermoresponsive (Figure 1). Disulfide terminal groups were employed to attach the polymers on gold. Disulfides, like thiols, are known for their strong affinity to nobel metal surfaces, thus enabling the self assembly of polymeric monolayers (*16*). The success and the kinetics of the grafting reaction can be conveniently followed by surface plasmon resonance (SPR).

EXPERIMENTAL SECTION

Instrumentation and Methods.

Polymers were grafted on gold surfaces by two methods : i) grafting of functionalized polymers from ethanol or from aqueous solutions (c=2g/L) during 23 h. ii) grafting of the disulfide functionalized initiator in DMSO (c=3g/L) during 16 h, and polymerization of the monomers in water (c= 40% w/v) by irradiation at 350 nm glass-filtered UV light for 1 h (2 lamps Sylvania, 6W, distance 4 cm), or by thermal initiation at 60°C for 23 h under argon atmosphere.

Infrared (GIR) spectra were taken with a Bruker Equinox 55 spectrometer. SPR measurements were recorded on a SPR-Prototype (BioTuL, Munich, Germany) with a diode laser (λ=784 nm) as light source, angular resolution f=0.005°. Gold coated prisms are made of BK7 glass (n=1.5168). The monochromatic light directed to a prism was reflected onto a photodiode. The intensity of the reflected light was measured as a function of external incidence angle. Every experiment was computer controlled. Thicknesses of the films were estimated by applying the standard Fresnel equations, assuming a refractive index of 1.478. Ellipsometry was performed on a commercial apparatus (Jobin-Yvon, Ellisel) operating in the fixed polarizer/rotating compensator/ fixed analyzer configuration. Every sample was measured 5 times on different places. The thickness was estimated by standard equations (*17a*) using a homemade software (*17b*), fitting the data with a model of a homogeneous film of a refractive index of 1.478, on a gold substrate of a refractive index $0.183 + j\ 2.82$ (j is the square root of -1).

Materials.

All solvents used were analytical grade. *N,N*-dimethylformamide was dried over molecular sieves (4 Å). CH_2Cl_2 was distilled over CaH_2. Water was purified by a Elgastat purification system (resistance 18 MΩ). Monomers *N*-[tris (hydroxymethyl)methylacrylamide] (Aldrich), and *N,N*-dimethyl-*N*-methacryl amido propyl-*N*-(3-sulfopropyl) ammonium betaine (Raschig, Ludwigshafen, Germany) were used as received. Gold supports (12 mm x 12 mm glass slides, thickness 0.5 mm recovered with 1 nm NiCr and 50 nm of gold) were provided by Dresdner Transferstelle für Vakuumtechnik e.V (Dresden, Germany).

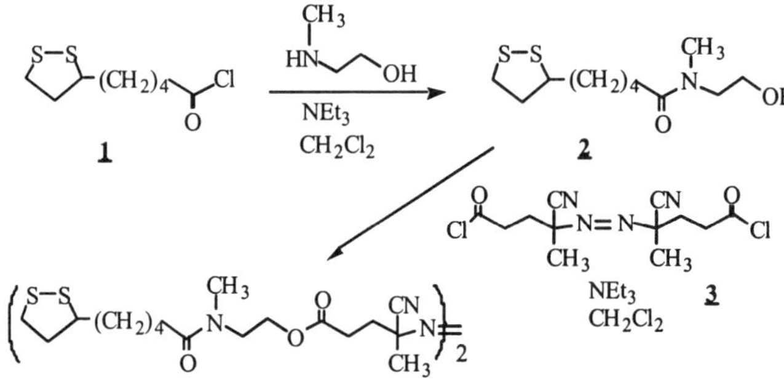

Figure 2. Synthesis of the disulfide functionalized azo-initiator $\underline{4}$.

Azo-disulfide initiator $\underline{4}$ (Figure 2): Thioctic acid chloride $\underline{1}$ (*18*) was dissolved in anhydrous CH_2Cl_2 and added dropwise under argon to methylaminoethanol in CH_2Cl_2 at -5°C. Triethylamine dissolved in CH_2Cl_2 was added slowly . The resulting mixture was stirred at room temperature for 2 days, was washed three times with 1M HCl and twice with saturated aqueous NaCl. The organic layer was dried over $MgSO_4$, evaporated in vacuo, and precipitated in diethyl ether to yield the functional alcohol $\underline{2}$ as yellowish viscous oil (67%). A solution of $\underline{2}$ in anhydrous CH_2Cl_2 was reacted with reagent $\underline{3}$ (*19*) at room temperature. When the solution became clear, triethylamine was added and the mixture was stirred for additional 36 h. Washing three times with 1M HCl, drying over $MgSO_4$ and evaporating under reduced pressure gives **4** in form of a yellowish viscous oil (89%).

End-group functionalized poly-[N-2-hydroxypropylmethacrylamide] (P-HPMA) **5**: N-2-hydroxypropylmethacrylamide (HPMA) (*20*) in absolute ethanol (conc=10% v/v) was polymerized using 2 mol% of the functional initiator **4** dissolved in a minimum of dimethylsulfoxide. The mixture was degassed in vacuo at the temperature of liquid nitrogen, and reacted at 60°C for 24h. The polymer was purified by repeated precipitation into acetone (72%).

End-group functionalized poly-[N-tris-(hydroxymethyl)methylacrylamide] (P-THMA) **7** was made from N-[tris-(hydroxymethyl)methylacrylamide] in 30v/7v ethanol/water (c=10% v/v) using initiator **4**, as described for **5** . Dialysis against water for 5 days and lyophilization gives **7** as white hygroscopic powder (85%).

End-group functionalized poly-N-[isopropylacrylamide-co-tris-(hydroxy-methyl)methyl acrylamide]) **8**: P-NIPAM-co-THMA was synthesized by free radical copolymerization of N-isopropylacrylamide (95%mol) and THMA (5%mol) in absolute ethanol (c=10% v/v) containing 2mol% of the functional initiator **4** dissolved in a minimum of dimethylsulfoxide. The mixture was degassed in vacuo at the temperature of liquid nitrogen, and reacted at 60°C for 24h. The polymer was purified by two precipitations in diethylether (yield 75%). The composition of the copolymer is P-NIPAM-co-THMA 95/5 according to the integration of the ^{1}H-NMR signals. The virtually identical composition of the monomer feed and of the copolymer obtained suggests reactivity ratios of both acrylamides close to 1. The statistical copolymer obtained has, therefore, probably a structure similar to a random copolymer.

The partial esterifications of P-HPMA and of P-THMA to obtain acetylated and cinnamoylated copolymers **6a** and **6b**, and **9** are described elsewhere (9).

RESULTS AND DISCUSSION

In order to fix the hydrogels to gold surfaces, we employed a disulfide functionalized initiator that is synthesized as shown in Figure 2. Polymers containing disulfide end-groups, as obtained by free radical polymerization using the azo-disulfide initiator **4**, were adsorbed via a reductive cleavage of the disulfide group followed by complexation with gold (Figure 3a). In the second strategy (Figure 3b), the initiator **4** is self-assembled on a gold surface, and the polymerization is started from this monolayer of initiator molecules, to form the polymer directly on the surface.

168

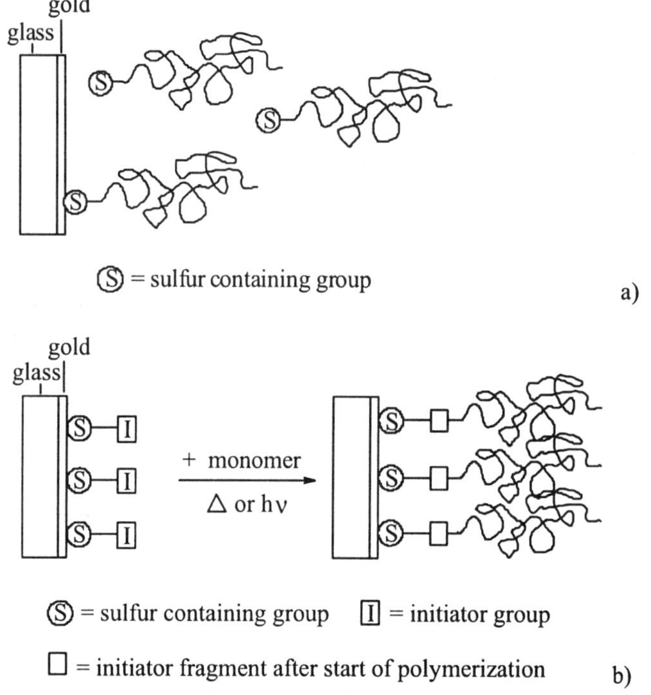

Figure 3. *Strategies to gold fixed hydrogels: a) "grafting to" of disulfide-functionalized polymers, b) "grafting from" in two steps: grafting of disulfide-functionalized initiator followed by polymerization in situ.*

For comparison and for evaluation of of grafting methods, we prepared hydrogel films of poly-[*N*-(tris-(hydroxymethyl)methyl) acrylamide] (P-THMA) by both strategies. We selected this polymer by virtue of its high hydrophilicity due to the three hydroxyl groups per repeat unit (*21*). Disulfide-functionalized P-THMA **7** was grafted on gold surfaces by immersion of clean gold plates in an aqueous solution of the polymer. The hydrophilicity of the gold plate increases markedly as evidenced by contact angle measurements (cf. Figure 14). The contact angle which originally for the unmodified gold surface is about 85 degrees, is reduced after the grafting to 25 degrees. These findings are corroborated by SPR measurements (Figures 4 and 5), which show a significant shift of the plasmon signal to a higher resonance angles for the grafted surfaces, due to the adsorption of the polymer. The thickness of the hydrogel in water was estimated to be in the range of 5 nm by SPR and about 6 nm by ellipsometry.

The kinetics of the grafting in Figure 5 show that most of the polymer is adsorbed rapidly within a few minutes, though the maximum of coverage was obtained only after about 180 min.

P-THMA hydrogels on gold surfaces were also prepared by the second strategy, i.e. by "grafting from" (Figures 6 - 8). In the first step, the disulfide containing initiator **4** was grafted on the gold surface by immersion of a clean gold plate in a solution of the initiator in DMSO during 16 h. To avoid the presence of residual initiator which was not fixed on the gold, the supports were washed extensively with DMSO and with water. In the second step, the grafted initiator was decomposed to induce the free radical polymerization from the gold surface. SPR allowed us to follow the two steps of the grafting process (Figures 7,8). After each step, the resonance angle is shifted to higher values, showing the binding of the initiator, or the grafting by the polymer, respectively

Thermal polymerization under argon atmosphere at 60°C during 23 h in a solution of monomer in water, gave ultra-thin hydrogel films of P-THMA with a thickness of about 3 nm (Figure 7), whereas the polymerization induced by irradiation at λ=350 nm for 1h provided hydrogel films with a thickness of about

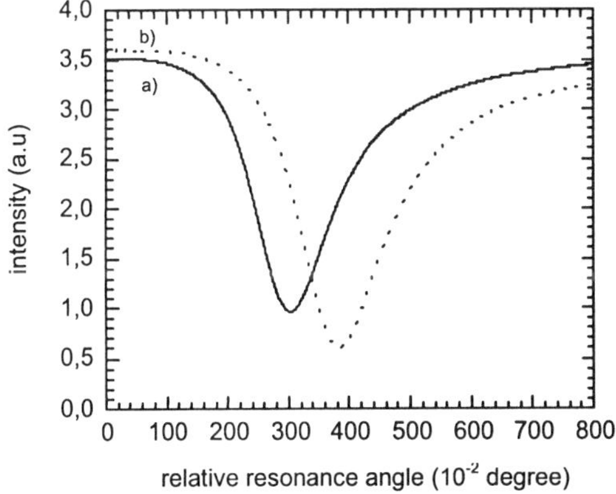

Figure 4. *SPR curves of gold surfaces exposed to aqueous solutions of P-THMA **7** ("grafting to"): a) bare gold before exposure, b) surface after exposure*

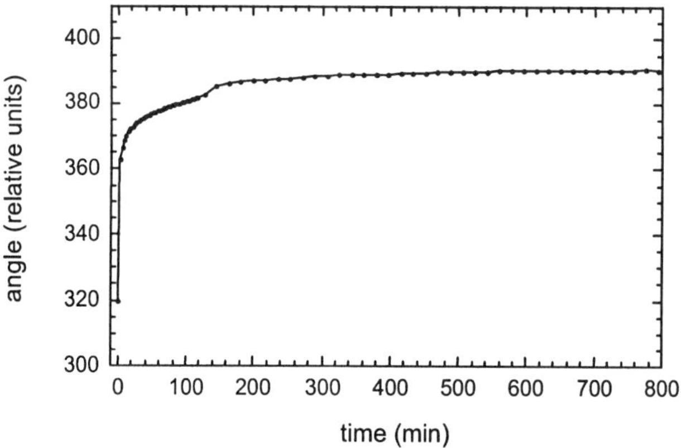

Figure 5. Kinetics of grafting of end-group functionalized P-THMA $\underline{7}$ on gold surfaces in water followed by SPR ("grafting to").

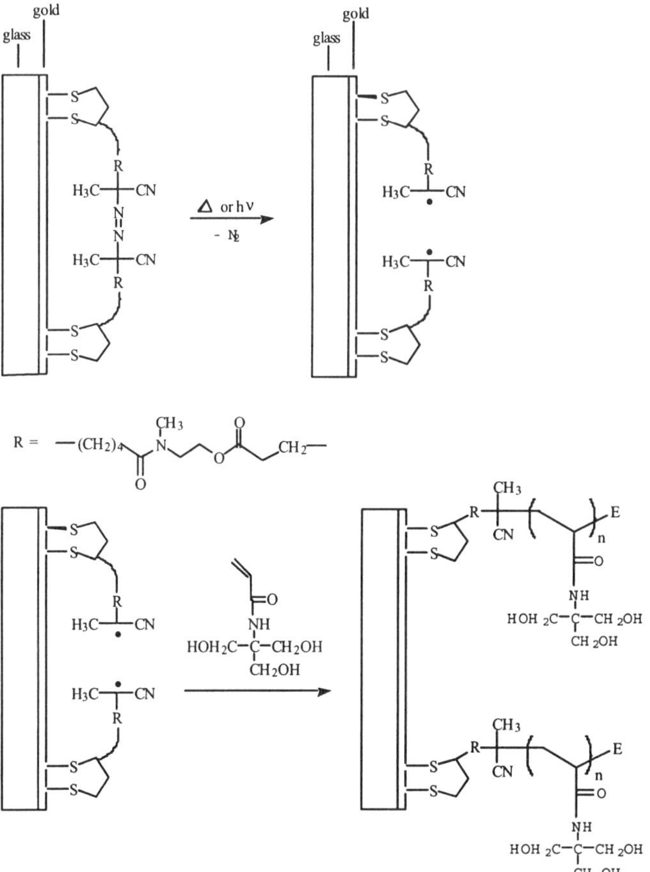

Figure 6. *Scheme of "grafting from" of P-THMA on gold surface: decomposition of grafted initiator followed by free radical polymerization.*

11 nm (Figure 8). Thus photopolymerization provides thicker hydrogel films within a shorter time than thermal polymerization does. Possibly, the thermal polymerization is handicapped by the weakness of the Au-S bond at 60°C (*16*), removing some polymeric material in the course of the reaction; this question remains to be clarified.

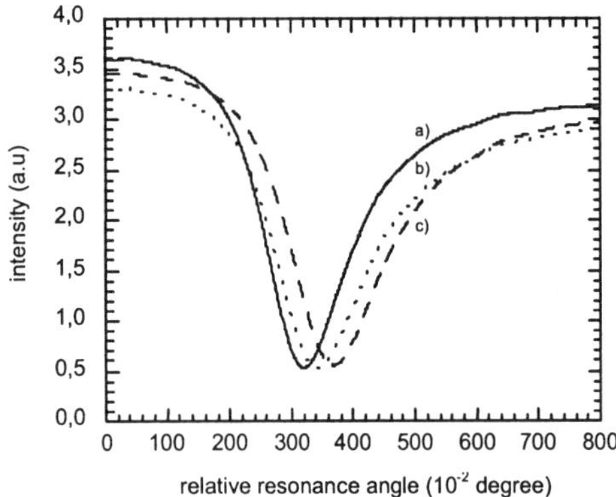

Figure 7. SPR curves of gold surfaces exposed to water solutions of P-THMA: a) bare gold before exposure, b) after initiator adsorption, c) after grafting with P-THMA via thermal initiation ("grafting from").

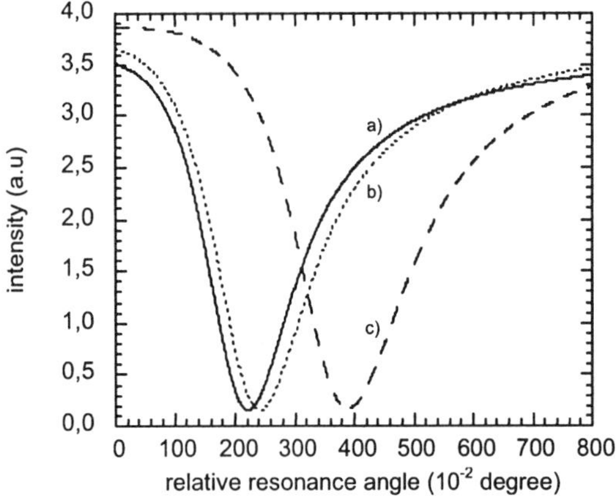

Figure 8. SPR curves of gold surfaces exposed to water solutions of P-THMA: a) bare gold before exposure, b) after initiator adsorption, c) after grafting with P-THMA via photoinitiation ("grafting from").

Having seen the better results obtained for P-THMA, we applied photoinitiation for the "grafting from" of the zwitterionic P-SPM **10** which typically exhibits a UCST (11), as an alternative stimuli-responsive hydrogel (Figure 9).

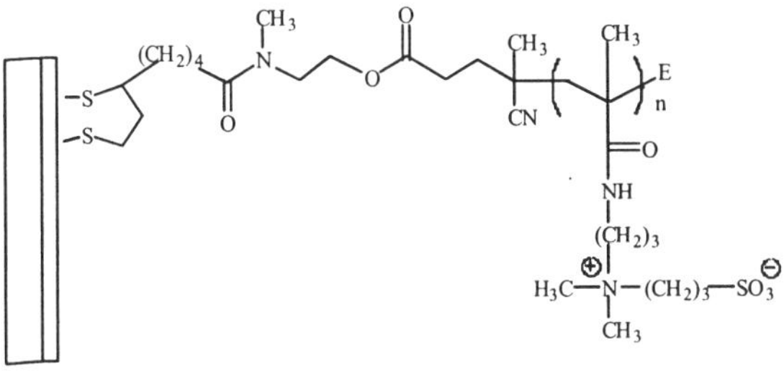

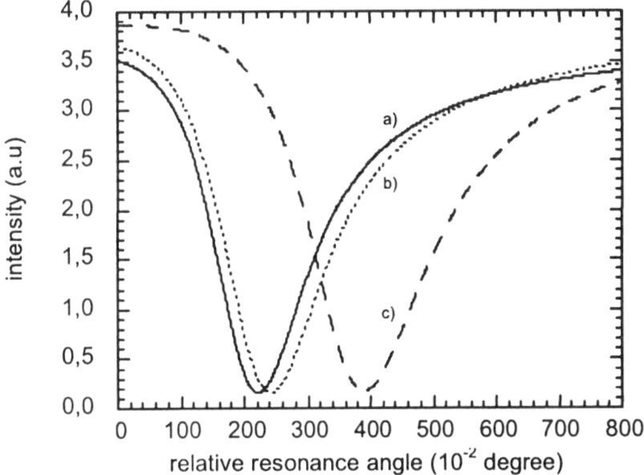

Figure 9. *"Grafting from" of P-SPM **10** on gold surface.*

Under the same reaction conditions as for P-THMA, gold surfaces covered by disulfide-azo initiator were immersed in a solution of monomer SPM in water, and films of P-SPM **10** with a thickness of about 5 nm were obtained using the "grafting from" method by irradiation at λ=350 nm for 1 h, as followed by SPR (Figure 10). The coatings obtained were stable towards washing with water. The thickness of the film was easily controlled by the time of irradiation.

Figure 10. *Photopolymerization. SPR curves of gold surfaces exposed to water solutions of P-SPM **10**: a) bare gold before exposure, b) initiator adsorption, c) surface grafted with P-SPM ("grafting from").*

The "grafting from" method is particularly attractive for homopolymers. For copolymers and modified homopolymers, however, this method seems to pose certain problems because it is difficult to obtain informations about the copolymers in the films, e.g. on their composition. Moreover, if thermoresponsive polymers are desired, this method poses difficulties to characterize the LCST behavior of the polymers in solution. Therefore, we chose the "grafting to" method to prepare thermoresponsive hydrogel films on surfaces (Figures 11-13).

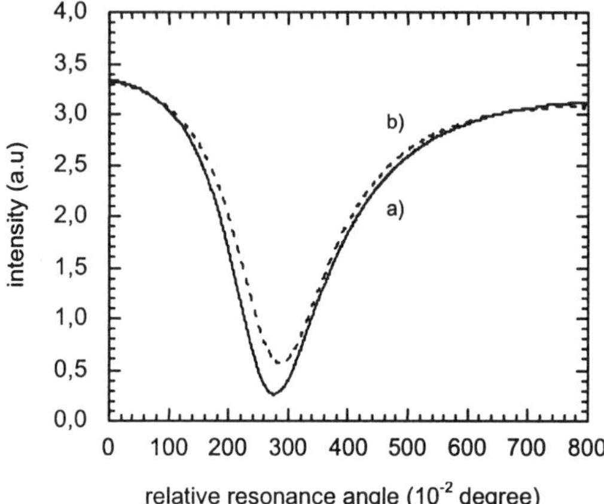

Figure 11. *SPR curves of gold surfaces exposed to ethanolic solutions of copolymer* **8** *(P-NIPAM-co-THMA:95/5): a) bare gold, b) surface after grafting ("grafting to").*

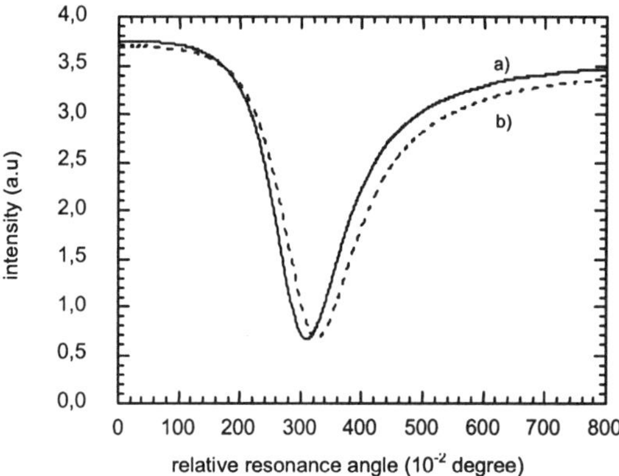

Figure 12. *SPR curves of gold surfaces exposed to water solutions of acetylated copolymer* **6a**: *a) bare gold, b) surface after grafting ("grafting to").*

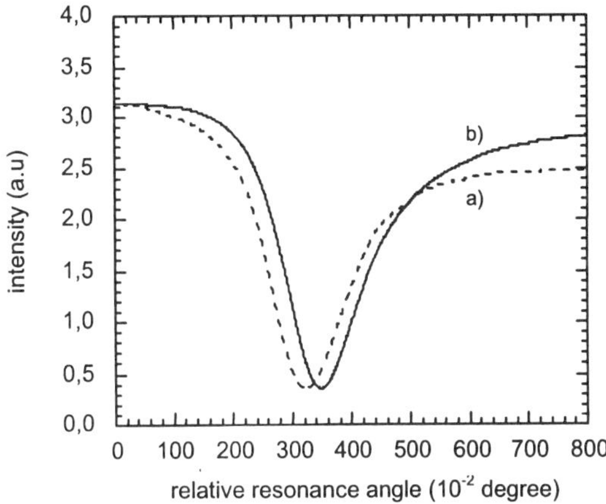

*Figure 13. SPR curves of gold surfaces exposed to water solutions of cinnamo-ylated copolymer **6b**: a) bare gold, b) surface grafted with **6b** ("grafting to").*

The end group functionalized, thermoresponsive copolymers studied comprised P-NIPAM-co-THMA:95/5 **8**, 80% acetylated P-HPMA **6a** and 21% cinnamoylated P-HPMA **6b**, as well as 40% acetylated P-THMA **9** (Figure 1). The cloud point of these copolymers in water were studied before the grafting, and were found to be 30°C for **8**, 35 °C for **6a**, 30°C for **6b**, and 28°C for **9** (cf also Ref.9).

The polymers were adsorbed by immersion of clean gold plates in an ethanolic or an aqueous solution for 23 h. Following the grafting by SPR, the plasmon signal shifts to higher resonance angles for all polymers, due to the adsorption of the polymers (Figures 11, 12, 13). Polymer adsorption was qualitatively corroborated by GIR-IR spectroscopy, taking advantage of the strong carbonyl bands in the IR. The SPR shifts observed are typical for hydrogel coatings with a thickness in the range of 1.5 nm for **6a** and 2 nm for **6b**. As for the "grafting from" method, the obtained coatings are stable towards washings with water.

The hydrophilicity of the gold plate increases significantly after the grafting as evidenced by contact angle measurements (Figure 14). Not surprisingly, most hydrophilic surfaces are obtained when employing the unmodified P-THMA. But note also that the widely used thermoresponsive copolymers based on *N*-isopropylacrylamide lead only to moderately hydrophilic surfaces. The partially esterified thermosensitive copolymers based on HPMA produce more hydrophilic coatings.

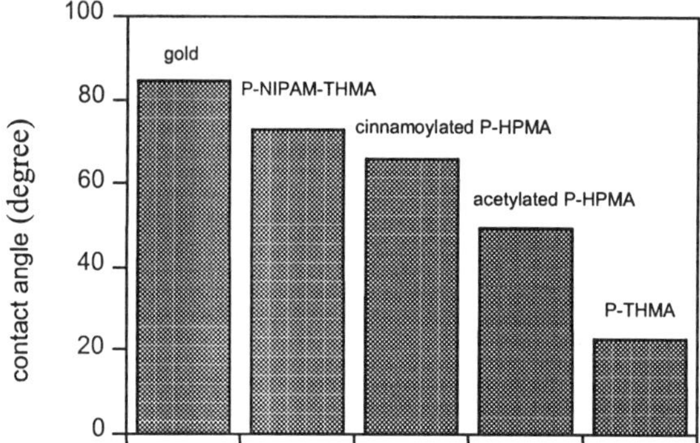

Figure 14. *Static contact angle measurements of ultrathin water soluble hydrogels on gold surfaces at room temperature, obtained by adsorption of disulfide end-group functionalized polymers.*

Remarkably, SPR-studies also demonstrate the thermoresponsive behavior of the hydrogels. When, for example, comparing the evolution of the resonance angle with increasing temperature, the simple hydrogel made from unmodified P-THMA **7**, which does not show a thermal transition in solution, exhibits a continuous, nearly linear decrease of the resonance angle due to the diminution of the refractive index upon heating. In contrast, the analogous measurements of thermoresponsive polymers, such as of the partially acetylated THMA **9**, exhibit a discontinuity at about 27°C (Figure 15).

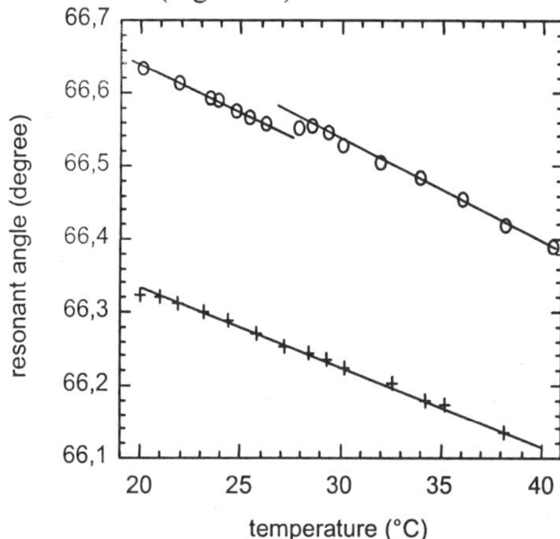

Figure 15. *Resonant angle of hydrogel films as function of the temperature: (+) = unmodified P-THMA **7**; (o) = partially acetylated P-THMA **9**. The lines are guides to the eye.*

The temperature at which this discontinuity appears matches well with the cloud point of the polymers observed in solution (see above and ref. 9). The discontinuity is explained by the collapse of the polymer coil, thus increasing its density in the close proximity of the gold surface. As the sensitivity of the SPR decreases exponentially with the distance from the surface, the thermal collapse of the water-swollen gel leads to a notable increase of the resonance angle

Hydrogel films on gold surfaces are known to reduce the nonspecific adsorption of proteins (*22*), and thus can improve the performance of SPR in biomedical analytics. First studies with bovine serum albumin (BSA) demonstrated that hydrogels of **7** indeed suppress efficiently the nonspecific adsorption of this protein: no change in the resonance angle could be detected by SPR upon exposure of the coated surfaces to aqueous solutions of BSA.

To study the influence of the nature of solvent on the kinetics of the grafting reaction, we grafted the P-NIPAM-co-THMA of **8** on gold from ethanolic solutions as well as from aqueous solutions. SPR-measurements show that the adsorption is considerably faster in water than in ethanol. As illustrated for films of P-NIPAM-co-THMA **8** in Figures 11 and 16, the hydrogel is twice as thick after 23 h of exposure when grafted from water (about 3 nm) compared to ethanol (about 1.5 nm). We hypothesize that the rather hydrophobic lipoic acid end-group fixed on the rather hydrophilic polymer favours the adsorption from the aqueous solution to the gold.

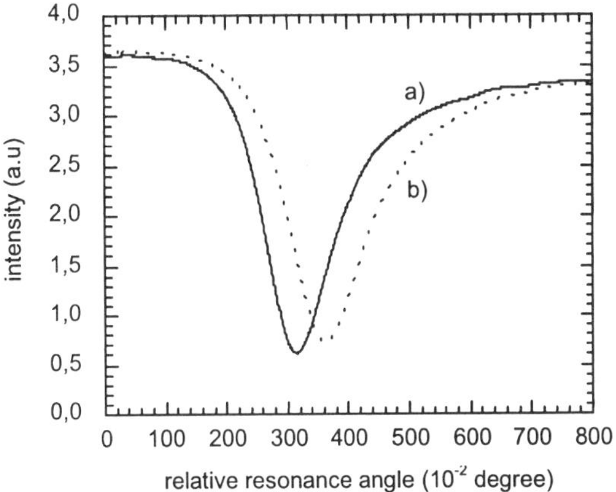

Figure 16. SPR curves of gold surfaces exposed to water solutions of copolymer **8** *(P-NIPAM-co-THMA:95/5) ("grafting to"): a) bare gold before exposure, b) surface exposed to the copolymer for 23h.*

CONCLUSION

Two different strategies to prepare stable, thin thermoresponsive hydrogel films on gold surfaces were investigated. The first one consists of the direct grafting of functionalized polymers ("grafting to"), and the second one is the polymerization in situ on functionalized gold surfaces ("grafting from"). The grafting was followed by Surface Plasmon Resonance which allows evaluation of the thickness of the hydrogels and to determine the kinetics of the grafting. By "grafting from", photopolymerization of the diazo system studied provides hydrogels with greater thicknesses in a shorter time than thermal polymerization at 60°C. The hydrogel films prepared efficiently suppress the nonspecific adsorption of bovine serum albumin. Furthermore, ultrathin hydrogels on gold could be prepared from polymers which exhibit a LCST or UCST in solution by the "grafting to" method. The hydrogels of this type studied exhibit stimuli-responsive behavior which matches the behavior in solution.

ACKNOWLEDGEMENT

E.D.Rekaï thanks the European Commission for a personal grant (CEE FMB1-CT98-2915). The work was supported by the European Commission (CEE BIO-CT97-962372). We are indebted to S.Bayoudh (Université catholique de Louvain), L.Jullien (ENS, Paris, France), J.Sadowski (VTT, Tampere, Finland) and D.W.Grainger (GAMMA-A, Technologies, Herndon,Virginia, USA) for stimulating discussions, to P.Rouxhet and Y.Adriensen (Université catholique de Louvain) for help in contact angle measurements, and to A.Jonas and X.Arys (Université catholique de Louvain) for ellipsometric measurements.

REFERENCES

1. Hoffman, A.S. *J. Controlled Release* **1987**, *6*, 297.
2. Peppas, N.A.; Langer, R. *Science* **1994**, *263*, 1715.
3. Taylor, L.D.; Cerankowski, L.D. *J. Polym. Sci., Polym. Chem. Ed.* **1975**, *13*, 2551.
4. a) Costas, M.; Patterson D. *J. Chem. Soc. Faraday Trans. I* **1985**, *81*, 2381; b) Goldstein, REJ. *J. Chem. Phys.* **1984**, *80*, 5340; c) Nakayama, H. *Bull. Chem. Soc. Jap.* **1970**, *43*, 1683.
5. Tager, A.A; Safronov, A.P.; Berezyuk, E.A.; Galaev, I.Y. *Colloid Polym. Sci.* **1994**, *272*, 1234.
6. a) Bae, Y.H.; Okano, T.; Kim, S. W. *J. Polym. Sci., Polym. Phys. Ed.* **1990**, *B 28*, 923; b) Ringsdorf, H.; Simon, J.; Winnik, F.M. *Macromolecules* **1992**,

25, 5353; c) Chen, G.; Hoffman, A.S. *Nature* **1995**, *373*, 49; d) Heskins, M.; Guillet, J.E. *J. Macromol. Sci. Chem.* **1968**, *2*, 1441.

7. a) Dong, L.C.; Hoffman, A.S. *J. Controlled Release* **1991**, *15*, 141; b) Deng, Y.L.; Pelton, R. *Macromolecules* **1995**, *28*, 4617; c) Hu Y.; Armentrout, R.S.; McCormick, C. L. *Polym. Prepr. Am. Chem. Soc. Polym. Chem. Div* **1996**, *37 (1)*, 661.

8. a) Ritter, H.; Stock, A. *Macromol. Rapid Commun.* **1994**, *15*, 271; b) Dieu, H. A., *J. Polym. Sci.* **1954**, *12*, 417.

9. a) Laschewsky, A.; Rekaï E.D.; Wischerhoff E. *Polym. Prepr. Am. Chem. Soc. Div. Polym. Chem.* **1999**, *40 (2)*, 189; b) Laschewsky A.; Rekaï, E.D.; Wischerhoff, E. *Macromol.Chem.Phys.*, submitted.

10. Schulz, D.N.; Peiffer, D.G.; Agarwal, P.K.; Larabee, J.; Kaladas, J.J.; Soni, L.; Handwerker B.; Garner R. T. *Polymer* **1986**, *27*, 1734.

11. Huglin, M.B.; Radwan, M.A. *Polym. Int.* **1991**, *26*, 97.

12. Köberle, P.; Laschewsky, A.; Lomax, T. D. *Makromol. Chem., Rapid Commun.* **1991**, *12*, 427.

13. Liaw, D.J.; Huang C.C.; Sang H.C.; Kang E.T. *Langmuir* **1998**, *14*, 3195.

14. a) Krenkler, K.P.; Laible, R.; Hamann, K. *Angew. Makromol. Chem.* **1976**, *53*, 101; b) Tsubokawa, N.; Hosoya, M.; Yanadori, K.; Sone,Y. *J. Macromol. Sci. Chem.* **1990**, *A27*, 445.

15. a) Prucker, O.; Rühe J. *Mater. Res. Soc. Symp. Proc.* **1993**, *304*, 167; b) Prucker, O.; Rühe, J. *Macromolecules* **1998**, *31*, 592 et ibid. 602; c) Weimer, M.W.; Chen, H.; Giannelis, E.P.; Sogah, D.Y. *J. Am. Chem. Soc.* **1999**, *121*, 1615; d) Nakayama, Y.; Takatsuka, M.; Matsuda, T. *Langmuir* **1999**, *15*, 1667; e) Jordan, R.; Ulman A. *J. Am. Chem. Soc.* **1998**, *120*, 243; f) Kratzmüller, T.; Appelhans, D.; Braun, H.G. *Adv. Mater.* **1999**, *11*, 555.

16. a) Nuzzo, R.G.; Fusco, F.A.; Allara, D.L. *J. Am. Chem. Soc* **1987**, *109*, 2358; b) Ulman A. *Chem. Rev.* **1996**, *96*, 1533; c) Wang, W.; Castner, D. G., Grainger, D. W. *Supramol. Sci.* **1997**, *4*, 83; d) Tsao, M.W.; Pfeifer K.H.; Rabolt, J.F. *Macromolecules* **1997**, *30*, 5913; e) Koutsos, V.; Van der Vegte, E.W.; Pelletier, E.; Stamouli, A.; Hadziioannou, G. *Macromolecules* **1997**, *30*, 4719; f) Erdelen, C.; Häussling, L.; Naumann, R.; Ringsdorf, H.; Wolf, H.; Yang, J.; Liley, M.; Spinke, J.; Knoll, W. *Langmuir* **1994**, *10*, 1246.

17. a) Azzam, R.M.A.; Bashara, N.M. In *Ellipsometry and Polarized Light*, North Holland, Amsterdam, Netherlands, 1977; b) Arys, X. PhD thesis, Université catholique de Louvain, Louvain-la-Neuve, Belgium, 2000.

18. Sabapathy, R.C.; Bhattacharyya, S.; Leavy, M.C; Cleland Jr, W.E.; Hussey, C.L. *Langmuir* **1998**, *14*, 124.

19. Smith, D.A. *Makromol. Chem.* **1967**, *103*, 301.

20. Strohalm, J.; Kopecek; J. *Angew. Makromol. Chem.* **1978**, *70*, 109.

21. a) Barthélemy, P.; Maurizis, J.C.; Lacombe, J.M.; Pucci, B. *Bioorg. Med. Chem. Lett.* **1998**, *8*, 1559;   b) Saito, N.; Sugawara, T.; Matsuda, T. *Macromolecules* **1996**, *29*, 313;   c) Köberle, P.; Laschewsky, A.; Van den Boogaard, D. *Polymer* **1992**, *33*, 4029.

22. a) Beyer, D.; Knoll, W.; Ringsdorf, H.; Wang, J.H., Timmons, R.B.; Sluka, P. *J. Biomed. Mater. Res.* **1997**, *36*, 181;   b) Seigel, R.R.; Harder, P.; Dahint, R.; Grunze, M.; Josse, F.; Mrksich, M.;   Whitesides, G.M. *Anal. Chem.* **1997**, *69*, 3321.

Chapter 11

Thermogelation in Aqueous Polymer Solutions

Alain Durand[1,2], Mélanie Hervé[1] and Dominique Hourdet[1,*]

**[1]Laboratoire de Physico-Chimie Macromoléculaire (UMR 7615)
ESPCI-CNRS-UPMC, 10 rue Vauquelin, 75231 Paris 05, France
[2]Current address: Laboratoire de Chimie-Physique Macromoléculaire,
UMR 7568 CNRS-INPL, BP 451, 54001 Nancy, France**

Physical gelation upon heating is reviewed in aqueous polymer solutions. Among the different systems, which basically rely on reversible associations between non-polar groups in water, we distinguish four different categories: cellulose derivatives, mixtures of amphiphilic systems, block copolymers and graft copolymers. For all of these the common features are described and the specific properties are underlined. More particularly we try to point out the relationship between the physical gelation mechanism and the primary structure of the copolymers, as well as the influence of external parameters like added salt, pH, surfactant, …

INTRODUCTION

Due to the ever-increasing development of water-based formulations, for both environmental and economic reasons, associating water-soluble polymers have received a lot of attention in the last two decades *(1-5)*. These amphiphilic macromolecules contain 1) a hydrophilic part (neutral or ionic) that maintains the solubility of the polymer in water and 2) a hydrophobic component

181

(alkyl, perfluoroalkyl, aromatic) which engenders the associative behavior. In water, these copolymers self-assemble forming hydrophobic clusters embedded in a sea of hydrophilic chains. In semi-dilute solution physical networks are formed. Thus, these polymers have found numerous applications as thickeners in aqueous-based fluids. The control of the rheological properties in these systems can be regulated at the molecular level by changing the primary structure of the copolymer. Moreover, for a given system, self-assembly can be modulated through external parameters like solvent quality, temperature, pH, ionic strength *(2)* or by favoring synergistic interactions with other amphiphiles: surfactant or polymers *(6,7)*. Nevertheless, for most of these systems it is rather difficult to maintain complete control over the association process, i.e. reversible switching of the hydrophobic aggregation using a single trigger. If temperature is the selected trigger a whole class of associative polymers can be defined that are commonly called thermoassociative, or in relation to their potential properties, thermothickening, thermoviscosifying or thermogelling. These terms are related to an association process driven by heating through hydrophobic interactions. This phenomenon is opposed to gelation upon cooling that has been described long ago with natural polymers like gelatin for instance. The first examples in literature of thermothickening behavior in aqueous solution were reported by Heymann in 1935 with cellulosic derivatives *(8)*. Until the beginning of the 70's this property was only attributed to cellulosic samples up to the development of block copolymers with poly(ethylene oxide) and poly(propylene oxide). It has been mainly during the last 10-15 years that this specific property has been reported for other systems *(9-11)* and clearly demonstrated for graft copolymers tailored with responsive side-chains *(12-15)*. Even though all these systems are chemically different, they follow the same rules of hydrophobic solvation. Indeed, water molecules reduce their entropy by forming ice-like structures around the non-polar groups. As the temperature is increased, the thermal motions progressively lead to a very unfavorable situation for the water shell, which can no longer subsist. This breakdown subsequently triggers the aggregation of the alkyl groups. A physical network is then created provided that the overall solubility of the polymer chain remains strong enough to counterbalance the attractive forces generated upon heating. More exactly, the interplay between attractions and repulsions is not uniform and must be considered as a spot-like distribution of hydrophobic clusters in a sea of highly swollen chains.

These thermoassociative systems are of great interest for various reasons. First of all, thermothickening is an original property compared to thermothinning which is characteristic of polymer systems or fluids in general. This property is also a technological answer to a set of problems found in industrial applications where aqueous based-fluids are submitted to heating stages during applications like cooking processes in the food industry *(16)*. The control of the rheology of drilling fluids in deep subterranean formations is another typical application *(12)*. Recently, similar systems called smart gels or intelligent gels, were also depicted *(17)* with potential applications as shoe inserts, or in the medical field for drug delivery or skin care applications. The associating behavior is conceptually reversible and this feature provides additional reason for technological applications where switching properties are required. From theoretical considerations, the control of hydrophobic associations with temperature provides an interesting model to follow step by step the formation of the micelles and to verify percolation theory or other recent developments based on associative systems.

Finally, as mentioned above, this property is not related to a given chemical structure but is a general concept that is exemplified by a broad spectrum of systems. In order to clarify the following discussion, we will classify the various thermothickening systems into four categories: 1) the cellulosic derivatives, 2) the mixture of amphiphilic systems, 3) the block copolymers and 4) the graft copolymers.

CELLULOSIC DERIVATIVES

General behavior in aqueous solution

Cellulose is basically an interesting example of a hydrophilic polymer, which is not soluble in water. Nevertheless, chemical modification of the glucose units offers an efficient way to solubilise these highly crystalline structures by substituting hydroxyl groups with ether functionalities *(18)*. Surprisingly, hydrophobic modification of the cellulosic backbone is often carried out to enhance the solubility of this polysaccharide in water. Various derivatives, named according to the chemical group of the ether functionality, are presented in Table 1.

Table 1. Abbreviated names and substituents for some cellulose derivatives

Cellulose derivatives	*Abbrev.*	*Substituent*
Methylcellulose	MC	$-CH_3$
Ethylcellulose	EC	$-CH_2-CH_3$
Hydroxyethylcellulose	HEC	$-(CH_2-CH_2-O)_x-H$
Ethyl(hydroxyethyl)cellulose	EHEC	$-(CH_2-CH_3)(-CH_2-CH_2-O)_x-H$
Hydroxypropylcellulose	HPC	$-(CH_2-CH(CH_3)-O)_x-H$
Carboxymethylcellulose	CMC	$-CH_2-COO^-Na^+$

The extent of the modification can be quantified by either of two parameters: the average number of substituted hydroxyl groups per glucose unit (DS) or the average number of alkyl oxide molecules reacting with one glucose unit (MS). The nature of the alkyl groups introduced into the backbone and the extent of the substitution have dramatic consequences on the water-solubility of the cellulosic derivative and consequently on the viscosity of semi-dilute aqueous solutions (Figure 1).

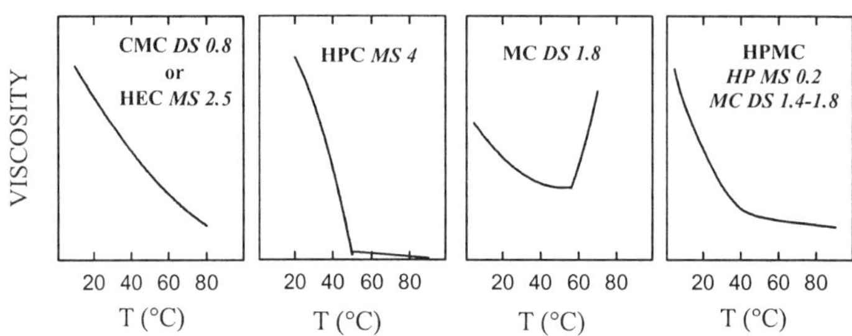

Figure 1. *Typical behavior of cellulosic derivatives in aqueous solutions. Reproduced with permission from reference 19. Copyright 1971 John Wiley & Sons, Inc.*

For hydrophilic modification (like sodium carboxylate groups in NaCMC) the polymer solution follows a typical Arrhenius behavior with a decrease in viscosity with increasing temperature. For derivatives with short alkyl substituents, like MC, the opposite, i.e. thermothickening behavior, is observed. In these systems, the alkyl groups have a very low affinity for water and their solubilisation can take place only at a very high cost of free energy due to the formation of ice-like structures around non-polar groups. When the temperature increases, the breakdown of water cages triggers the aggregation of the alkyl tails. A physical network is then created provided that the overall solubility of the polymer chain remains strong enough to counterbalance the attractive forces generated upon heating. If the hydrophobic forces are very strong, the phase separation process expands at the macroscopic scale as with HPC solutions (Figure 1). In that case the viscosity drops rapidly when the temperature exceeds a critical value since the phase separation gives rise to a highly concentrated polymer phase that cannot percolate through the total volume of the solution. To answer this problem, one can boost the osmotic pressure of the HPC chains by introducing some ionic groups along the backbone (like an amine in the ionized salt form for example). As a result, phase separation is cancelled out and the expected thermothickening behavior is observed *(19)*.

The origin of thermal gelation and its characteristics

Since the beginning of the first studies on thermal gelation, authors have attributed this phenomenon to a local desolvation of the chains upon heating[8]. Some years later, Rees used the concept of the hydrophobic effect and described these systems in terms of physical gels cross-linked by micelles *(20,21)*. An original study based on the influence of the pressure on aqueous solutions of MC allowed Suzuki et al. *(22)* to demonstrate that hydrophobic interactions were effectively the driving force of the gelation process. Using the same system, Sarkar *(23)* was able to determine by differential scanning calorimetry (DSC.) the enthalpy of dehydration which is characteristic of the hydrophobic associations triggered upon heating. The transition temperature was shown to decrease as the concentration of MC was increased. More recently Haque et al. *(24)* studied in detail MC solutions by comparing the results obtained from different techniques: DSC, ^{1}H NMR and rheology. They clearly show that above a temperature threshold the solution develops an important increase of the elastic modulus that is closely correlated to the release of the hydration shell and to the loss of mobility of the cellulosic units. Since the solution becomes very sensitive to shear rate upon gelation, this parameter must be carefully controlled in order to obtain reproducible results. Another important parameter is the distribution of the alkyl substituents along the cellulosic backbone *(25-28)*. This is due to the marked hydrophobic character of tri-substituted cellulose units. The more heterogeneous is the distribution of the substituents, the higher is the proportion of the tri-substituted units and the more stable are the physical cross-links. Desbrières et al. *(25)* have studied these aspects in detail by preparing homogeneously modified MC. They showed that for a given DS, the heterogeneously substituted MC always form stronger physical gels compared with the homogeneous sample. Also, if the nature of the alkyl group is slightly changed (methyl / propyl) the hydrophobic interactions will be greatly modified as well as the temperature responsiveness.

The influence of co-solutes on thermal gelation

In addition to the primary structure of cellulose derivatives, modification of the aqueous environment provides another way to control the association phenomenon. This was used very early with MC solutions for which authors noticed that the addition of salts modified the association temperature *(8,29)*. The nature of the salt is very important and especially that of the anion. While sulfate, characterized by a high affinity for water molecules, strongly decreases the association temperature (T_{ass}), nitrate has a very limited effect and thiocyanate slightly increases the temperature threshold. Considering the effect

186

of salts on the gelation temperature, we can see that they follow exactly the well-known Hofmeister series reported earlier in the case of protein denaturation *(30,31)*. Non-ionic species like sucrose, sorbitol, glycerine or co-solvent (ethanol, propylene glycol...) were also reported to influence the association temperature either by decreasing or increasing the quality of the solvent *(32,33)*.

As we can see, additives can strongly modify the association process by disturbing the interactions between solvent molecules and non-polar groups of the macromolecular backbone. Ionic solutes mainly decrease the polymer / solvent affinity, as the cellulosic backbone becomes more hydrophobic towards the salty aqueous medium. Nevertheless the opposite is also true and one can switch a solvophobic polymer into a more solvophilic one by improving the solvent quality or by playing with specific interactions. This second point will be detailed in the upcoming section.

MIXTURES OF AMPHIPHILIC SYSTEMS

Several ternary mixtures have been reported for their thermothickening properties. Even if they are chemically different, their common feature is a mixture of amphiphilic molecules and the rules for the thermally induced association remain the same with hydrophobic interactions as the driving force. We will classify the various examples into three categories.

Mixtures of an ionic surfactant and a hydrophobically modified water-soluble polymer.

This type of system is directly connected to the previous discussion about cellulosic polymers taking into account some of the more hydrophobic derivatives like EHEC *(34-37)* or HPC *(38)*. As is known, EHEC (or HPC) gives aqueous solutions that phase-separate upon heating. When EHEC is mixed with sodium dodecyl sulfate (SDS), hydrophobic associations take place between the chains and the surfactant molecules, provided that the concentration of surfactant is higher than the critical association concentration (c.a.c.). Mixed aggregates are thus formed and the (initially) neutral polymer backbone is converted into a polyelectrolyte chain decorated with surfactant micelles. This phenomenon has two important consequences on the behavior of the aqueous solution upon heating. First of all, the phase separation mechanism is delayed towards higher temperatures since the polymer is now less hydrophobic. Furthermore, above the temperature threshold of the complex (EHEC/SDS), the phase separation remains confined at the local scale due to the presence of ionic surfactant (Figure 2). The hydrophobic units of the polymer chains self-assemble into mesoscopic aggregates stabilized by adsorbed surfactant molecules *(35)*. This formation of mixed micelles is responsible for the thermothickening properties *(38)* and for the clear gel pattern observed in the low temperature range.

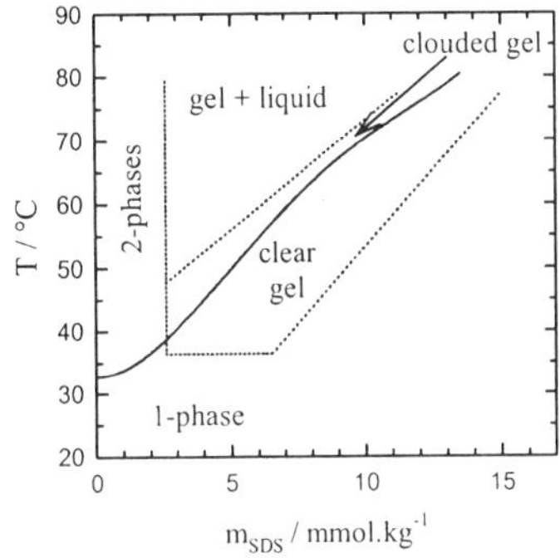

Figure 2. Phase diagram of an aqueous solution of 1% EHEC with SDS. The solid line represents the cloud point curve and the dashed lines the borders of the high viscosity solutions. Reproduced from reference 34. Copyright 1997 American Chemical Society

If the temperature is increased, the size of the heterogeneities grow up and a clouded gel situation is reached in an intermediate range, just before a macroscopic phase separation takes place (gel + liquid). As we can easily imagine, this problem is quite general and not limited to some cellulosic polymers. It deals with the stabilization of hydrophobic aggregates at a local scale. If the macromolecular backbone itself cannot fulfill this condition, amphiphilic molecules, like surfactants, are a typical answer to that problem.

Mixtures of a thermoresponsive polymer and an associative polymer

Surfactant molecules are known to develop hydrophobic interactions with non-polar groups carried by water-soluble polymers. This is the case for SDS molecules which can form micelle necklaces with POE or poly(N-isopropylacrylamide) [PNIPA] chains. What will happen if one substitutes a hydrophobically modified polyelectrolyte (a kind of polysoap) for SDS ?
Bokias et al *(39)* have answered this question studying the behavior in aqueous solution of mixtures containing 1) a poly(sodium acrylate) backbone (Mw=150 000g/mole) randomly modified with 3% (molar) of octadecyl groups [3C18] and 2) a poly(N-isopropylacrylamide) (M_w=650,000g/mole) which exhibits a Lower Critical Solution Temperature (LCST) in water around 32°C. Due to the amphiphilic nature of PNIPA, hydrophobic interactions take place between the two polymers at low temperature and give rise to a strong enhancement of the viscosity level. The initial physical network of 3C18 is now "over-cross-linked" by the long PNIPA chains (Figure 3). This synergy is very sensitive to the

composition of the mixture and to the temperature. As a matter of fact, when the temperature is raised the viscosity increases around ten times its initial value in the range 10 - 25°C.

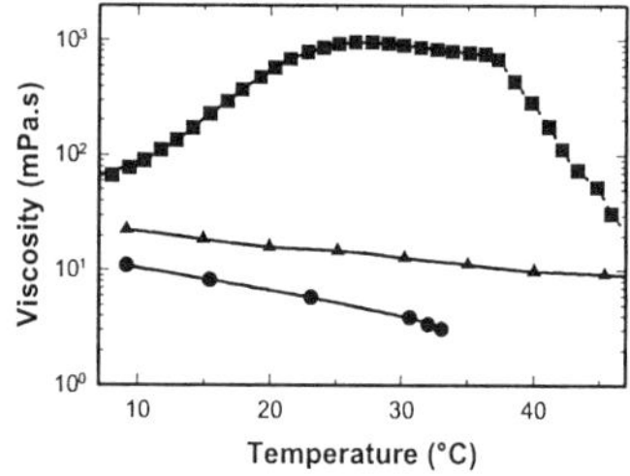

Figure 3. Rheological behavior of aqueous solutions (1.3% w/w): ● PNIPA; ▲ 3C18; ■ a mixture of 3C18/PNIPA (90/10 in weight). Reproduced from reference 39. Copyright 1997 American Chemical Society

This thermothickening effect is ascribed to the strengthening of the hydrophobic associations between the C18 tails of the polymer and non-polar groups of PNIPA upon heating. Nevertheless above 35°C these interactions cannot further avoid the coil-globule transition of PNIPA and its phase separation. PNIPA chains are then progressively released from the network and, after a huge decrease in viscosity, the situation becomes comparable at high temperature to that obtained for a 3C18 aqueous solution. This example clearly illustrates the synergy that can be developed between an associative polymer and a thermoresponsive compound able to interact with the former. Here the coupling between the two species is lost as soon as the phase transition occurs. However we will see now that other mixtures have been developed to trigger a thermothickening effect when crossing the phase transition.

Mixtures of a thermoresponsive surfactant and an associative polymer

This type of ternary system has been exemplified by Iliopoulos and co-workers using 1) a similar associative water-soluble polymer to that mentioned above (here they use a [1C18]) and 2) a nonionic surfactant of the oligoethylene glycol alkyl ether type *(40,41)*. Hydrophobically modified polymers (HMP) are known to associate strongly with surfactants, forming mixed micelles and their rheological behavior is strongly dependent on the molar ratio R=[micelle]/[alkyl group]. At high R values (R > 0.2), when there are only a few alkyl grafts per micelle belonging mainly to the same polymer chain, there is no effective cross-links in the system (see Figure 4a). On the other hand oligoethylene glycol derivatives are known to self-assemble under different morphologies according to a phase diagram (temperature vs concentration) that strongly depends on the length of the hydrocarbon tail. For instance, the surfactant denoted $C_{12}E_4$ [$CH_3(CH_2)_{11}$-$(OCH_2CH_2)_4$-OH] exhibits a phase transition from micellar (L1) to lamellar phase (Lα) when the temperature is increased above T_L=23°C.

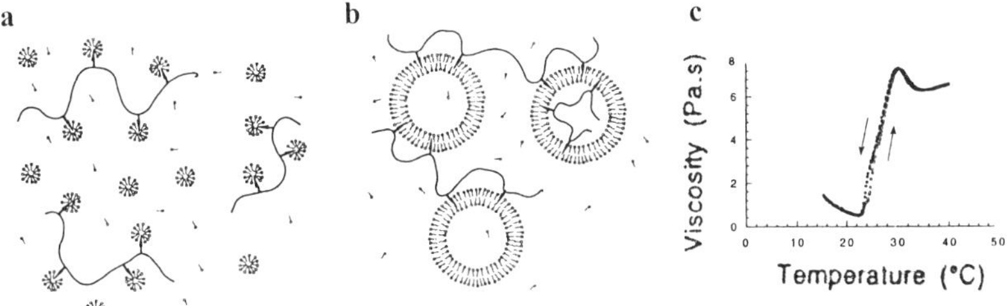

Figure 4. *Thermal gelation of 1C18/C12E4 mixture in aqueous solution. a and b illustrate the gelation mechanism at $T<T_L$ and $T>T_L$, respectively. C is the temperature dependence of the viscosity for a mixture of 1C18 (1%) and C12E4 (2%). Reproduced from reference 41. Copyright 1995 American Chemical Society*

Now when an aqueous mixed solution of $C_{12}E_4$ and 1C18 is considered at low temperature ($T<T_L$), with convenient concentrations of surfactant and polymer (Cs =2%, Cp=1% w/w), it behaves as a viscous liquid according to the picture drawn in Figure 4a. Upon heating ($T > T_L$), the surfactant starts to form bilayers which are stabilized as large vesicles. The volume fraction of surfactant aggregates becomes very large and these vesicles can form bridges between the polymer chains (Figure 4b). The resulting thermal gelation reported on Figure 4c is clearly driven by the phase diagram of the surfactant and one can control the network formation by changing either the oligoethylene glycol derivative (T_L=52°C for $C_{12}E_5$ for example) or mixing various surfactants *(41)*.

In this example the responsive amphiphile (here the surfactant) is not able to develop by itself a physical network at low concentrations and an "amplifier" (i.e. the associative polymer) is necessary to obtain thermothickening properties. Nevertheless there are some examples where a binary system «surfactant + water» can exhibit a thermothickening behavior at moderate concentration. This was reported by Greenhill-Hooper et al.[42] with oligomeric surfactants tailored with a sodium sulfonate head, a thermosensitive spacer, combining ethylene oxide (EO) and propylene oxide (PO) units, and a hydrophobic tail containing an aromatic ring terminated by an octadecyl group. The viscosity behavior of a low concentration solution (C = 0,01 mole/l ~ 1% w/w) of surfactant denoted [C18-ϕ-P5E11-S] is shown in Figure 5. The thermothickening behavior was studied by light scattering and clearly identified as a consequence of the surfactant phase transition: here from micelle to rod-like aggregates. This modification of the surfactant organization is related to the dehydration of the PO units upon heating. The viscosity enhancement observed above 45°C and the shear-thinning behavior is thus ascribed to the entanglements between the rod-like aggregates.

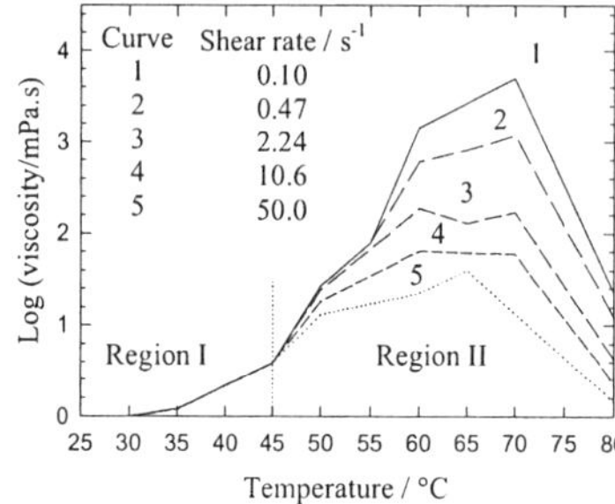

Figure 5. *Temperature and shear-rate dependence of the viscosity of a 0.01 mol/L solution of [C18-φ-P5E11-S] in 3.5% saline solution. Reproduced with permission from reference 42. Copyright 1988 Academic Press Inc.*

At higher temperatures, above 65 to 70°C, a macroscopic phase separation occurs, indicating that the hydrophobicity of the C18-φ-P5E11 block overcomes the hydrophilicity of the ionic head. This example clearly demonstrates that an increase of viscosity with temperature does not necessary require the formation of a connected physical network. As we will see now, this duality between entanglements and bridging is the basis of the thermothickening properties in block copolymer solutions.

BLOCK COPOLYMERS

In this section, we deal with di- or triblock copolymers that combine hydrophobic (or at least thermoresponsive) blocks (B) and blocks (A) which could be permanently hydrophilic or thermoresponsive but in a temperature range higher than (B). Three cases will be distinguished in what follows: 1) the diblocks copolymers [AB], 2) the triblocks with a central hydrophobic segment [ABA] and 3) the triblocks with a central hydrophilic segment [BAB]. While the BAB copolymers are expected to form a physically cross-linked network, the properties of AB and ABA gels will be mainly dominated at high concentrations by the entanglements between the A block coronas of the close-packed micelles *(43)*.

Thermoassociative diblock copolymers

With diblock copolymers, two strategies can be explored in order to obtain thermothickening systems: 1) the coupling between a thermosensitive sequence and a permanent hydrophobic block, which is similar to the behavior described for an oligomeric surfactant *(42)* or 2) the combination of a thermosensitive block with a hydrophilic one *(43,44)*. In the first case, the diblock macromolecules form micelles with a B-core above their c.a.c. . Upon heating, the affinity of the hydrophilic corona (A) for water molecules progressively decreases and this induces drastic changes in the organization of the molecules in water. On the contrary, in the second case, the molecules are initially isolated (unimers) and the desolvation of the thermoresponsive block gives rise to the formation of micelles. In each case, the thermothickening effect is entirely due to the entanglements of the macromolecular aggregates since no bridging between micelles is expected with diblock copolymers. The main consequence is that the copolymer concentrations required to obtain thermothickening properties are generally very high compared to the previous systems. This can be exemplified with the results reported by Deng et al. *(43)* in the case of diblock copolymers tailored with a random ethylene oxide/propylene oxide [EO/PO] copolymer (hydrophobic block) coupled with a hydrophilic poly(ethylene oxide) [PEO] block. They show that the formation of micelles upon heating requires a minimum content of PO units (20 mol%) in the hydrophobic block as well as a minimum length of the PEO block. On this basis, a further increase of temperature above the micellisation threshold leads either to a phase separation at low concentration or to a thermothickening behavior above 30% (w/w). For more hydrophobic diblocks, containing a high PO content (> 70 mol%), the copolymer solution exhibits only a phase separation upon heating, without micellisation.

Thermoassociative triblock copolymers

Copolymers with a central hydrophobic block [ABA]

In this section, we will focus on triblock copolymers containing a central hydrophobic block (usually poly(propylene oxide) [PPO]) and two hydrophilic blocks at each end (like PEO). In fact both the hydrophobic and the hydrophilic blocks are sensitive to the temperature (PEO as well as PPO exhibit a LCST behavior). Nevertheless, the temperature at which the dehydration of PPO block takes place is much lower compared to that of PEO. The thermothickening properties of aqueous solutions of such triblock copolymers proceeds in a similar way to the one reported with diblock copolymers. Once again the triggering of elastic properties in the solution rely upon the formation of an entangled network between the macromolecular assemblies. The symmetrical triblocks of general formula: EO_x-PO_y-EO_x are obviously the most widespread example in the ABA family. They are usually called "Poloxamers" or "Pluronics" but other trade names are also used *(45,46)*. These copolymers have been studied in aqueous solution with a wide variety of techniques, which encompass rheology *(47,48)*, radiation scattering *(49-51)* as well as differential scanning calorimetry *(46,52)* and interfacial tension measurements *(53)*. The results, which have been reviewed extensively *(54)*, generally try to correlate the morphology of the aggregates which could be quite complex to the rheological behavior of the solution. For aqueous solutions of P85 ($EO_{27}PO_{39}EO_{27}$), it was shown *(48)* that unimers (single chains), micelles and clusters (large aggregates) coexist initially at lower concentrations (and temperatures) in relative proportions which depend on temperature. As the temperature is increased, the coexistence between unimers and micelles remains over a broad range of concentration while at still higher temperatures only micelles are present. Further increase in concentration and temperature causes a pronounced increase in micellar mass and disymmetry. At high polymer concentrations (above 20% w/w), a thermogelation can be observed as a result of the close packing of the micelles with a high degree of entanglement of the PEO chains in the micellar mantle (see Figure 6).

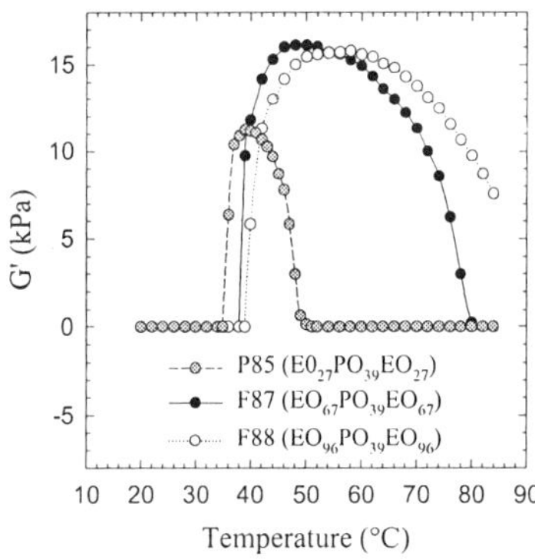

Figure 6. *Comparison of temperature-dependent gelation of 0.25g/ml solutions of P85, F87 and F88. The figure shows G' at 0.05 Hz versus temperature during heating scan (0.3°C/min). Reproduced from reference 48. Copyright 1992 American Chemical Society.*

At still higher temperatures, the dehydration of the PEO blocks gives rise to a macroscopic phase separation with a loss of the viscoelastic properties. It is worth mentioning that if the PPO blocks are responsible for the initial micellisation in the solution, the further changes in the shape of the aggregates that occur upon heating are associated with the temperature sensitivity of the PEO blocks.

Copolymers with a central hydrophilic block [BAB]

These BAB thermoassociative copolymers can be roughly compared to the telechelic polymers, like HEUR[4]. The hydrophilic central block is generally PEO while several terminal thermoresponsive moieties have been proposed like PPO *(55)* or PNIPA *(56,57)*. The main difference with the ABA triblocks is obviously the possibility of bridge formation between aggregates. Mortensen et al. *(55)* studied the behavior of symmetrical copolymers with the general formula (PPO)y-(PEO)x-(PPO)y. For aqueous solutions of the copolymer 15/156/15 (x=156 and y=15) a rather complex phase diagram (temperature vs concentration) was evidenced with two kinds of physical networks built up upon heating. These are called "random network" and "micellar network" by the authors. The "random network" contains association domains with low aggregation numbers spread among PEO hydrated coils. It could be obtained in a rather broad range of concentrations and, interestingly, in the low concentration domain with gelation temperature around 30-40°C for C=1-2% w/w. On the contrary, the "micellar network" is formed by micelle-like aggregates involving more PPO sequences with PEO bridges between them. Because of the structural difference the random network is weaker compared to the micellar network that exhibits a high viscosity (higher number of elastically active chains). Nevertheless, concentrations as high as 50% (w/w) are necessary for the micellar network to percolate through the entire volume of the solution. Other copolymers were prepared by Yoshioka et al. *(56,57)* on the basis of a triblock with a central PEO sequence and PNIPA blocks at each end. Even if the procedure does not ensure that all the copolymer chains have the expected triblock structure, a thermothickening was observed in semi-dilute solutions and ascribed to the micro-phase separation of the PNIPA blocks. By incorporating hydrophobic comonomers inside the PNIPA sequence (butyl methacrylate for instance) the gelation temperature was modulated from 32°C down to 0°C. The addition of salt in the solution was also reported to decrease the gelation temperature similarly to the cloud point of PNIPA solutions.

These experimental results can be compared with the Monte Carlo simulation performed by Nguyen-Misra et al. *(58)* on BAB triblocks in a selective solvent. The solvent is assumed to be poor for the B blocks and athermal for the A block, which is assumed to have similar properties to the solvent. As a result, solutions of BAB are reduced to a pseudo-binary mixture which can be described using the degree of incompatibility: $x = 2N_B\beta\epsilon$; where N_B is the number of B units, $\beta = (k_BT)^{-1}$ and ϵ the positive interaction energy ($\epsilon = \Delta\epsilon_{BA} = \Delta\epsilon_{BS} > 0$). On this basis, thermoassociative triblocks can be considered taking into account that the degree of incompatibility increases with the temperature. The main results are that both the critical micelle concentration (c.m.c.) and the gel transition point (ϕ_{gel}) scale with x as:

$$c.m.c. \sim x^{-0.4} \qquad \phi_{gel} \sim x^{-a} \qquad \text{(with } 0.1 < a < 0.2\text{)}$$

The c.m.c., like ϕ_{gel}, is not affected by the length of the middle block (N_A) since only small variations are evidenced by the calculation. Nevertheless, N_A influences the chain distribution functions (fraction of free chains, dangling ends, loops, bridges). The results can be summarized by plotting the degree of incompatibility (similar to the temperature for thermoassociative copolymers) versus the copolymer volume fraction (Figure 7).

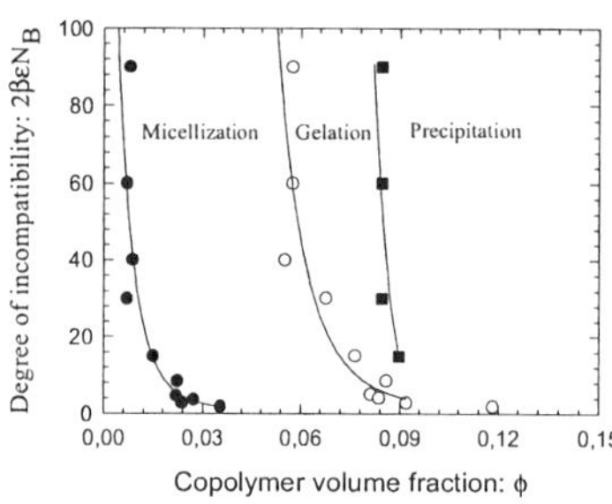

Figure 7. *Phase diagram of a triblock copolymer system BAB (N_A=10) with insoluble ends. Reproduced from reference 58. Copyright 1995 American Chemical Society.*

Four regions are distinguished on this phase diagram: homogeneously dispersed chains, micellization, gelation and precipitation. Obviously, the degree of incompatibility governs the association behavior of the copolymers.

GRAFT COPOLYMERS

Thermoassociative graft copolymers were first developed at the beginning of the 90's *(12-15)* as a technological answer to a set of industrial problems arising when aqueous-based fluids undergo a heating stage during their application. The control of the viscoelastic properties of drilling fluids is a typical problem, for instance at temperature as high as 200°C *(12)*. Clearly these copolymers resulted from the wide knowledge obtained during the preceding decade on associative water-soluble polymers. They were tailored on the basis of hydrophobically modified water-soluble polymers (HMP), taking the opportunity to reversibly switch hydrophobic interactions through LCST polymeric stickers. As reported with the other thermogelling systems, the most important ingredients in the development of responsive graft copolymer networks are:

- The thermodynamic properties of both the water-soluble backbone and the responsive stickers,
- The primary structure of the copolymers.

These are the two main aspects that we will try to concentrate on in the following sections.

Design of thermoassociative graft copolymers

Starting from an ideal picture of a graft copolymer, with a water-soluble backbone supporting LCST side-chains, several grafting pathways can be followed:

1) **Grafting onto**. This synthesis includes a) the synthesis of a LCST polymer (or copolymer) with a specific functionality at one end (telomerization) and b) the coupling reaction between this reactive end-group (amine for example) and one of the chemical functionalities carried on the backbone (carboxylic acid for instance) *(59-61)*.

2) **Grafting through**. This can be carried out by copolymerization between a water-soluble monomer and a thermoresponsive macromonomer obtained by modification of the former telomer *(14,59)*.

3) **Grafting from**. This can be realized by polymerization of a water-soluble monomer in the presence of a thermoresponsive polymer able to promote chain transfer reactions *(59,62)*.

The different methods display their own advantages and drawbacks, but the "grafting onto" method is definitely the best way to achieve well-controlled primary structures. Some of the most relevant structures are given in Table 2.

Table 2. Primary structure of thermoassociative graft copolymers.

Name	*Backbone*[a]	*Side chains*[a]	*W (%)*[b]	*Grafting*	*Ref*
PAA-g-PEO	PAA 10^2-10^3 kg/mol	PEO 1-30 kg/mol	15-50	*Onto*	*13*
PAA-g-PPO		PPO (0.5 kg/mol)	20-60		*59*
PAA-g-PNIPA		PNIPA 2-10 kg/mol	15-40	*Onto*	*60*
PAM-g-PNIPA	PAM 10^3 kg/mol			*Through*	*63*
PAA-g-pluronic	PAA 4.10^3 kg/mol	EO_{100}-PO_{65}-EO_{100} 13 kg/mol	43	*From*	*64*
PAM-g-PEO	PAM 8.10^2 kg/mol	PEO 0.4 kg/mol	5	*Through*	*14*

[a]) PAA: poly(sodium acrylate); PAM: poly(acrylamide); PEO: poly(ethylene oxide); PPO: poly(propylene oxide); PNIPA: poly(N-isopropylacrylamide), [b]) W_{LCST}: weight fraction of LCST side chains.

As we can see, the side chains could be very long (with molecular weight up to 25000 g/mol) compared to the "short" alkyl tails (C8 to C18) used in HMP. One of the main consequences is that the molar grafting ratio must remain very low (generally under 1%) to preserve the water-solubility of the copolymer during association. Actually mainly polyethers (PEO, PPO or copolymers) and PNIPA derivatives have been used as responsive stickers but of course the number of candidates is very large, especially if copolymers are considered.

We will see now how they work and what are the main correlations between the primary structure and their properties in aqueous solutions.

General behavior of thermoassociative graft copolymer solutions

Grafted structures with polyether stickers (12-15,59,61,62,64-67)

Typical thermothickening behavior is presented in Figure 8 with semi-dilute solutions of PAA-g-PEO25 (25 kg/mol is the molar mass of the PEO) *(65)*.

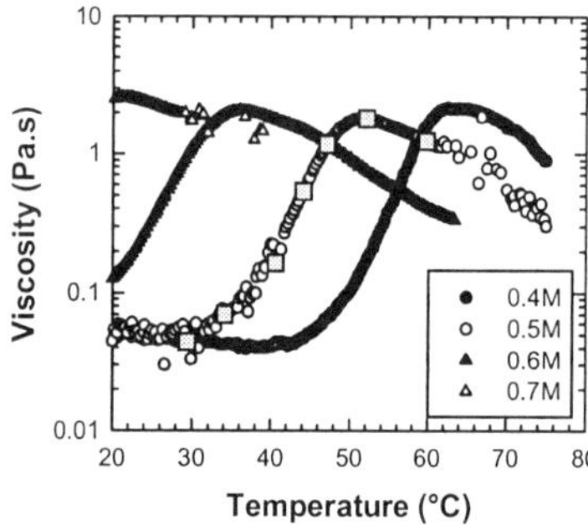

***Figure 8.** Influence of K_2CO_3 concentration on the thermothickening behavior of PAA-g-PEO25 aqueous solutions: Cp= 2.7%; shear rate=50s⁻¹ Reproduced from reference 65. Copyright 1998 American Chemical Society.*

As for cellulosic derivatives, salt is added to the solution in order to decrease the association of PEO25, initially around 110°C in pure water. For a given salt concentration, say $[K_2CO_3]$ = 0.5M (O), three different temperature regimes are distinguished.

1) Below T ≈ 30°C, the PEO side-chains are still soluble in the solvent and the copolymer is hydrophilic as a whole. The decrease of the viscosity with T is well described by the Andrade law ($\eta = A.e^{E/kT}$) with an activation energy for viscous flow E_η ≈ 20KJ/mole. This value is close to the one calculated for the solvent itself as one expects at this copolymer concentration.

2) Above 30°C, the association behavior is switched on and the viscosity starts to increase with temperature. The thermothickening process, i.e.

the formation of "hydrophobic" microdomains, is clearly observed above 30°C, but at the same time the copolymer solution becomes shear-thinning, so the viscosity increase finally levels off around 50°C under these experimental conditions (shear rate=50s^{-1}).

3) Beyond this point the viscosity starts to decrease again, but now with a higher activation energy ($E_\eta \approx$ 70KJ/mole), as generally reported with associative polymer solutions *(4)*.

In these experiments, the temperature of association ($T_{ass} \approx$ 30°C) is very close to the cloud point of the PEO25 precursor, determined under the same conditions (Figure 9).

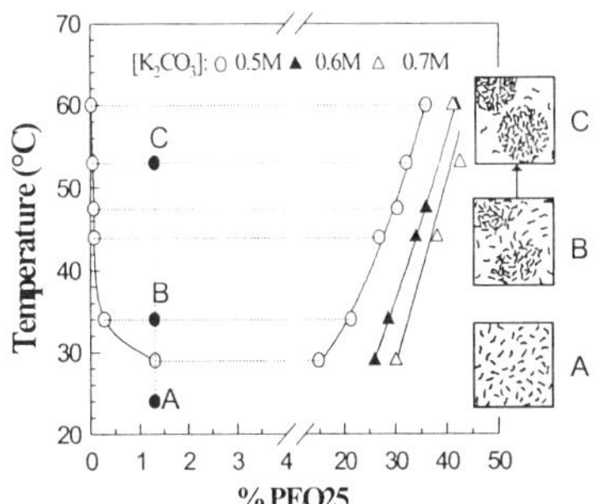

Figure 9. *Phase diagram of the pseudobinary system PEO25 / water for various concentrations of K_2CO_3. Reproduced from reference 65. Copyright 1998 American Chemical Society*

This general result *(15,66)* indicates that the thermodynamics of ⌣ST grafts drives the association process and that consequently, a good knowledge of the phase diagram of PEO is a useful tool to predict the temperature of transient network formation. This basic principle is typically applied in the experiments depicted in Figure 8 for which potassium carbonate is added to decrease the switching temperature of the copolymer solution. Looking at the local scale now, by using small angle neutron scattering [SANS] experiments (Figure 10), an interesting parallel can be drawn with the previous results.

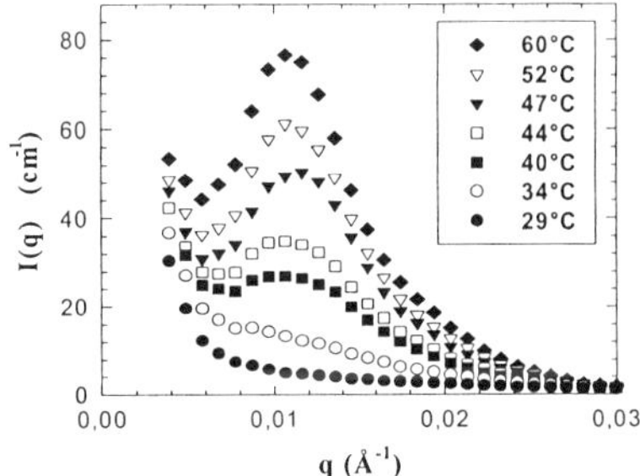

Figure 10. *Variation of SANS intensities with temperature of PAA-g-PEO25 solutions: Cp=2.7%; [K_2CO_3]=0.5M Reproduced from reference 65. Copyright 1998 American Chemical Society.*

As the temperature exceeds 30°C, a sharp scattering peak arising from the correlation between the PEO domains starts to grow. Beyond these conditions, the magnitude of the fluctuation keeps on increasing while the wavelength, or the periodicity of the microdomains, remains constant. A detailed analysis of the asymptotic behavior indicates that PEO side-chains undergo a continuous transition from a homogeneous solution of random coils to a two-phase structure stabilized at the nanoscopic scale. According to this analysis, the PEO microdomains can be pictured in the strong segregation regime (high temperature and/or salt concentration) by polydisperse spherical aggregates (radius of ~ 10nm). They display a sharp interface and an average local concentration of 40-45%, which is finally very similar to the one obtained from the phase diagram (Figure 9).

Similar systems were developed recently by Bromberg *(64)*, on the basis of pluronic stickers grafted on a poly(sodium acrylate) backbone (see Table 2). In agreement with percolation theory, it was shown that the zero-shear viscosity (η_0) and the equilibrium modulus (G_0) could be scaled with the distance from the gelation threshold (ε) :

$$\eta_0 \sim \varepsilon^s \text{ and } G_0 \sim \varepsilon^t \text{ with } \varepsilon = \varepsilon_T = T/T_g - 1 \text{ or } \varepsilon = \varepsilon_C = C/C_g - 1$$

where s and t are the transient exponents and T_g and C_g are the critical gelation temperature and concentration, respectively. As with the PAA-g-POE25 aqueous solutions, a decrease of the viscoelastic properties (G') with increasing temperature was also noticed around 20°C after the beginning of the thermothickening process.

Grafted structures with PNIPA stickers (60,63,68,69)

Going back to the original grafted PAA and replacing PEO25 by PNIPA10 (10kg/mol), we observe a rheological behavior (Figure 11) in agreement with the phase diagram PNIPA10 (Figure 12).
As for pluronic derivatives, the association process can be triggered now at low temperature in pure water. Whatever the copolymer concentration is, the viscosity starts to deviate from the initial Andrade law around the LCST of PNIPA. More precisely, T_{ass} decreases from 36 to 26°C by increasing the copolymer concentration as it could be predicted taking into account the salting-out effect of the polyelectrolyte backbone *(60,68)*. There are also some differences between the concentration profiles that can be attributed: 1) to the connectivity of the transient network, with a sol-gel transition which is not far below Cp=0.8% and 2) to the competition between hydrophobic attraction and electrostatic repulsion, the result being strongly dependent of the concentration (or ionic strength). Contrary to POE derivatives, the aggregation process of PNIPA can be followed by NMR (^{1}H or ^{13}C), as the responsive stickers tend to form solid-like aggregates above their LCST. The strong decrease of the molecular mobility of the side-chains is clearly evidenced in Figure 13 where the ^{1}H signals of PNIPA (isopropyl protons are shown in this figure) progressively disappear upon heating, leaving only the protons of the PAA backbone (between 1.2 and 2.5 ppm).

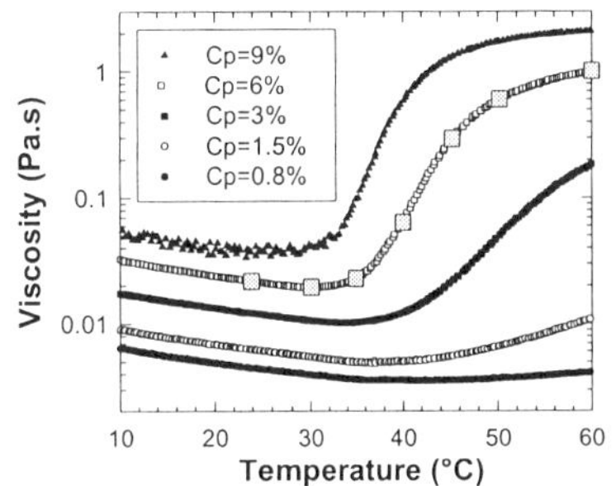

Figure 11. *Thermogelation of PAA-g-PNIPA10 in pure water; Shear rate = 100s⁻¹*

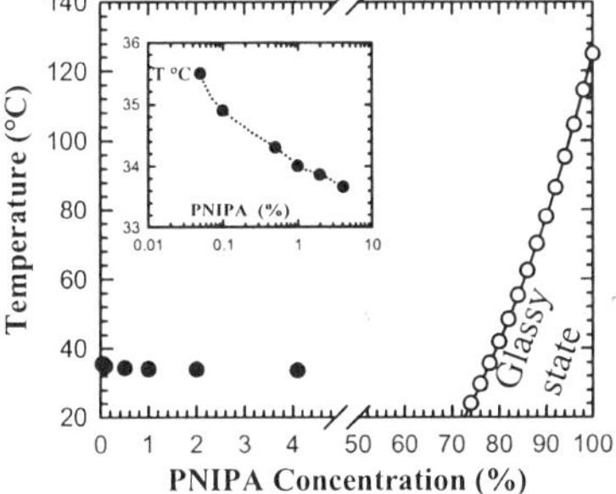

Figure 12. *Cloud point curve ● and calculated glass transition ○ of the binary system PNIPA10 / water*

200

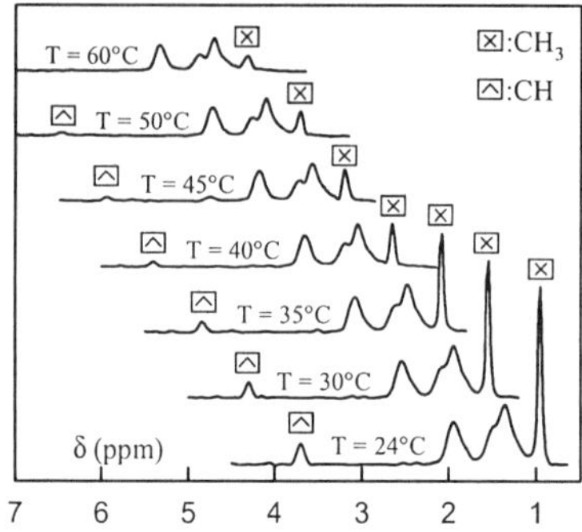
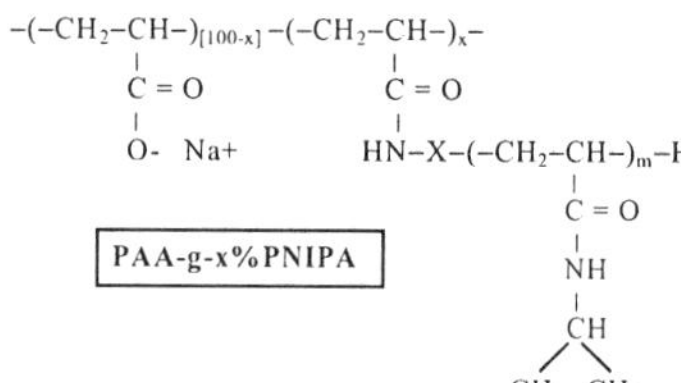

Figure 13. NMR spectroscopy of a PAA-g-PNIPA10 aqueous solution (Cp=6% in D2O) during gelation.

At 60°C for instance, one can estimate that less than 30% of the PNIPA remains observable, i.e. at least 70% of the side-chains are embedded in solid hydrophobic clusters. This loss of mobility evidenced by NMR is at least qualitatively in good agreement with the phase diagram of PNIPA10 (Figure 12), where the calculated glass transition was plotted against the PNIPA/water composition. These experiments can be completed by DSC, as the PNIPA phase separation gives rise to a clear endothermic peak attributed to the dehydration process (like for HPC samples). Such experiments demonstrate that 1) the PNIPA grafts behave similarly whether they are grafted on the PAA backbone or are free in the solution and 2) that the whole thermal process is achieved in less than 20°C after the beginning of the association (T_{ass}). Again if we come back to the thermothickening profile of PAA-g-PNIPA, we can see that the main difference compared to polyether derivatives is the relative independence of the viscosity in the high temperature range. This can be put in line with the difference of morphology between the liquid-like PEO aggregates and the solid-like PNIPA clusters.

Control of the association process

Due to the versatility of their chemistry, graft copolymers with responsive side-chains afford the best technological answer to the main problem "How to control the association process and the rheological properties of aqueous based fluids under a temperature gradient?" At least two opposite ways can be outlined:

- by modifying the primary structure of the copolymer in regard to a given environment,
- by modifying the environmental conditions on the basis of a given primary structure.

Modification of the primary structure

Two kinds of structural parameters have to be distinguished: 1) those related to the side chains (length, grafting rate, chemical nature, distribution, ...) and 2) those concerning the water-soluble backbone (length, degree of ionization, hydrophilicity...).

Concerning the length of the themoassociative side chains and their grafting ratio, an extensive study has been carried out on PAA-g-PEO derivatives *(65,67)*. Their influence on the characteristics of the thermally induced physical network are very similar to what was previously reported with HMP. The longer the side chains, the higher is the viscosity increase above T_{ass}. The same holds for the grafting ratio but these conclusions are restricted to a small range of composition. As a matter of fact true phase separation is expected and observed when the weight fraction of thermoresponsive moieties becomes too high in respect to the hydrophilicity of the backbone. Generally the weight fraction of PEO or PNIPA in graft copolymers never exceeds 50% (w/w). Moreover, the control of the chemical composition of the side chains, especially by copolymerization, affords a very convenient way to adjust the association temperature *(70)*. A good knowledge of the thermodynamic properties of the side chain precursors is consequently the best guideline for designing polymers with well-controlled associative properties. The structure of the backbone is also very important for designing a convenient thermoassociative system. First of all the length of the backbone is closely related to the overlap concentration of the copolymer solution. Consequently it gives the concentration threshold of the gelation process and the ratio between intermolecular cross-links and intramolecular cross-links (degree of overlapping), at a given copolymer concentration. In the case of graft copolymers with a low content of very long side chains (see PEO25) it is very important to use water-soluble backbones with high molecular weight in order to keep enough responsive side-chains (more than 2) in the copolymer structure. The chemical composition of the backbone is also very important in the design of these systems. Polyelectrolytes are good candidates as they are very soluble in water and in some applications sodium sulfonate units will be preferred to sodium acrylate which are sensitive to divalent cations *(67)*. Non-ionic backbones can also provide very good thermothickening properties and various [poly(acrylamide) derivatives]–g–PNIPA were recently developed and patented for applications in the field of capillary electrophoresis *(63)*. However, if the hydrophilicity of the backbone is hardly decreased, the tendency of the copolymer chain will be to uncoil and to reduce the connectivity of the physical network *(71)*. From a general point of view, the formation of physical cross-links is favored when the incompatibility between the side chains and the backbone is high. Nevertheless, there are some cases where specific interactions between the grafts and the backbone give rise to a sharper thermothickening behavior. These properties were observed with PAA-g-PNIPA aqueous solutions at low pH (around 4-5), in a situation close to the phase separation (complex coacervation) *(69)*.

At first thought, the prediction of associative behavior from the primary structure may seem rather complex. However, this can be accomplished using the following relation established from neutron scattering studies performed on a series of PAA-g-PEO derivatives:

$$N_{ag} \sim N_B^{0.9} / N_A^{\alpha}$$

From the primary structure of the copolymer, it is then possible to extrapolate the aggregation number (N_{ag}) which is a key-parameter for the viscoelastic properties, from:

- N_A, the average number of water-soluble units between two grafts,
- N_B, the average number of ethylene oxide units in the side chains,
- $\alpha = 0.5$-0.7, an adjusting parameter accounting for the affinity between the hydrophilic block A and the solvent.

Modification of the environmental conditions

Salts have been extensively used to modify the association temperature. This is especially the case for potassium carbonate, which is known to strongly decrease the LCST of polymers (see Figure 8). As reported *(65)* for PAA-g-PEO25, the effect of K_2CO_3 on T_{ass} is similar to that observed on the cloud point of PEO25 under the same conditions. The same conclusion also holds for other LCST grafts *(15,59,60,69)*. Many other salts have also been tested in PAA-g-PNIPA aqueous solutions *(72)*, and the main conclusions are qualitatively the same as those reported early with cellulosic derivatives, which are based on the Hofmeister series. Nevertheless if the main effect of salt, and especially K_2CO_3, is to lower the solvent quality for the side chains (or to increase the degree of incompatibility x as defined in the section devoted to block copolymers), it also increases the ionic strength of the solution and reduces the electrostatic repulsions inside the hydrophilic coronas. Consequently, added salts not only shift T_{ass} but also modify the characteristics of the aggregates (size and shape) and the resulting viscoelastic properties. If the ionic strength is a critical variable, salt can be replaced by non-ionic molecules like sugar or organic liquids which are known to also modify (and generally decrease) the LCST of the side-chains.

As previously reported, surfactants are known to interact specifically with the thermosensitive polymers through the formation of mixed hydrophobic aggregates. This has been particularly studied in the case of aqueous mixtures of PAA-g-PNIPA and SDS *(72)*. Contrary to the positive synergy reported between HMP and SDS in aqueous solution *(6)*, the main contribution of SDS above T_{ass}

of PAA-g-PNIPA is a general weakening of the physical network due to the solubility enhancement of the SDS/PNIPA micellar necklace. Nevertheless this behavior was used to turn an associative polymer into a thermoassociative one.

This is exemplified in Figure 14 using a PAA derivative grafted with short side-chains (M < 5 kg/mole) of poly(tertiobutylacrylamide) [PAA-g-PTBA] *(73)*.

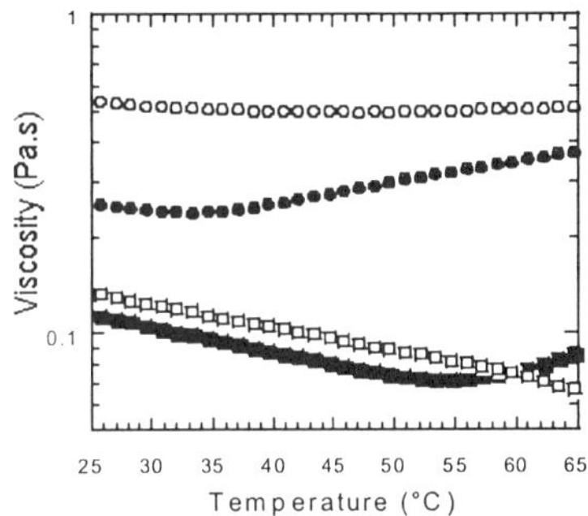

Figure 14. Viscosity of PAA-g-PTBA aqueous solutions (2%) with added SDS :

○ 0mM; ● 4mM; ■ 15mM. Comparison is given with the unmodified PAA precursor ❑. Shear rate=100s⁻¹.

Since PTBA is not soluble in water PAA-g-PTBA are physically cross-linked in semi-dilute aqueous solutions whatever the temperature. By adding SDS, the formation of a soluble complex SDS/PTBA leads to a decrease of the network connectivity at lower temperatures. When the saturation of the side-chains by SDS micelles is reached (~15mM), the viscosity of the solution becomes very close to the viscosity of an unmodified PAA solution. However, when the temperature is increased, the complex becomes less stable and PTBA side-chains start to self-assemble again while SDS micelles are progressively released. This example based on graft copolymers is analogous to those reported in the section on mixtures of amphiphilic systems.

CONCLUSION

Thermal gelation in aqueous macromolecular solutions has been discussed and exemplified through a wide selection of data. It appears that the gelation pathways can be quite different from one system to another. However they initially proceed through the same driving force which is an increase in hydrophobic interactions upon heating. The associative properties of these systems are well understood. Taking into account the various examples which were discussed, one can find or adapt one to fit a specific need, including any toxicity requirements or environmental conditions. As a general rule, an aqueous thermothickening system must combine:

204

- a responsive component (B) that undergoes a phase transition in water upon heating within a given temperature range,
- a hydrophilic component (A) that stabilize the hydrophobic aggregates in water at a local scale in order to avoid a macroscopic phase separation,
- the formation of a physical network through the bridging of elastic chains A between hydrophobic cross-links B, or an increase of entanglements between the A coronas of growing macromolecular objects.

For a given system, the association temperature can be modified quite easily on the basis of the thermodynamic properties of the ternary system (A/B/S). We have seen that salts generally lower the solubility of B in water in the same way that sugar or organic liquids which can share favorable interactions with water. On the other hand, surfactant can enhance the solubility of the responsive stickers in aqueous media and retard the temperature of phase separation and the strength of the transition.

Nevertheless, behind these well-known aspects related to phase transition itself, a lot of questions remain unanswered concerning the viscoelastic properties and their relation to molecular organization. Of course, some interesting answers were afforded with the development of graft copolymers, which were specifically tailored for the thermogelation purpose. As members of the amphiphilic polymers family, the thermoassociative macromolecules will certainly benefit from the increasing body of knowledge in the associative thickener area. However a lot of work remains before one is quantitatively able to control the viscosity profile of the macromolecular solution from the primary structure of the system.

REFERENCES

1. Water-soluble Polymers, Synthesis, Solution Properties and Applications; Shalaby, S.W.; McCormick, C.L.; Butler, G.G., Eds.; ACS Symposium Series 467; American Chemical Society: Washington DC, **1991**.
2. Iliopoulos, I.; Wang, T.K.; Audebert, R.; *Langmuir* **1991**, *7*, 617.
3. Hill, A.; Candau, F.; Selb, J.; *Prog. Colloid Polym. Sci.* **1991**, *84*, 61.
4. Annable, T.; Buscall, R.; Ettelaie, R.; Whittlestone, D.J.; *J. Rheol.* **1993**, *37*, 695.
5. Yekta, B.Xu.; Duhamel, J.; Adiwidjaja, H.; Winnik, M.A.; *Macromolecules* **1995**, *28*, 956.
6. Audebert, R.; Iliopoulos, I.; Hourdet, D.; *Polimery*; **1997**, *42*, 237.
7. Thuresson, S.; Nilsson, B.; Lindman, B.; *Langmuir*; **1996**, *12*, 530.
8. Heymann, E., *Trans. Farad. Soc.* **1935**, *31*, 846.
9. Taylor, L. D., Kolesinski, *J. Polym. Sci: Polym. Letters* **1986**, *24*, 287.

10. Holme, K. R., Hall, L. D., *Macromolecules* **1991**, *24*, 3828.

11. Tanigami, T., Ono, T., Suda, N., Sakamaki, Y., Yamaura, K., Matsuzawa, S., *Macromolecules* **1989**, *22*, 1397.

12. Maroy, P., Hourdet, D., L'Alloret, F., Audebert, R.; *Eur. Patent* **1993**, 0 583 814 A1 and Maroy, P., L'Alloret, F., Hendriks, H., Hourdet, D.; *Eur. Patent* **1994**, 0 629 649 A1.

13. Hourdet, D.; L'alloret, F.; Audebert, R.; *Polym. Prepr.*; **1993**, *34*, 972.

14. De Vos, S. C., Moller, M., *Makromol. Chem., Macromol. Symp.* **1993**, *75*, 223.

15. Hourdet, D.; L'alloret, F.; Audebert, R.; *Polymer* **1994**, *35*, 2624.

16. Bell, D. A., Steinke, L. W., *Cereal Foods World* **1991**, *36*, 941.

17. Dagani, R., *Chemical and Engineering news* **09/06/1997**, *vol. 75 n°23*, ACS publication, Washington.

18. Doelker, E. "*Cellulose Derivatives* " In *Advances in Polymer Science*; Springer-Verlag, Ed.; **1993**; *Vol. 107.*

19. Klug, E. D., *J. Polym. Sci., Part C* **1971**, *36*, 491.

20. Rees, D. A., *Adv. Carbohydr. Chem. Biochem.* **1969**, *24*, 267.

21. Rees, D. A., *Chem. Ind.* **1972**, 630.

22. Suzuki, K., Taniguchi, Y., Enomoto, T., *Bull. Chem. Soc. Jpn.* **1972**, *45*, 336.

23. Sarkar, N., *J. Appl. Polym. Sci.* **1979**, *24*, 1073.

24. Haque, A., Morris, E. R., *Carbohydrate Polymers* **1993**, *22*, 161.

25. Desbrieres, J., Hirrien, M., Rinaudo, M., *Carbohydrate Polymers.* **1998**, *37*, 145.

26. Vigouret, M., Rinaudo, M. and Desbrieres, J., *J. Chim. Phys.* **1996**, *93*, 858.

27. Takahashi, S., Fujimoto, T., Miyamoto, T. and Inagaki, H., *J. Polym. Sci., Polym. Chem.* **1987**, *25*, 987.

28. Savage, A. B., *Ind. Eng. Chem.* **1957**, *49*, 99.

29. Heymann, E., Bleakley, H. G., Docking, R., *J. Phys. Chem.* **1938**, *42*, 353.

30. Collins, K. D., Washabaugh, M. W., *Quaterly Rev. Biophys.* **1985**, *18*, 323.

31. Collins, K. D., *Biophys. J.* **1997**, *72*, 65.

32. Iso, Y., Yamamoto, D., *Agr. Biol. Chem.* **1970**, *34*, 1867.

33. Levy, G., SchwarzZ, T. W., *J. Am. Pharm. Ass.* **1958**, *47*, 44.

34. Wang, G., Lindell, K., Olofsson, G., *Macromolecules* **1997**, *30*, 105.

35. Cabane, B., Lindell, K., Engstrom, S., Lindman, B., *Macromolecules* **1996**, *29*, 3188.

36. Nystrom, B., Kjoniksen, A.-L., Lindman, B., *Langmuir* **1996**, *12*, 3233.

37. Nystrom, B., Lindman, B., *Macromolecules* **1995**, *28*, 967.

38. Carlsson, A., Karlstrom, G., Lindman, B., *Colloids Surf.* **1990**, *47*, 147.

39. Bokias, G.; Hourdet, D.; Iliopoulos, I.; Staikos, G.; Audebert, R.; *Macromolecules* **1997**, *30*, 8293.

206

40. Sarrazin-Cartalas, A., Iliopoulos, I., Audebert, R., Olsson, U., *Langmuir* **1994**, *10*, 1421.

41. Loyen, K., Iliopoulos, I.; Audebert, R.; Olsson, U.; *Langmuir* **1995**, *11*, 1053.

42. Greenhill-Hooper, M. J., O'Sullivan, T. P., Wheele, P. A., *J. Colloid and Int. Sci.* **1988**, *124*, 77.

43. Deng, Y., Ding, J., Stubbersfield, R. B., Heatley, F., Attwood, D., Price, C., Booth, C., *Polymer* **1992**, *33*, 1963.

44. Topp, M. D. C., Dijkstra, P. J., Talsma, H., Feijen, J., *Macromolecules* **1997**, *30*, 8518.

45. Schmolka, R., *J. Biomed. Mater. Res.* **1972**, *6*, 571.

46. Hvidt, S., Jorgensen, E. B., Brown, W., Schillen, K., *J. Phys. Chem.* **1994**, *98*, 12320.

47. Nystrom, B., Walderhaug, H., Hansen, K., *Farad. Discus.* **1995**, *101*, 335.

48. Brown, W., Schillen, K., Hvidt, S., *J. Phys. Chem.* **1992**, *96*, 6038.

49. Wu, G., Chu, B., Schneider, D., *J. Phys. Chem.* **1995**, *99*, 5094.

50. Brown, W., Schillen, K., Almgren, M., Hvidt, S., Bahadur, P., *J. Phys. Chem.* **1991**, *95*, 1850.

51. Mortensen, K., Brown, W., Jorgensen, E., *Macromolecules* **1995**, *28*, 1458.

52. Hvidt, S., *Colloids Surf. A* **1995**, *112*, 201.

53. Prasad, K. N., Luong, T. T., Florence, A. T., Paris, J., Vaution, C., Seiller, M., Puisieux, F., *J. Colloid and Interface Sci.* **1970**, *69*, 225.

54. Alexandridis, P., Hatton, T. A., *Colloid Surf. A* **1995**, *96*, 1.

55. Mortensen, K., Brown, W., Jorgensen, E., *Macromolecules* **1994**, *27*, 5654.

56. Yoshioka, H., Mikami, M., Mori, Y., *J. M. S. Pure Appl. Chem.* **1994**, *A31*, 113.

57. Yoshioka, H., Mikami, M., Mori, Y., Tsuchida, E., *J. M. S. Pure Appl. Chem.* **1994**, *A31*, 121.

58. Nguyen-Misra, M., Mattice, W. L.; *Macromolecules* **1995**, *28*, 1444.

59. Hourdet, D.; L'alloret, F.; Audebert, R.; *Polymer* **1997**, *38*, 2535.

60. Durand, A., Hourdet, D., *Polymer* **1999**, *40*, 4941.

61. De Vos, S. C., Moller, M., Visscher, K., Mijnlieff, P. F., *Polymer* **1994**, *35*, 2644.

62. Bromberg, L., *Ind. Eng. Chem. Res.* **1998**, *37*, 4267.

63. Viovy, J.-L.; Hourdet, D.; Sudor, J.; *French Patent applied for capillary electrophoresis* **12/98**, (*n° 98 16676*).

64. Bromberg, L., *Macromolecules* **1998**, *31*, 6148.

65. Hourdet, D., L'Alloret, F., Durand, A., Lafuma, F., Audebert, R., Cotton, J.-P., *Macromolecules* **1998**, *31*, 5323.

66. L'alloret, F.; Hourdet, D.; Audebert, R.; *Colloid Polym. Sci.* **1995**, *273*, 1163.

67. L'alloret, F.; Maroy, P.; Hourdet, D.; Audebert, R.; *Rev. de l'Inst. Français du Pétrole* **1997**, *52*, 117.
68. Durand, A., Hourdet, D., *Polymer* **2000**, *41*, 545.
69. Durand, A.; Hourdet, D.; *Macromolecular Chem. Phys.* **2000**, in press.
70. Taylor, L. D., Cerankowski, L. D., *J. Polym. Sci.* **1975**, *13*, 2551.
71. L'Alloret, F.; Ph. D. thesis, University of Paris 6, Paris, France, 1996.
72. Durand, A., Ph. D. thesis, University of Paris 6, Paris, France, 1998.
73. Lagarrigue, S. and Hourdet, D., *unpublished results*.

Chapter 12

Fluorescence Studies of the Thermoresponsive Behavior of Aqueous Dispersions of Microgels Based upon Poly(N-isopropylacrylamide)

N.J. Flint, S. Gardebrecht, I. Soutar* and L. Swanson

The Polymer Centre, School of Physics and Chemistry, Lancaster University, Lancaster, LA1 4YA, UK

Fluorescence measurements, involving both pyrene (as a solubilized probe) and an acenaphthenyl species (as covalently bound label), have been used to study the thermoresponsive behavior of a series of microgels based on N-isopropylacrylamide) (NIPAM). The gels all exhibit cloud points which constitute their lower critical solution temperatures (LCST). Incorporation of dimethylacrylamide (DMAC) as a hydrophilic modifier raises the LCST of the microgel. However, as the DMAC content of the gel is increased, the capacity of the nanoparticles to solubilize organic guests is reduced. Time-resolved anisotropy measurements (TRAMS) on ACE-labeled PNIPAM nanoparticles have confirmed the proposal [Hirotsu et al. *J. Phys. Soc. Jpn.* 1989, *58*, 210; Fujishige et al. *J. Phys. Chem.* 1989, *93*, 3311] that the rapid reversibility of the smart thermal response of such microgels involves a 2-stage mechanism. Furthermore, the TRAMS data have allowed the onset temperatures of the two distinct steps in this mechanism to be identified in dispersions containing 0.13 wt % microgel, in total.

Introduction

Microgels, cross-linked polymer colloids of dimensions of the order of microns, have become important to technologies dependent upon solubilization and lubrication effects (*1*). For example, they find application in paints, oil products, drug release systems and food products. The microgels which have attracted most attention to date are those based upon poly(N-isopropylacrylamide) PNIPAM, which confers a "smart", thermoreversible responsivity to the colloid. The dimensions of the nanoparticles reduce dramatically as the temperature is raised above a critical value, the lower critical solution temperature (LCST).

LCST behavior, rarely encountered in mixtures consisting solely of low molar mass species, is expected (*2*) to be a much more common feature of polymer/solvent systems. Solutions of high molar mass, linear PNIPAM exhibit an LCST of *ca.* 32°C (*3,4*). The rapidity of the thermoreversible changes in condition which can be effected has led to the proposal (*5-8*) that the phase separation observed above the LCST occurs in 2 stages. Conformational collapse of individual polymer chains is followed by aggregation of the collapsed coils to form the scattering centres responsible for the turbidity which is observable, visually, in moderately concentrated solutions (at PNIPAM contents of about 10^{-2} wt% and greater) above the LCST.

Hirotsu et al (*5*) and Fujishige et al. (*6,8*) have observed the changes in dimension which accompany each stage of this 2-step process using quasi-elastic light scattering. Winnik (7), on the other hand, has illustrated the plausibility of the mechanism proposed (*5-8*) to rationalize the rapidity of the thermally reversible transition through fluorescence energy transfer studies of the coil collapse which constitutes its first stage. More recently, we have used time-resolved anisotropy measurements (TRAMS) on an acenaphthylene (ACE) tagged PNIPAM, both to demonstrate the occurrence of the coil-globule transition which precedes inter-globular aggregation and to quantify the effects of this transition upon the rate of intramolecular segmental motion of the polymer (*9*).

In the current investigation, we have employed a range of fluorescence techniques which demonstrate conclusively the validity of Hirotsu's proposals (*5*). In particular, we present data based upon time-resolved (fluorescence) anisotropy measurements (TRAMS) upon ACE-labeled microgels which support the findings of Hirotsu et al. (*5*).

The TRAMS technique, discussed briefly below (see "Results and Discussion"), involves examination of the polarization characteristics of the luminescence emitted by a chromophore dispersed within the medium under investigation. Studies of the molecular dynamics of polymers in fluid solutions involve the use of fluorescent labels which are covalently bound to the macromolecule in such a fashion as to reflect the particular relaxation process (e.g. that of a main chain segment, terminus or substituent) of interest.

Previously, we have used fluorescent labels, resultant upon incorporation of acenaphthylene (ACE) residues (at low concentrations) into the polymer backbone *via* copolymerization, to study the segmental dynamics of various polymers in organic media (*10,11*), conformational changes in aqueous solutions of polyacids (*12-14*) and the thermoresponsive behavior of PNIPAM in dilute aqueous solution (*9*).

Here, we employ the ACE label as an interrogator of changes in mobility of segments of the cross-linked PNIPAM matrix which constitutes the microgel particles. We also report upon the effects of incorporation of the comonomer, N,N-dimethylacrylamide (DMAC), upon the thermoresponsivity of NIPAM-based microgel particles.

Experimental Section

Materials

N-Isopropylacrylamide (NIPAM; Aldrich; 97%) was purified by multiple crystallization from a mixture (60/40%) of toluene and hexane (both spectroscopic grade; Aldrich).

Ammonium persulfate (APS; Aldrich; 99%) sodium lauryl sulfate (SDS; Aldrich; 99%), N,N'-methylenebisacrylamide (MBA; BDH; electrophoresis grade), and N,N,N'N'-tetramethylethylenediamine (TMED; Lancaster Synthesis; 99%) were used as received.

Dimethylacrylamide (DMAC; Aldrich; 99%) was vacuum distilled prior to use. Pyrene (Aldrich; 99%) was recrystallized five times from spectroscopic-grade toluene.

Microgels containing varying amounts and NIPAM and DMAC were synthesized *via* a single-step emulsion polymerization process in doubly-distilled, degassed water at 25°C, in the presence of MBA as cross-linker[15]. The total aqueous charge of the system was 600 cm^3 and contained a monomer content of 8.25g. Microgels prepared for TRAMS investigations contained, in addition, varying amounts (0.1 $\rightarrow$ 1 mol% of total monomer) of acenaphthylene (ACE). APS (0.6g)/TMED (0.3g) was used to initiate polymerization. HCl (0.5M, 20 cm^3) was added to terminate reaction after 55 min. All microgels were purified by dialysis.

Characterization

The LCSTs of the various microgels were determined using visual observation of the onset of turbidity to establish their cloud points. Dispersions for spectroscopic investigations of solubilization behavior contained *ca.* 0.13 wt% microgel and 10^{-6}M pyrene. Dispersions for TRAMS contained varying amounts of microgel, as discussed below.

Steady state-fluorescence spectra (uncorrected for wavelength dependence of excitation source intensity and instrument response) were measured on a Perkin-Elmer LS50 spectrometer.

Fluorescence lifetime data were acquired on an IBH System 5000 time-correlated, single-photon counter, employing D_2 as a discharge medium in the nanosecond flashlamp used as the pulsed excitation source.

Time-resolved fluorescence anisotropy measurements (TRAMS) were made upon ACE-labelled microgels using time-correlated, single photon counting (TPSPC) following excitation of the ACE label by vertically polarized synchrotron radiation from the SRS, Daresbury. The SRS, and its use in the study of macromolecular dynamics *via* TRAMS, has been described elsewhere (*16*).

TRAMS, involving pyrene dispersed as a probe/guest species, were performed using a specially modified form of an Edinburgh Instrument 199 lifetime spectrometer using TPSPC. The intensities of luminescence from the pyrene probe were analyzed, following photoselection using polarized radiation (from a flashlamp containing D_2 as discharge medium) in planes parallel $[I_{||}(t)]$ and perpendicular $[I_{\perp}(t)]$ to that of the vertically polarized excitation. An automated "toggling procedure" was adopted, under computer control, in collation of $I_{||}(t)$ and $I_{\perp}(t)$. In this procedure, the position of the analyzing polarizer was altered in a regular and sequential fashion, as appropriate segments of the memory of the multichannel analyzer recording the temporal profiles of $I_{||}(t)$ and $I_{\perp}(t)$ were switched. Collection runs involved, typically, about 40 cycles of the switching procedure, resulting in data sets containing a minimum of 20,000 counts in the channel of minimum population of the memory recording $I_{||}(t)$.

Results and Discussion

Details of typical microgel compositions and their resultant LCSTs are presented in Table 1. Incorporation of ACE as fluorescent label (at levels between 0.1 and 1 mol% of total monomer) had negligible effect upon the gross physical properties of aqueous dispersions of the resultant microgels, relative to their unlabeled counterparts.

Reference to Table 1 reveals that the presence of increasing amounts of DMAC within the microgel serves to increase its LCST (as determined by "cloud point" measurements). This observation is consistent with the predictions of Taylor and Cerankowski (2) who reasoned that increases in the hydrophilic/hydrophobic balance of a thermoresponsive polymer would raise the LCST of water-soluble, uncrosslinked systems. The data also reinforce the findings of other studies involving NIPAM/DMAC linear polymers (17) and gels (18).

Table 1. Details of Microgel Compositions and LCSTs

System	NIPAM (wt %)[a]	DMAC (wt %)[a]	wt % solids[b]	LCST (°C)[c]
PNIPAM	90.9	–	1.32	34
NIPAM-DMAC(4)	87.3	3.6	1.35	40
NIPAM-DMAC(16)	74.5	16.4	1.33	45
NIPAM-DMAC(31)	59.9	31.0	1.34	51

[a]Feed content with respect to the total monomer.

[b]Solids content following dialysis.

[c]Determined by cloud-point measurements.

Fluorescence Spectroscopic Investigations

We have employed two types of fluorescent systems in these studies of the thermoreversible behavior of the PNIPAM microgels. In the first instance, pyrene has been dispersed as a fluorescent probe which can report, through its spectroscopic characteristics (including its fluorescence lifetime), upon the nature of the environment which it inhabits as the microgel responds to the temperature of its surroundings. The second approach involves the use of acenaphthylene (ACE) labeled nanoparticles. This allows us to study, through time-resolved (fluorescence) anisotropy measurements (TRAMS) the relaxation of segments of PNIPAM within the cross-linked interiors of the nanoparticles which constitute the microgel. In addition, the TRAMS data provide incontrovertible evidence that at longer times/higher temperatures above the LCST of the microgel, inter-particle aggregation occurs.

Fluorescence Studies Involving Pyrene Probes

Spectroscopic studies.

Both the fluorescence emission and excitation spectra of pyrene (dispersed at an overall concentration within the system of 10^{-6}M) reveal that, below the LCST of the PNIPAM microgel, the fluorescent probe is located predominantly in the aqueous phase (*15*). Above the LCST, the hydrophobic domains which are created within the nanoparticles provide a solubilizing environment which can act as a host medium for the fluorescent probe (*15*). Location of the pyrene within these hydrophobic domains is evident both in the vibronic structure of its emission spectrum (*19,20*) and in the small, but significant, red-shifts apparent in its fluorescence emission (*14*) and excitation spectra (*14*), which become evident at *ca.* 32°C as the microgel collapses during its thermally-induced transition. This estimate of the LCST of the PNIPAM microgel from these spectroscopic data agrees well with that of *ca.* 34°C from cloud-point measurements (*15*). It also agrees with that which is obtained (*4*) for the LCST of high molar mass solutions of PNIPAM in water.

As the hydrophilic/hydrophobic ratio of the microgel is increased, through incorporation of varying proportions of DMAC to NIPAM, the temperature marking the onset of change in the fluorescence spectrum of the pyrene probe is progressively raised. In addition, the magnitude of change in I_3/I_1 (*20*) ratio is reduced until at 31 wt % DMAC, there is no evidence of the LCST (at 51°C, from cloud point measurements) in the spectroscopic data. At such DMAC concentrations within the microgel, the nanoparticles have lost their capacity to solubilize organic guests, such as pyrene, at any temperature.

Fluorescence lifetime measurements.

The lifetime of the singlet excited state of pyrene is sensitive to the polarity of the medium in which the probe is dispersed (*21*). Previous studies have shown (*15*) that the average fluorescence lifetime of pyrene (10^{-6}M) dispersed in aqueous PNIPAM microgels can be used to monitor the LCST of the system. We have confirmed these observations in the current work, as shown in Figure 1.

It can be noted that, as the temperature of the system is raised, the average lifetime, τ_f, of the pyrene increases abruptly and significantly at *ca.* 35°C as the probe becomes solubilized within the hydrophobic interiors of the collapsed nanoparticles which exist above the microgel's LCST. A single transition is apparent in the τ_f data. τ_f maximizes at *ca.* 45°C.

As the DMAC content of the microgel is increased, the transition in τ_f is observed to shift to higher temperatures in a similar fashion to that characteristic of I_3/I_1 as described in (i) above. In addition, again the intensity of the transition is reduced and the trend in the values of τ_f of the probe confirm the implications of the I_3/I_1 data that pyrene is not solubilized at any temperature in the NIPAM/DMAC(31) nanoparticles.

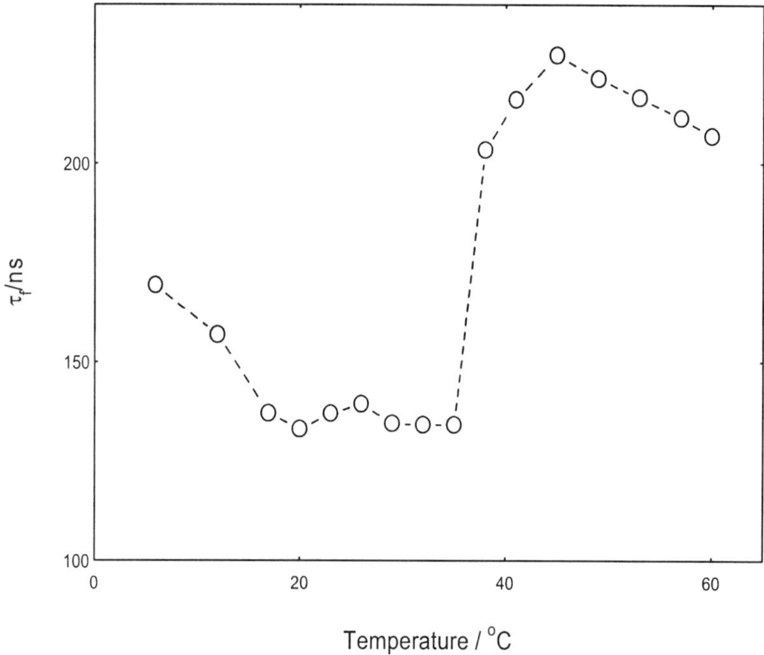

Figure 1. *Variation of pyrene probe fluorescence lifetime with temperature in a PNIPAM microgel dispersion.*

Fluorescence quenching measurements.

Fluorescence quenching, with CH_3NO_2 as a dynamic water-borne quencher has been used to confirm that, above the LCST of the microgel and at DMAC contents of 16 wt % and less, the pyrene probes are located in hydrophobic domains within the nanoparticles.

The quenching data conform to Stern-Volmer kinetics whereby

$$\tau^{o}/\tau = 1 + k_q \tau^{o} [Q] \tag{1}$$

τ^{o} and τ being the fluorescence lifetimes of the pyrene probe in the absence and presence of quencher, respectively, at a particular concentration [Q] of quencher. For the PNIPAM and NIPAM/DMAC(4) microgels, values of k_q, obtained from the slopes of these plots decrease by almost two orders of magnitude above the LCST (*15*) when compared to those obtained at temperatures below the LCST. This is likely to be due to a combination of two effects which might come into operation as the probe becomes occluded within the collapsed PNIPAM particles:

1) CH_3NO_2 may be less soluble within the hydrophobic domains created within the NIPAM-based particles than it is in the bulk aqueous phase;
2) mutual diffusion of the quencher and pyrene will be severely restricted by the high microviscosities which characterize the crowded interiors of the collapsed colloids.

In either eventuality, the data show that the pyrene is solubilized in environment within the interiors of the PNIPAM and NIPAM/DMAC(4) particles which afford a degree of protection to its singlet state from the quenching effects of the CH_3NO_2. This observation supports the evidence provided earlier by the spectroscopic and lifetime measurements.

In the case of the NIPAM/DMAC(16) and NIPAM/DMAC(31) microgels, on the other hand, the pyrene fluorescence is quenched with much greater efficiency above the gel's LCST than is observed for either the PNIPAM or the NIPAM/DMAC(4) system. At 330K, for example, the values of k_q for NIPAM/DMAC(16) and NIPAM/DMAC(31) are more than an order of magnitude greater than that for the PNIPAM microgel at the same temperature *cf.* Table 2. That for NIPAM/DMAC(31) is reduced only by a factor of 1.5 relative to that obtained for the same system at 298K, below its LCST. Clearly, this gel does not form highly protective solubilizing domains for the pyrene guest above its LCST when compared to those of lower DMAC content.

Table 2.   Bimolecular Quenching Constants Obtained for the Various Microgels

	$k_q \times 10^{-9}$ $(M^{-1} s^{-1})$	
Sample	*298 K*	*330 K*
PNIPAM	8.8	0.1
NIPAM-DMAC(4)	6.3	0.2
NIPAM-DMAC(16)	7.2	1.1
NIPAM-DMAC(31)	6.7	4.4

Reference to Table 2 reveals that similar values of k_q are obtained for each of the various microgels at 198K, below their respective LCSTs. Furthermore, the observed quenching efficiency is of the order which might be expected for the interaction between freely mobile fluorescent and quenching species in a solvent of the viscosity of water at this temperature. This observation is consistent with the conclusion, drawn from the spectroscopic and fluorescence lifetime data [discussed in (i) and (ii) above], that none of the nanoparticles examined in the current work, create protective, hydrophobic domains for organic guests below their LCSTs.

Time-resolved anisotropy measurements (TRAMS).

TRAMS can be used to determine the rate of rotational motion of a chromophore tumbling in a fluid medium. The method is described briefly as follows.

The observed anisotropy, R(t), is constructed from its components according to equation 2

$$R(t) = \frac{I_{\parallel}(t) - I_{\perp}(t)}{I_{\parallel}(t) + 2I_{\perp}(t)} = \frac{D(t)}{S(t)} \tag{2}$$

Problems attend the recovery of reliable kinetic data concerning reorientation of a fluorescent probe (or label) when either its fluorescence lifetime and/or the correlation time characterizing its rotational mobility are comparable to the width of the perturbation produced in the observed time-resolved fluorescences [$I_{\parallel}(t)$ and $I_{\perp}(t)$] by the finite width of the excitation pulse. The problem, discussed in some detail elsewhere (*16*), concerns the fact that reconvolution routines (regularly adopted in routine analyses of individual fluorescence decays) do not commute through the process of division of the difference function, D(t), by the sum function, S(t). R(t) data can **not** be deconvoluted by "simple" reconvolution of anisotropy [R(t)] data with those of the "excitation prompt".

In the present case, such considerations are unimportant. The long-lived singlet excited state of the pyrene probe [*cf.* Figure 1], allied to its slow reorientation when solubilized in the microgels, allows analysis of the TRAMS data from a channel beyond that at which the excitation pulse exercizes a perturbing influence. This validity of this approach to analysis of the TRAMS data obtained for the pyrene probe in the systems described in the current report, was confirmed. Direct analyses of the various R(t) data sets gave concordant (within experimental error) estimates of the rate of molecular reorientation of the pyrene probe to those furnished by both the impulse reconvolution (*22,23*) and the autoreconvolution (*16,24*) techniques. (Both latter forms of analysis compensate for the perturbation induced in the probe's fluorescence, by the excitation "prompt". Data sets, at selected temperatures, were analyzed across the *entire* temporal profile of the emission observed from the fluorescent probe for comparison with "direct analyses" performed as described above).

If the microBrownian rotation of the fluorophore occurs as a single relaxation process, the true (i.e. unperturbed) emission anisotropy, r(t), will decay exponentially;

$$\text{i.e.} \qquad\qquad r(t) \ = \ r_o \exp(-t/\tau_c) \qquad\qquad (3)$$

where r_o is the intrinsic anisotropy of the fluoresence probe and τ_c is the rate parameter characterizing the molecular tumbling. Although such an approach can prove inadequate under certain circumstances (*25*) in resolving the dynamics of a probe molecularly dispersed within an aqueous polymer solution, it can usually provide a reasonably quantitative description of the probe's dynamics. Consequently, it has been adopted in the current study.

At the signal/noise ratios afforded by flashlamp excitation, analysis of each data set characterizing the fluorescence from the pyrene probe was justified upon statistical grounds ($\chi^2 < 1.3$, random distribution of residuals, examination of autocorrelation function) when a single exponential function was chosen to describe the decay of fluorescence anisotropy of the pyrene probe.

As reported previously (*15*), no anisotropy was evident in the fluorescence of the pyrene probe, on nanosecond timescales below the LCST of the PNIPAM microgel. This is to be expected since, as discussed above, the PNIPAM microgel particles exist as large, open, water-swollen structures below the LCST. The pyrene molecules exist for the most part in the bulk, aqueous phase. For probes of molar volume comparable to that of pyrene in solutions of fluidity equal to that of water at ambient temperature rotational relaxation would occur within tens of picoseconds.

Above the LCST, on the other hand, finite anisotropies are obtained which decay slowly. At 51°C, for example, a value of τ_c of the order of 1 μs was obtained from analysis of R(t) using equation (3). This long correlation time constitutes incontrovertible evidence that above its LCST the microgel contains hydrophobic "pockets" which can accommodate pyrene guests. It is also evident from the TRAMS data that virtually all of the pyrene probes are solubilized into the nanoparticles above the LCST.

The longevity of the singlet excited state of the pyrene, solubilized in the interior of the microgel above its LCST ($\tau_f > 0.2$ μs, *cf.* Figure 1), means that the TRAMS estimate of τ_c for reorientational relaxation of the probe is reasonably reliable. Modeling the relaxing moiety as a spherical entity results (*15*) in an estimate of the order of 20 nm for the diameter of the tumbling sphere. This is considerably less than that afforded by dynamic light scattering for microgels (synthesized at the higher temperature of 70°C[1]). This difference between estimates of the dimensions of the effective rotor (afforded by fluorescence) and microgel particles could have several origins (*15*).

Perhaps the most plausible explanation of this apparent discrepancy is that the microviscosity of the solubilizing domains created within the collapsed forms of the microgels is such as to allow rotation of the pyrene probe independent of that of the microgel particle as a whole. Among other possible rationales (*15*) is the potential for energy migration between solubilized pyrene guests, concentrated within the contracted nanoparticles, to produce a pathway for enhanced decay of R(t). This mechanism for increasing the rate of loss in orientation of transition vectors governing the emission of radiation from photoselected chromophores is discussed further below (with respect to ACE-labeled microgels). However, it is not expected to make a significant contribution to the overall loss of anisotropy at the relative concentration levels of host:guest employed here.

Fluorescence Studies of ACE-labeled Microgels

Spectroscopic studies.

Both emission and excitation spectral profiles of the fluorescence emitted from the ACE-labeled microgels were (as expected) relatively insensitive to the occurrence of the LCST of each microgel. [*Minor* changes were observed in both the wavelengths of maximum fluorescence intensity and the relative intensities of specific vibronic bands of the ACE fluorescence spectra as the hydrophobicity of the label's local environment changed, at the LCST of each particular host].

The fluorescence lifetime of the emission from the ACE-labeled microgels varied according to the state of its polymeric host: above the LCST, the ACE label occupied a "more protective", hydrophobic environment, resulting in a longer lifetime of its fluorescence relative to that observed below the LCST of the aqueously-dispersed matrix.

TRAMS studies of macromolecular mobility within the microgel particles.

The ACE label is equipped to report upon the segmental motions of the PNIPAM chains within the cross-linked nanoparticles. Below the LCST, segmental motion within the microgel is much slower than that of linear PNIPAM (*9*) itself under similar conditions. The resultant slow decay of fluorescence anisotropy observed at 20°C, for example, is shown in Figure 2.

The long correlation time obtained (*ca.* 100 ns) reflects the considerable constraints exerted upon the polymer segments by the cross-links present in the system, compared to the freedom of motion ($\tau_c \sim 4$ ns) (*9*) enjoyed by PNIPAM

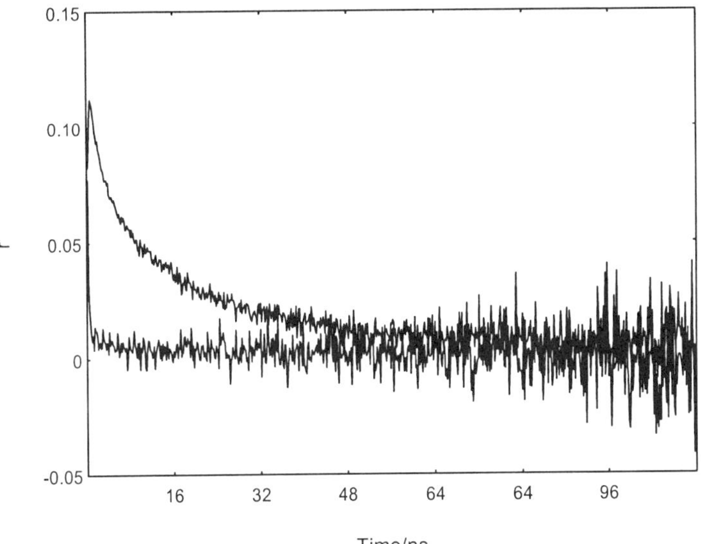

Figure 2. *Time-resolved fluorescence anisotropy of PNIPAM microgel at 20 °C (upper curve) and 48 °C (lower curve).*

segments in the isolated coils of the linear polymer, which exist in ultradilute aqueous solutions below the LCST.

Above the LCST, coil collapse will be accompanied by a much closer packing of the polymer chains. A large increase in τ_c would be anticipated due to the considerable restrictions that will be imposed upon segmental motion in the shrunken interiors of the nanoparticles. Reference to Figure 2, however, reveals that at 48°C the anisotropy of the fluorescence from a microgel containing 1 mol % of the ACE label decays almost instantaneously. At first sight, it would appear that the interiors of the microgel have become very fluid such that the PNIPAM segments enjoy similar degrees of rotational freedom to those located in isolated, water-swollen coils of the linear polymer *below* its LCST! However, the effect is illusory.

Initially, we were tempted to ascribe the enhanced decays in anisotropy at high temperatures (>40°C) to the effects of energy migration between ACE labels. The local chromophore concentration would increase as the collapsed PNIPAM particles formed aggregates. We imagined that energy migration, between probes located within different microgel globules, would occur leading to loss of emission transition vector orientation and decreased polarization of the observed fluorescence. This explanation seemed all the more plausible since we had observed previously energy migration between ACE species bound, at similar levels of labeling, to acrylic latices (*26*). However, flaws in our reasoning became apparent upon further testing (*27*) of this proposed mechanism:

The effect was not diminished when the ACE content of the microgel was reduced to 0.5 and 0.1 mol %, respectively.

The effect *was* reduced when the concentration of fluorescently-labeled globule was lowered [while adding untagged polymer to keep the overall PNIPAM concentration (and the rate of aggregation) constant].

It would appear that the principal cause of the rapid loss in emission anisotropy observed at T>40°C is the turbidity of the aggregated system: light scattering will depolarize both excitation and emission.

Closer examination of the temperature dependence of the TRAMS data from the ACE-labeled microgel reveals the processes involved in its thermal response. As the temperature of the system is raised through the LCST, a marked increase in τ_c is observed. This is to be expected, in view of the restrictions upon segmental mobility that will be imposed by collapse of the nanoparticles. As the temperature and "residence time" above the LCST is further increased, and under the concentration conditions employed here, interparticle aggregation occurs. This is evident in the *apparent* dramatic reduction in τ_c at temperatures greater than *ca.* 38 °C. Aggregation of the nanoparticles leads to increased turbidity resulting in a rapid loss of anisotropy [*cf.* Figure 2]. The data confirm the reports of Hirotsu et al. (*5*) and Fujishige et al. (*6*). It would appear that aggregation between collapsed globules occurs at temperatures greater than that marking the onset of conformational collapse of the nanoparticle. The results of a continued program of research involving TRAMS upon such systems will be published in due course (*27*).

Conclusions

1. Fluorescence spectroscopic and lifetime measurements, involving a pyrene probe, have shown the controlled solubilization/release properties of PNIPAM microgels.
2. Increasing the hydrophilic:hydrophobic balance of the microgels through incorporation of DMAC raises the LCST, as expected[2]. However, such modification reduces the magnitude of the particle collapse associated with criticality: the gels lose their solubilization capacity for organic guests at higher DMAC contents.
3. TRAMS on ACE labeled PNIPAM nanoparticles have provided unambiguous evidence that the thermoresponsive behavior of such microgels involves two distinct stages: (a) collapse of individual nanoparticles, followed by (b) inter-particle aggregation. In moderately concentrated solutions (0.13 wt % in microgel) TRAMS allow identification of the temperatures at which each phase of this dual mechanism comes into play.

Acknowledgments

Financial support from the Leverhulme Trust (fellowship NJF) and EPSRC (beamtime at the SRS) is noted, with gratitude.

References

1. Snowden, M.J.; Chowdry, B.Z. *Chem. Brit.* **1995**, *1995*, 943.
2. Taylor, L.D.; Cerankowski, L.D. *J. Polym. Sci. Polym. Chem. Ed.* **1975**, *13*, 2551.
3. Heskins, M.; Guillet, J.E. *J. Macromol. Sci., Chem.* **1968**, *A2*, 1441.
4. Schild, H.G. *Prog. Polym. Sci.* **1992**, *17*, 163.
5. Yamamoto, I.; Iwasaki, K.; Hirotsu, S. *J. Phys. Soc. Jpn.* **1989**, *58*, 210.
6. Fujishige, S.; Kubota, K; Ando, I. *J. Phys. Chem.* **1989**, *93*, 3311.
7. Winnik, F.M. *Polymer*, **1990**, *31*, 2125.

8. Kubota, K; Fujishige, S.; Ando, I. *J. Phys. Chem.* **1990**, *94*, 5154.
9. Chee, C.K.; Rimmer, S.; Soutar, I.; Swanson, L. *Polymer* **1996**, *38*, 483.
10. Soutar, I.; Swanson, L. *Macromol. Symp.* **1995**, *90*, 267.
11. Soutar, I.; Swanson, L.; Christensen, R.L; Drake, R.C.; Phillips, D. *Macromolecules* **1996**, *29*, 4931.
12. Soutar, I.; Swanson, L. *Macromolecules* **1994**, *27*, 4304.
13. Soutar, I.; Swanson, L.; Thorpe, F.G.; Zhu, C. *Macromolecules* **1996**, *29*, 918.
14. Ebdon, J.R.; Hunt, B.J.; Lucas, D.M.; Soutar, I.; Swanson, L.; Lane, A.R. *Can. J. Chem.* **1995**, *73*, 1982.
15. Flint, N.J.; Gardebrecht, S.; Swanson, L. *J. Fluor.* **1998**, *8*, 343.
16. Soutar, I.; Swanson, L.; Imhof, R.E.; Rumbles, G. *Macromolecules* **1992**, *25*, 4399.
17. Barker, I.C.; Soutar, I.; Swanson, L.; Cowie, J.M.G. *to be published.*
18. Shibayama, M.; Mizutani, S.; Nomura, S. *Macromolecules* **1996**, *29* , 2019.
19. Nakajima, A. *Bull. Chem. Soc. Jpn.* **1970**, *43*, 967.
20. Kalyanasunderam, K.; Thomas, J.K. *J. Amer. Chem. Soc.* **1997**, *108*, 6270.
21. Birks, J.B. *Photophysics of Aromatic Molecules,* Wiley-Interscience, London, **1970**.
22. Wahl, P. *Chem. Phys.* **1975**, *7*, 220.
23. Barkley, M.D.; Kowalczyk, A.A.; Brand, L. *J. Chem. Phys.* **1981**, *75*, 3581.
24. Marsh, A.J.; Rumbles, G.; Soutar, I.; Swanson, L. *Chem. Phys. Lett.* **1992**, *195*, 31.
25. Ghiggino, K.P.; Haines, D.A.; Smith, T.A.; Soutar, I.; Swanson, L. *to be published.*
26. Soutar, I.; Swanson, L.; Annable, T.; Padget, J.C.; Satgurunathan, R. *to be published.*
27. Flint, N.J.; Gardebrecht, S.; Soutar, I.; Swanson, L. *to be published.*

Chapter 13

Manipulating the Thermoresponsive Behavior of Poly(N-isopropylacrylamide)

C.K. Chee, S. Rimmer, I. Soutar and L. Swanson*,

The Polymer Center, School of Physics and Chemistry, Lancaster University, Lancaster LA1 4YA UK

Fluorescence techniques, including quenching and time-resolved anisotropy measurements have been used to follow the temperature induced conformational transition from an open coil to globular structure, of poly(N-isopropylacrylamide), PNIPAM. The onset of the coil collapse occurs at 32°C, the lower critical solution temperature (LCST). In view of the potential of such polymers to act as carriers in controlled release applications, it would be attractive if ways could be found to manipulate the LCST of the polymeric host through chemical modification. In this paper, we present the results of initial attempts to achieve this aim by copolymerization of NIPAM with varying amounts of styrene. Fluorescence spectroscopy has shown that alteration of the hydrophobic/hydrophilic balance of NIPAM-based polymers, through random copolymerization, changes the LCST of the thermoresponsive polymer. Unfortunately, the magnitude of the transition, is also reduced in such cases. Manipulation of the polymer topography, on the other hand, seems to offer an alternative route whereby the LCST of smart systems may be controlled.

Introduction

Polymers which exhibit "smart" behavior (i.e. respond to external stimuli such as pH and/or temperature) have attracted much interest in both the academic and industrial communities over a number of years (*1,2*). Much of this attention stems from the fact that these systems are fundamentally interesting, and also are industrially important. They are involved in a diverse range of technologies such as controlled-release systems, agrochemicals, adhesives, coatings, enhancers for oil recovery, foodstuffs, rheology modifiers, personal care products, superabsorbents, catalysis, inks and coding systems. The use of fluorescence spectroscopy (*3,4*) has been particularly prominent in investigation of smart polymers since it allows examination of *ultra*-dilute solutions, permitting examination of purely *intra*-molecular effects. Indeed, luminescence techniques have proved invaluable in confirming (*5,6*) that poly(methacrylic acid), PMAA, undergoes a pH-induced conformational transition from an uncoiled (at high pH) to a *hyper*coiled structure at pH 4. With poly(acrylic acid) (*7*) this transition is much less dramatic: the polymer coil essentially adopts an open chain conformation at all values of pH.

Poly(N-isopropylacrylamide), PNIPAM, undergoes a similar conformational transition to that of PMAA, except that contraction and expansion of the coil is controlled by temperature. The onset of the coil collapse occurs at $32°C$, the lower critical solution temperature (LCST) (*8,9*). Under semi-dilute conditions, the polymer forms a turbid solution upon heating above the LCST, which *rapidly* turns clear again upon cooling. In this respect, we have recently confirmed (*10*) (*via* time-resolved anisotropy measurements, TRAMS) that the LCST behavior in linear PNIPAM is governed by a 2-stage mechanism (*11,12*). The first step involves intramolecular coil collapse. This is followed by intermolecular aggregation between collapsed coils. The ability of "smart" polymers to expand and contract "on demand" could lead to such systems being used as carriers with controlled release capabilities. For example, in the compact form, above its LCST, PNIPAM can solubilize low molar mass organic species (*13*). The solute can subsequently be released into the aqueous phase by simply lowering the temperature of the dispersion below $32°C$. In view of this potential, it would be attractive if ways could be found to manipulate the LCST of the polymeric host through chemical modification. We have recently modified PNIPAM (*14,15*) and synthesized microgels (*16*) based upon PNIPAM so that we can control the conformational switch of the polymer over a wide temperature range (e.g. from 4-100 $°C$) including the physiological temperature of $37°C$. The resultant modified polymers might then find application in a much more extended range of industrial and medical activities.

In this paper, we present the results of initial attempts to achieve this aim.

Experimental

Materials

N-isopropylacrylamide, NIPAM, (Aldrich; 97%) was purified by multiple recrystallization from a mixture (60/40 %) of toluene and hexane (both spectroscopic grade; Aldrich).

Styrene was purified by washing with an aqueous solution of NaOH (5 wt%) to remove inhibitor, washed with distilled water until the washings were neutral to litmus and fractionally distilled under high vacuum.

Acenaphthylene (ACE) was purified by multiple recrystallization from ethanol followed by vacuum sublimation.

Nitromethane (Aldrich; Gold Label) was used as received.

Acenaphthylene labeled poly(N-isopropylacrylamide) [ACE-PNIPAM] was prepared by copolymerisation of NIPAM with a trace amount (*ca.* 0.5 mole %) of ACE in dioxane solution (80% by weight of solvent) at 60°C using AIBN as initiator. Fluorescently labeled styrene-NIPAM copolymers were prepared in a similar manner to that of the homopolymer. Two samples were synthesized: one contained 8.9 weight % styrene [ACE-(STY8.9)-NIPAM] while the second contained 16.9 weight % of hydrophobe [ACE-(STY16.9)-NIPAM].

All polymers were purified by multiple reprecipitation from dioxane into diethylether (May and Baker).

Graft copolymers containing a styrene backbone and NIPAM side chains were prepared by a macromonomer (*17*) technique and purified by ultra-filtration. Details will be published (*14*) in due course.

Contents of all copolymer samples were obtained by proton NMR and elemental analyses. Molecular weights were determined by aqueous GPC (*14*). The data are listed in Table 1.

Table 1. Physical characteristics of the NIPAM based polymers.

Sample	M_n g mol^{-1} (x1000)	Styrene content (%)	LCST (°C)
ACE-PNIPAM	21	- -	32
ACE-STY(8.9)-NIPAM	19	8.9	20
ACE-STY(16.9)-NIPAM	21	16.9	9
ACE-STY(16)-gNIPAM	1800	16.0	37
STY(14)-gNIPAM-ACE	1700	14.0	37

Instrumentation

Optical density measurements were made on a Hitachi U-2010 spectrophotometer.

Steady state fluorescence spectra were measured on a Perkin-Elmer LS50 spectrometer.

Fluorescence lifetime data were acquired on an Edinburgh Instruments 199 time-correlated single photon counter.

Time-resolved anisotropy measurements (TRAMS) were made at the synchrotron radiation source, SRS, Daresbury, UK. A complete description of the experimental set-up and a detailed discussion of analysis of anisotropy data can be found elsewhere (*18*).

Results and Discussion

(i) Behavior in methanol

Fluorescence quenching

The accessibility of a fluorescent species, F, to a low molar mass species, Q, (which is capable of dynamic quenching of an excited state) can be estimated from quenching experiments.

This technique follows the general principle outlined below. The quenching process may be represented by equation (1).

$$F^* + Q \longrightarrow F + Q^* \tag{1}$$

The efficiency with which a dynamic quencher accesses the excited state is described by the Stern-Volmer equation:

$$\tau^o/\tau = 1 + k_q \tau^o [Q] \tag{2}$$

where τ is the excited state lifetime at some concentration Q, τ^o is that in the absence of Q and k_q is the bimolecular quenching constant. (Consequently, k_q can be considered as a measure of the "ease of access" of Q to F*). If a fluorescently labeled polymer is used, information regarding the "openness" or compactness of the chain can be accrued from such experiments (*19,20*).

Stern-Volmer experiments were carried out on each of the ACE-labeled NIPAM based polymers in methanol at various temperatures using nitromethane as a quencher. The resultant k_q values which were derived from excited state lifetime data are listed in Table 2.

Table 2. Bimolecular quenching constants for the various ACE-labeled polymers in methanol.

Sample	Temperature (°C)	k_q (x 10^{-9})/ $M^{-1}s^{-1}$
Ace-NIPAM	25	7.8
	42	9.3
Ace-STY(8.9)-NIPAM	15	6.3
	35	7.6
Ace-STY(16.9)-NIPAM	4	4.9
	25	6.5

Examination of the k_q values listed in Table 2 reveals that they are close to that which would be expected for a diffusion controlled process at each temperature in a solvent of viscosity corresponding to that of the methanol solvent at that temperature. This infers that each polymer sample adopts essentially an open chain conformation at each temperature accessed. In addition, a "normal" thermal response is observed in that the bimolecular quenching constant increases with increasing temperature. This is consistent with the fact that these samples do not show an LCST in methanol and consequently do not exhibit the conformational change which would accompany such behavior.

Time-resolved anisotropy measurements (TRAMS)

Time-resolved (fluorescence) anisotropy measurements, TRAMS, upon suitably labeled samples, allows estimation of the degree of macromolecular mobility within a polymer. Briefly, the experiment involves the use of vertically polarized excitation radiation. Detection of the 2 time-dependent orthogonal components of the intensity of fluorescence emitted from the label, $i_{\parallel}(t)$ and $i_{\perp}(t)$, respectively, is achieved by use of a rotatable polarizer (*18*). $i_{\parallel}(t)$ and $i_{\perp}(t)$ are related to the anisotropy r(t) via equation 3

$$r(t) \; = \; \frac{i_{\parallel}(t) - i_{\perp}(t)}{i_{\parallel}(t) + 2i_{\perp}(t)} \tag{3}$$

For a simple relaxation process the time-dependent anisotropy r(t) can be described by a single exponential function of the form of equation 4

$$r(t) = r_0 \exp(-t/\tau_c) \tag{4}$$

where r_0 is the intrinsic anisotropy and τ_c, is the correlation time characteristic of the motion under study.

There are several methods of analysis which can be used to recover relaxation information from anisotropy data. The merits of the various approaches have been discussed at length elsewhere (*18*). For the current decay sets, we have favored the impulse reconvolution technique (*21*) which is particularly useful (*18*) when the timescale of the motion under investigation is close to that of the duration of the excitation pulse.

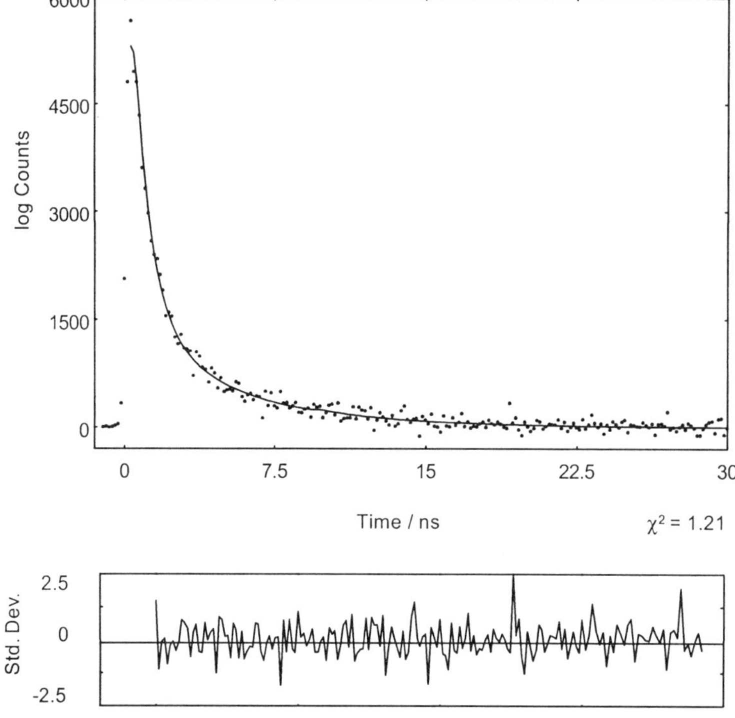

Figure 1. Difference function, D(t), and impulse reconvolution fit for ACE-PNIPAM (10^{-2}wt%) in methanol at 298K.

Figure 1 shows an example of an IR fit to the difference function, D(t), of ACE-PNIPAM in methanol, assuming a single exponential model for r(t).

The low value of χ^2 (χ^2 should be close to unity for a good fit) and the random distribution of the residuals provide statistical confidence in the quality of fit. A τ_c of *ca.* 1.9 ns results from this form of analysis. This value is similar to that of poly(methyl methacrylate) in dichloromethane (*22*) and implies that PNIPAM exists as a flexible coil under these conditions. Copolymerization with styrene (at 8.9 and 16.9%, respectively) has no significant effect on the dynamics of the resulting polymers: correlation times of *ca.* 2ns for both ACE-STY(8.9)-NIPAM and ACE-STY(16.9)-NIPAM, respectively, were observed at 298K.

The thermal dependence of the rate of macromolecular motion, k_c (where $k_c = 1/\tau_c$) for ACE-PNIPAM, ACE-STY(8.9)-NIPAM and ACE-STY(16.9)-NIPAM was also investigated for these solutions in methanol. The resulting Arrhenius plots are shown in Figure 2.

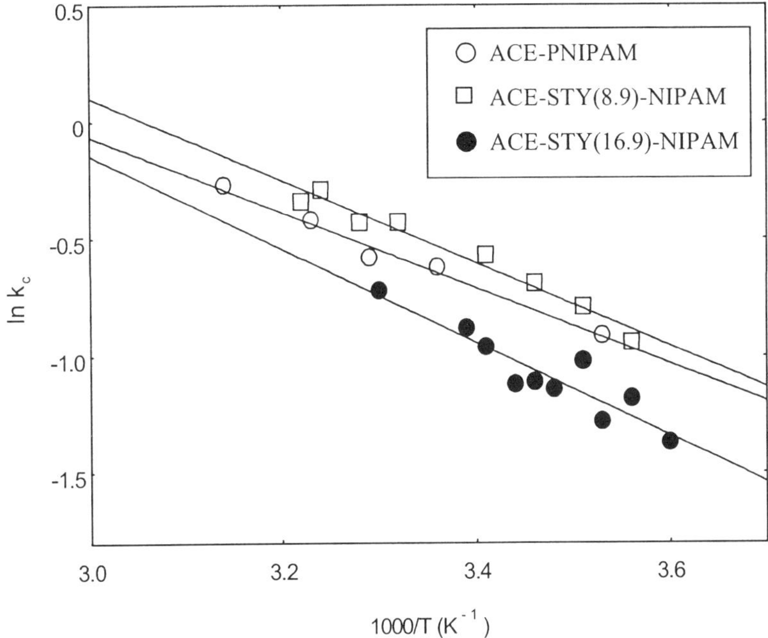

Figure 2. Arrhenius representation of the rate of macromolecular motion, k_c ($k_c = \tau_c^{-1}$) for for the NIPAM based polymers (all $10^{-2} wt\%$) in methanol

Clearly, by reference to Figure 2, it can be concluded that each of the current NIPAM based polymers undergo normal Arrhenius type behavior in methanol. The resultant activation energy, E_a, was calculated to be *ca.* 15±1.5 kJmol^{-1} for each polymer sample.

(II) Behavior in aqueous media

Cloud point measurements

LCSTs (*cf.* Table 1) were determined for each of the linear polymer samples in aqueous media by estimating cloud points from optical density measurements. Clearly, it can be concluded that increasing the styrene content within the macromolecule serves to shift the LCST to lower temperature. Variation of the relative amounts of copolymerized STY and NIPAM has the effect of changing the hydrophobic to hydrophilic balance within the polymer. (The resultant effect upon the LCST of polymers which have been hydrophobically modified in this fashion has been predicted previously (*23*) in the literature and observed experimentally) (*9*).

230

Steady state spectra

The steady state excitation spectra of each labeled polymer sample are shown in Figure 3 at temperatures below and above their respective LCSTs. For ACE-PNIPAM, the spectral profile remained unchanged when the temperature was raised from 23 °C (below the LCST) to 42 °C (above the LCST). This contrasts with the behavior of ACE-STY(8.9)-NIPAM. At 15 °C the excitation spectrum is similar to that of ACE-PNIPAM. However, as the temperature is raised to 42 °C a shoulder to the main peak (between 260 and 275 nm) is apparent. This reflects the occurrence of energy transfer (ET) from the styrene units to the ACE label as the polymer undergoes a conformational transition from a loose coil to a compact structure. As a result, the styryl groups come into closer proximity to the naphthyl species, which promotes ET to the label. Further examination of Figure 3 reveals that ET is apparent from STY to ACE in the excitation spectrum of ACE-STY(16.9)-NIPAM even at low temperature (9 °C). This constitutes an early indication of the existence of intramoleular hyrophobic cavity formation, even below the LCST in this copolymer.

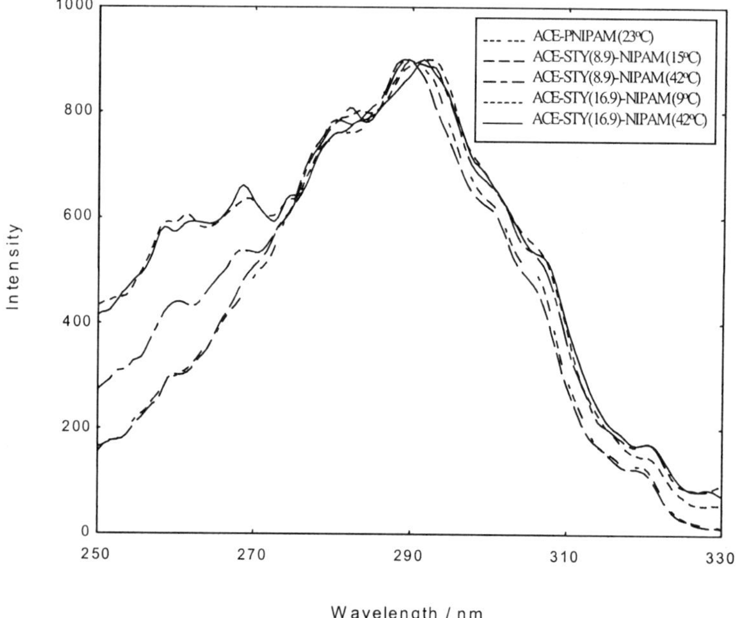

Figure 3. *Steady state excitation spectra for the various polymers analysed at 340 nm*

Presumably a critical concentration of styryl units is exceeded in this polymer which promotes intramolecular aggregation even in its expanded form. As the temperature is raised above that of the copolymer's LCST, the spectral profile of ACE-STY(16.9)-NIPAM remains similar to that observed at 9 °C.

Fluorescence quenching

Stern-Volmer experiments were carried out on each of the NIPAM based polymer samples above and below their respective LCSTs using CH_3NO_2 as a quencher. The fluorescence decays were complex at each temperature and CH_3NO_2 concentration accessed. In order to treat the data (for Stern-Volmer analysis), a triple exponential function of the form of equation 5 was required to adequately describe the curves on statistical grounds.

$$I(t) = A_1 \exp(-t / \tau_1) + A_2 \exp(-t / \tau_2) + A_3 \exp(-t / \tau_3) \tag{5}$$

An average lifetime $<\tau>$ was subsequently calculated from equation 6.

$$<\tau> = \frac{\sum A_i \tau^2_i}{\sum A_i \tau_i} \tag{6}$$

Linear Stern-Volmer plots were obtained for each system at each temperature accessed. The resulting bimolecular quenching constants (derived from the τ data) are listed in Table 3. For ACE-PNIPAM at 22°C (below its LCST) a k_q value of 7.0×10^9 $M^{-1}s^{-1}$ was obtained. This is close to that expected for a diffusion controlled process and infers that the polymer adopts an open chain conformation under these conditions. The quenching efficiency decreases markedly when the temperature is increased to 45°C. This is consistent with collapse of the PNIPAM into a globular structure: this would result in hindered access by the quencher to the label. The resultant k_q value is comparable to that of poly(methacrylic acid) at low pH (*24*) in its hypercoiled conformation.

Similar trends are apparent in the quenching data for the two hydrophobically modified NIPAM copolymers at low and high temperature: above the LCST the value of k_q is almost an order of magnitude less than that obtained at temperatures below that of the conformational collapse of the polymer.

Further examination of Table 3 reveals that the estimates of k_q obtained for the 2 copolymers below their respective LCSTs are less than that for PNIPAM itself. This will, in part, derive from the fact that the data have been obtained at different temperatures. However, the differences in quenching efficiency are likely also to reflect formation of hydrophobic domains between styryl residues, at temperatures below the LCST of each copolymer sample. Indeed, these observations reinforce the spectroscopic evidence (*cf.* excitation spectra in

232

Table 3. Bimolecular quenching constants for the various ACE-labeled polymers in aqueous solution.

Sample	Temperature ($^{\circ}C$)	k_q (x 10^{-9}) $M^{-1}s^{-1}$
ACE-PNIPAM	22	7.00
	45	0.18
ACE-STY(8.9)-NIPAM	15	3.63
	35	0.37
ACE-STY(16.9)-NIPAM	5	1.28
	25	0.25
ACE-STY(16)g-NIPAM	20	0.22
	45	0.06
STY(14)g-NIPAM-ACE	20	3.9
	45	0.26

Figure 3) for such intramolecular aggregates. Consistent with this inference is the fact that the quenching is inhibited to a greater extent, the higher the styrene content of the macromolecule. In addition, solubilization studies (*13*) also support the proposition that hydrophobic domains exist below the LCSTs of these polymers.

TRAMS

TRAMS were performed, as a function of temperature on ACE-PNIPAM, ACE-STY(8.9)-NIPAM and ACE-STY(16.9)-NIPAM, respectively. The anisotropy decays were observed to be complex at temperatures in excess of the LCSTs of each sample, requiring a double exponential model to adequately describe the data on statistical grounds (*14*). Such behavior has been observed previously in water-soluble systems (*6,7*) and presumably reflects the existence of a heterogeneous environment in the compact form of these NIPAM polymers. However, valuable information can still be derived by modeling the data by a single exponential function of the form of equation 4, *via* IR analysis, in order to examine the *relative* change in macromolecular mobility as a function of temperature.

The correlation time, τ_c, derived from such experiments, which characterizes the average degree of macromolecular motion, is plotted in Arrhenius form in Figure 4. Several features are apparent in consideration of the plot and are worthy of note:

(a) at temperatures lower than *ca.* 32 $^{\circ}$C, the dynamic behavior of ACE-PNIPAM is much as expected; the rate of macromolecular motion increases with temperature.

(b) the onset of the LCST for ACE-PNIPAM is accompanied by a marked decrease in k_c. [Clearly, the thermoreversible coil to globule transition of PNIPAM results in a dramatic reduction in the macromolecule's segmental mobility. Indeed, this relative change in macromolecular motion is similar to that observed for the pH dependent conformational change of poly(methacrylic acid) (*6*)].

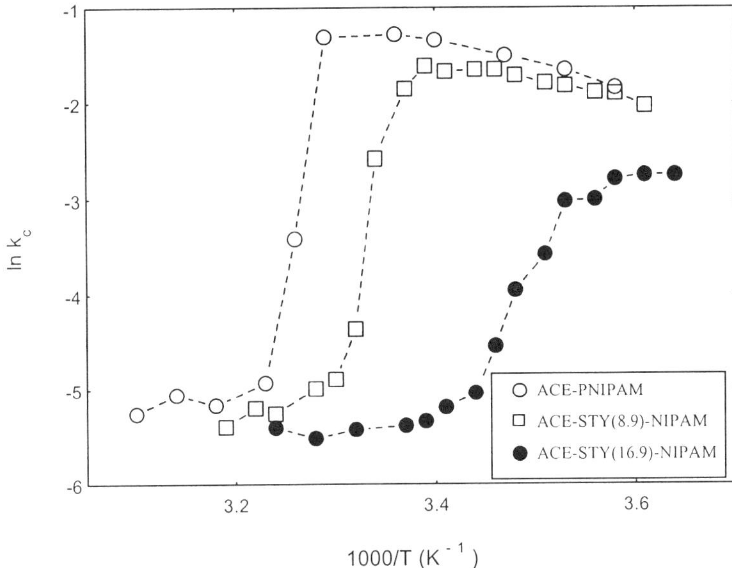

Figure 4. *Arrhenius representation of the rate of macromolecular motion, k_c ($k_c = \tau_c^{-1}$) for the statistical copolymer samples in aqueous media.*

(c) as the styrene content of the macromolecule increases, the magnitude of the observed transition decreases. This behavior appears to be a general phenomenon when PNIPAM is modified as has been observed by calorimetry measurements (*25*).

(d) the onset temperature of the coil-collapse is reduced at higher degrees of hydrophobic modification. In this case, the hydrophobic to hydrophilic balance within the polymer changes, which serves to shift the transition to lower temperature. Such behavior has been predicted previously in the literature (*23*) and observed experimentally (*9*).

Figure 4 also highlights the fact that, as the LCST of the NIPAM-STY copolymers is progressively lowered, the apparent "intensity" of the transition is also reduced. This is consequent upon the fact that the TRAMS reflect the creation of pockets of enhanced solubilization capacity both above and below the LCST (*13*), above a certain styrene content (5 wt%). This is not the case when the hydrophilic content of the polymer is increased, for example, by incorporation of dimethylacrylamide (DMAC) (*15*). In this instance, the LCST is elevated as the DMAC content of the polymer is increased (*15*). Unfortunately, in the latter instance, the solubilization capacity of the copolymer is also reduced (*15*). In an effort to overcome this drawback, we have recently initiated investigation of the effects of polymer topography on manipulation of the LCST. Preliminary results are reported below.

(III) Graft copolymer samples.

Copolymers containing *ca.* 15% styrene backbone with NIPAM grafts (5000 gmol^{-1}) exhibit LCSTs of 37 $^{\circ}$C (*cf.* Table 1). (One sample, ACE-STY(16)-gNIPAM, has an ACE label in its backbone. The other, STY(14)-gNIPAM-ACE, contains the label in its NIPAM arms). Despite the fact that a hydrophobic modifier (in the form of styrene) is present in the system, a higher LCST than that of PNIPAM results.

Fluorescence quenching measurements

Quenching experiments were carried out on each of the labeled graft copolymer samples at low and high temperature. The Stern-Volmer plots (using τ data, treated in a similar manner to that of the random polymers, see earlier) were linear. The resultant bimolecular quenching constants are listed in Table 3.

When the ACE label is located in the styryl backbone [ACE-STY(16)g-NIPAM], very inefficient quenching is observed at 20°C (below its LCST). This implies that hindered access of the quencher to the fluorescence label results under these conditions. Indeed, the k_q value is smaller than that observed for the same STY content linear sample at a similar temperature. This indicates that a more tightly coiled backbone conformation exists in the graft copolymer sample below its LCST compared to that in the random sample. At 45°C, the quenching efficiency is further reduced which implies that a yet more compact conformation exists under these conditions.

Examination of the bimolecular quenching data for STY(14)g-NIPAM-ACE at 20°C (below the LCST) reveals that a k_q value close to that for a diffusion controlled process results (*cf.* Table 3). This implies that the NIPAM arms adopt a very open structure at this temperature. A dramatic reduction in quenching efficiency occurs when the experiment is carried out at 45°C. Indeed, a k_q value close to that obtained for ACE-PNIPAM itself results. This infers that a conformational change has occurred in the NIPAM grafts from an open structure to a compact coil.

TRAMS

Further information can be obtained by examination of the dynamic behavior of each fluorescently tagged site within the graft coplymer. TRAMS were performed on the two ACE labeled samples as a function of temperature. The data are plotted in Arrhenius form in Figure 5. The thermoreversible transition in the PNIPAM "arms" is of a similar magnitude to that evident in high molecular weight PNIPAM itself (*10*). In this respect, the implications of the data are similar to those drawn from quenching data discussed in the previous section. The ACE label incorporated into the NIPAM arms of the graft copolymer, exists in a relatively unrestricted, open environment. Furthermore, attachment of these chains to the styrene backbone has little effect upon their mobility.

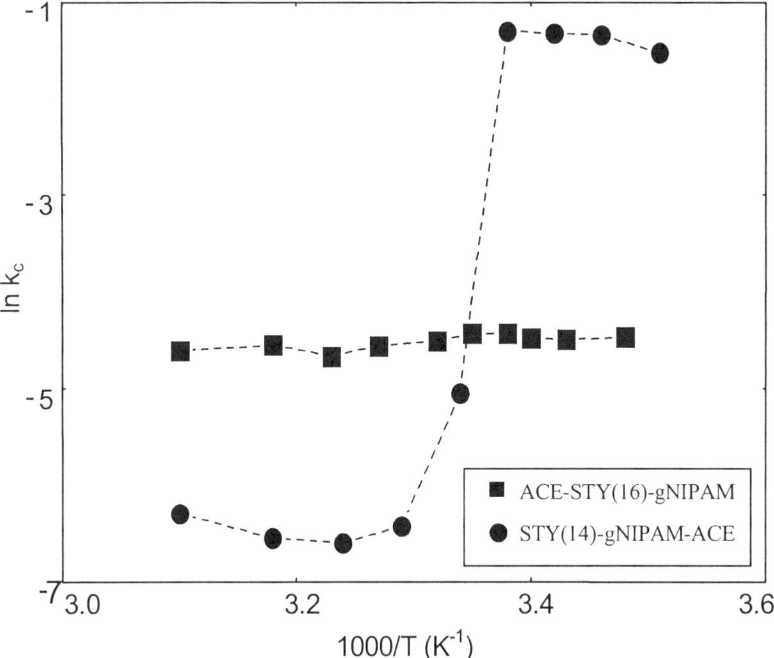

Figure 5. Arrhenius representation of the rate of macromolecular motion, k_c, $(k_c = \tau_c^{-1})$ for the graft copolymer samples in aqueous media.

Consideration of the styrene backbone, on the other hand, below the LCST of the graft copolymer, reveals that the value of τ_c is considerably greater than that of the NIPAM arms over the same temperature range. This implies that a degree of coiling exists under these conditions, presumably due to intramolecular aggregation involving the styryl residues. In addition, τ_c remains invariant with temperature. Interestingly, the value of τ_c obtained for the backbone motion above the LCST of the acrylamide arms, is less than that observed for the labeled grafts. This means that the label is detecting some motion within the core, even under these conditions, rather than motion of the entire globule. Consideration of the quenching and TRAMS data together leads to the following conclusions regarding the conformational behavior of the graft copolymer:

(i) the NIPAM grafts form open extended arms at low temperature, below the LCST of the graft copolymer. At 37°C, these fronds undergo a conformational transition, collapsing upon the styryl core to form a compact globular structure.

(ii) the styrene backbone forms a compact core-type structure at all temperatures. (These observations are further supported by pyrene solubilization data.) (*13*) In addition, there is evidence for increased hindrence of access of the ACE label by CH_3NO_2 following collapse of the NIPAM grafts.

A diagramatic representation of the conformational behavior of the graft copolymer sample, deduced from the fluorescence data, is depicted in Scheme 1.

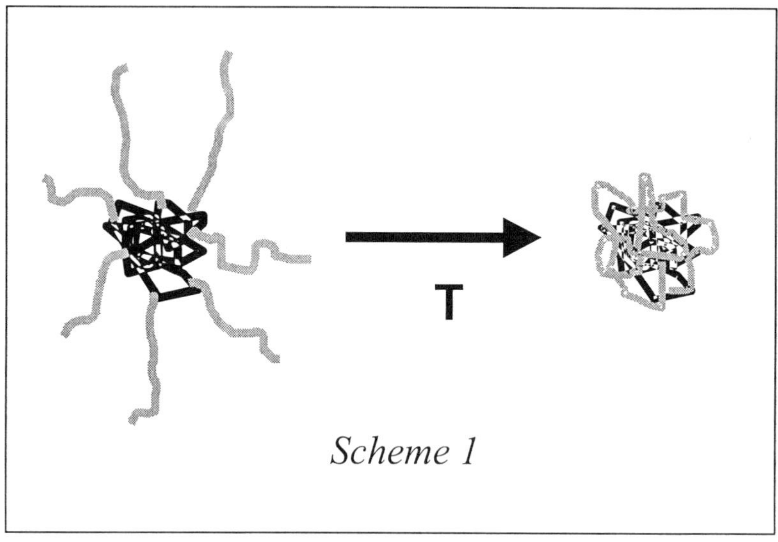

Scheme 1

One obvious limitation of the use of the styrene-based graft copolymer and other species of this type, as far as controlled release applications are concerned, is that the macromolecule retains solubilization of organic species (*13*) at all temperatures. One possible route to overcome this problem, to alter the LCST *via* "entropic control", would be to use a more hydrophilic comonomer thus reducing the tendency to form intramolecular aggregates in aqueous media.

Studies of such modifications in graft copolymers of varying composition continue.

Conclusions

1. Luminescence techniques are a useful means of probing the effects of chemical modification upon "smart" polymer behavior.
2. Alteration of the hydrophobic/hydrophilic balance of NIPAM-based polymers, through random copolymerization changes the LCST of the thermoresponsive polymer. Unfortunately, this stratagem also serves to reduce the magnitude of the transition, in many cases.
3. Control of the LCST can also be effected through manipulation of polymer topography.

References

1. *Polymers in aqueous media: performance through association*; Glass, J. E., Ed.; ACS Adv. Chem. Ser.; ACS, Washington, D.C., 1989; Vol 223.
2. *Water Soluble Polymers: Synthesis, Solution Properties and Applications*; Shalaby, S.W.; McCormick, C.L.; Butler, G.B., Eds.; ACS Symp. Ser.; ACS, Washington, D.C., 1997; Vol 467.
3. Ghiggino, K.P.; Tan, K.L. In *Polymer Photophysics*; Phillips, D., Ed; Chapman and Hall, London, 1985.
4. Winnik, F.M. *Polymer,* **1990,** *31,* 2125.
5. Bednar, B.; Trnena, J.; Svoboda, P.; Vajda, S.; Fidler, V. and Prochazka, K. *Macromolecules* **1991,** *24,* 2054.
6. Soutar, I.; Swanson, L. *Macromolecules,* **1994,** *27,* 4304.
7. Soutar, I.; Swanson, L., *to be published.*
8. Heskins, M.; Guillet, J.E. *Macromol. Sci. Chem.* **1968,** *A2,* 1441.
9. Schild, H.G. *Prog. Polym. Sci.,* **1992,** *17,* 163.
10. Chee, C.K.; Rimmer, S; Soutar, I.; Swanson, L. *Polymer,* **1997,** *38,* 483.
11. Yamamoto, I.; Iwasaki, K.; Hirotsu, S. *J. Phys. Soc. Jpn.,* **1989,** *58,* 210.
12. Winnik, F.M. *Polymer,* **1990,** *31,* 2125.
13. Chee, C.K.; Rimmer, S.; Soutar I.; Swanson, L. *Polym. Prep. ACS Div. Polym. Chem.,* **1999,** *40,* 232.
14. Chee, C.K.; Rimmer, S; Soutar, I.; Swanson, L., *to be published.*
15. Barker, I.C. ; Cowie, J.M.G.; Soutar, I.; Swanson, L., *to be published.*
16. Flint, N.J.; Gardebrecht, S.; Swanson, L. *J. Fluorescence,* **1998,** *8,* 343.
17. Rimmer, S; Mohd Ramli, A.N.; Lefèvre, S. *Polym*er, **1996,** *37,* 4135.
18. Soutar, I.; Swanson, L.; Imhof, R.E.; Rumbles, G. *Macromolecules,* **1992,** *25,* 4399.
19. Soutar, I.; Swanson, L. In *Current trends in polymer photochemistry.* Allen, N.S., Ed; Ellis Horwood Publishers, 1995.
20. Soutar, I.; Swanson, L. *ACS Symp. Ser.,* **1995,** *598,* 388.
21. Barkley, M. D.; Kowalczyk, A. A.; Brand, L. *J. Chem. Phys.,* **1981,** *75,* 3581.
22. Soutar, I.; Swanson, L.; Christensen, R. L.; Drake, R. C.; Phillips, D. *Macromolecules,* **1996,** *29,* 4931.
23. Taylor, L.D.; Cerankowski, L.D. *J. Polym. Sci., Polym. Chem. Edn.,* **1975,** *13,* 2551.
24. Soutar, I.; Swanson, L. *Eur. Polym. J.,* **1993,** *29,* 371.
25. Shibayama, M; Mitzutani, S.; Nomura, S. *Macromolecules* **1996,** *29,* 2019.

Chapter 14

Non-Ionic Water-Soluble Polysilanes and Polysilynes, Chameleons in Solution

Thomas J. Cleij, Jennifer K. King, and Leonardus W. Jenneskens

Debye Institute, Department of Physical Organic Chemistry, Utrecht University, Padualaan 8, 3584 CH Utrecht, The Netherlands

By incorporating non-ionic oligo(oxyethylene) and crown ether containing substituents, previously inaccessible water soluble linear polysilanes $[SiRR']_n$ and branched polysilynes $[SiR]_n$ have been prepared. Their remarkable σ-conjugation related properties are a sensitive probe for silicon backbone conformation. In aqueous solution at ambient temperatures the polymers appear to be dissolved in a highly folded random coil. At the lower critical solution temperature (LCST; 45-50°C) the solution becomes turbid and the polymers aggregate in a more extended conformation. This temperature-induced formation of ordered particles shows remarkable metal cation responsiveness.

Polysilanes $-(SiR_1R_2)_n-$ are one-dimensional polymers with an *all*-silicon backbone and organic substituents (R_1, R_2) (*1*). They attract considerable attention due to their remarkable optical and electronic properties, which differ considerably from their related *all*-carbon analogues, as a result of the occurrence of σ-conjugation (*2*). The degree of σ-conjugation, which is strongly dependent on the conformation of the silicon backbone, markedly affects these properties. As a result, a variety of useful electrical and optical properties are found. Examples of these properties include hole- and photoconduction, electroluminescence, microlithography, thermo-, piezo-, solvato- and surface-mediated chromism as well as thermoresponsive behavior (*1,3*). Introducing branching points in the silicon backbone branched polysilanes, *viz.* polysilynes - (SiR)$_n$- are accessible, which exhibit typical semiconductor-like properties (*4-6*).

After the discovery of soluble polysilanes in the 1980's, a variety of alkyl and aryl substituted representatives became available (*1,2,7*). Polymerization was accomplished using the Wurtz-type coupling reaction (Na, toluene, 110 °C). Unfortunately, the conventional Wurtz-type coupling reaction has disadvantages, such as its incompatibility with many functional groups, the low and often unpredictable yields as well as the fact that multi-modal weight distributions are obtained (*1,8*). Hence, it is not surprising that in the past decade a considerable effort has been directed towards the development of improved preparation methods. Despite the fact that several of these alternative procedures offer advantages over the reductive coupling method, concomitantly other disadvantages appear, *viz.* the procedures are complex and not (yet) universally applicable. Hence, reductive coupling still remains the preferred method.

Polar Functionalized Polysilanes

An interesting class of polysilanes, which has been developed in the past decade, consists of polar functionalized polysilanes. Interest in these materials originated from the observation that polysilanes with phenol moieties were able to form Langmuir monolayers at the air/water interface (*9*). Such layers provide an opportunity to organize and subsequently deposit polysilanes in a well-defined manner, which is of interest for their potential applications. In addition, these materials are expected to have enhanced solubility properties and good wettabillity towards polar inorganic substrates (*10*).

The most widely used synthetic approach of polar polysilanes is to introduce ether functionalities in the alkyl side chains. The first polysilane with such substituents was poly(methyl-3-methoxypropylsilane) **1** (Scheme 1), which, besides being soluble in common apolar organic solvents, is also soluble in alcohols (*11*). The improved solubilities attracted substantial attention, and soon related examples with either one or two 4-propoxybutyl, 5-ethoxypentyl or 6-methoxyhexyl substituents (**2-5**, Scheme 1) were prepared (*12*). As speculated, these polysilanes form excellent Langmuir-Blodgett monolayers and are (moderately) soluble in alcohols (*12*). Notwithstanding, the substituents incorporated in **1-5** were not sufficient to induce water-solubility. This was unfortunate, since polar non-ionic water-soluble polysilanes were expected to have remarkable (photo)physical and physico-chemical properties. Hence, we were prompted to enhance their water solubility by the introduction of extra ether units in the side chains.

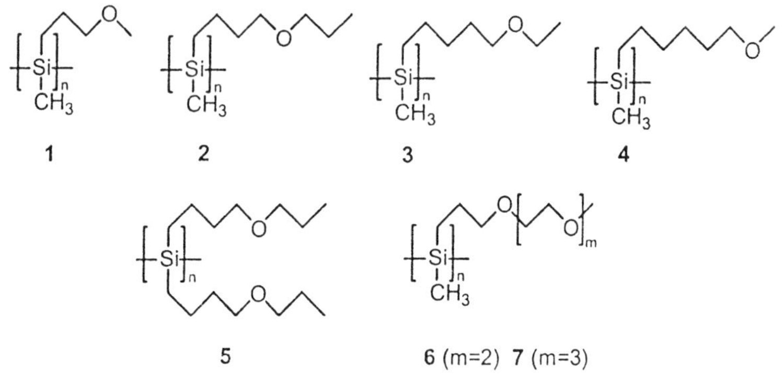

1   **2**   **3**   **4**

5   **6** (m=2) **7** (m=3)

Scheme 1. *Examples of polysilanes with ether units in the side chains.*

Poly(4,7,10-trioxaundecylmethylsilane) **6** and poly(4,7,10,13-tetraoxatetra-decylmethylsilane) **7** (Scheme 1) indeed have a sufficient number of oxygen atoms in the side chains to be soluble in water, *i.e.* they represent the first non-ionic water-soluble polysilanes (*13,14*). However, in the case of **7** prepared using the conventional Wurtz type coupling (Na, toluene, 110 °C), isolated yields were low and varied considerably [**7**; <<1.0% (*14*) up to 10% (*13*)].

Polar water-soluble polysilanes can also be prepared by chemical modification after polymerization. In this method poly(β-phenethylsilane) is fully chloromethylated, followed by a quaternization using a trialkylamine (*15,16*). Using this method, polysilanes can be prepared with 95% of the phenethyl substituents being quaternized. Even though interesting materials are obtained, the occurrence of an additional long wavelength band in the fluorescence emission spectrum represents an inherent disadvantage (*16*). This band is likely to originate from defects that are introduced during the chloromethylation step. Hence, at this moment the reductive coupling method is the only viable general method to prepare high purity polar polysilanes. This observation instigated our quest for alternative reducing agents.

C_8K, an Alternative Reducing Agent

To establish the suitability of graphite potassium, C_8K, for the preparation of **7**, 4,7,10,13-tetraoxatetradecylmethyldichlorosilane (*13*) was polymerized under various conditions (Scheme 2). Initial experiments revealed that both the formation of **7** as well as its final yield are strongly dependent on the *ratio* C_8K:monomer, reaction temperature and duration of polymerization. In tetrahydrofuran (THF) high molecular weight **7** was only obtained with a *ratio* C_8K:monomer of exactly 2, a reaction temperature of 0 °C and a reaction time of 2-2.5 h (*17*). After preparative SEC **7** was isolated with a considerable enhanced yield of 20% to 35%. ^{29}Si NMR analysis only gave the characteristic silicon backbone resonance at δ -31.4 ppm. The absence of resonances attributable to end-groups and/or siloxane moieties indicates that no low-molecular weight material is present. In marked contrast, for aryl substituted polysilanes C_8K gave ambiguous results. Although poly(methylphenylsilane) is accessible using C_8K its (polar) analogue poly(4-dimethylaminophenylmethyl-silane) cannot be prepared (*18-20*).

Hence, the following questions arise: What is the scope of the novel C_8K based polysilane polymerization and to what extent does the successful use of C_8K depend on the type and physico-chemical properties of the monomers? These topics were addressed by preparing a series of copolymers poly(dimethyl-*co*-4,7,10,13-tetraoxatetradecylmethylsilanes) by reductive coupling of polar 4,7,10,13-tetraoxatetradecylmethyldichlorosilane and apolar dimethyldichloro-silane in various ratios using either C_8K or Na (*21*). The polymer yields and optical characteristics show that, when using C_8K, concomitant with an increase in feed ratio of the apolar monomer, the copolymer yield is markedly reduced. At high feed ratios of the apolar monomer, no high molecular weight polysilane is formed. In marked contrast to this decrease in copolymer yield with high dimethyldichlorosilane feed ratios, the yields when using Na appear not to be strongly correlated to the feed ratio. The superior results of the use of C_8K in the polymerization of **7** presumably originates from the fact that specific (polar) monomers can intercalate into the graphite layers (*21*). Hence, C_8K is a promising reducing agent for the preparation of comparable non-ionic water-soluble polysilanes and polysilynes which cannot be prepared using Na.

Novel Polysilanes and Polysilynes

It has been demonstrated that polysilanes with crown ether substituents are not accessible *via* the conventional Wurtz type coupling of functionalized dichlorosilane monomers (*22*). Since this is likely due to the polar character of the crown ethers, the polymerizations were attempted using C_8K (*23,24*). Indeed poly[methyl(methylene-13-crown-4)silane] **8**, poly[methyl(methylene-16-crown-5)silane] **9** and poly[methyl(methylene-19-crown-6)silane] **10** (Scheme 3) can be prepared in 6-28% yield using C_8K from the respective dichloromethyl(methylene-n-crown-m)silane monomers, although a higher

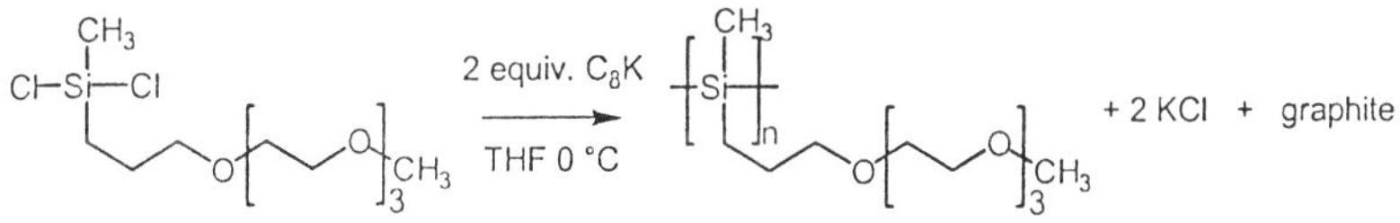

Scheme 2. Synthesis of poly(4,7,10,13-tetraoxatetradecylmethylsilane) 7.

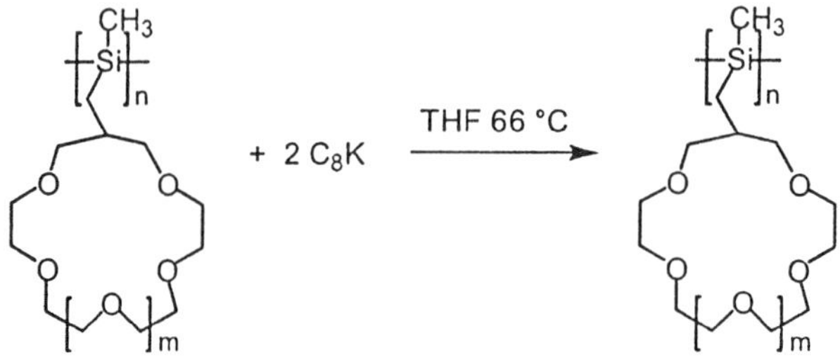

Scheme 3. *Polysilanes with crown ether substituents.*

temperature is necessary, *i.e.* 66 °C (*23*). The molecular weight distributions (analytical SEC, polystyrene standards) of **8-10** range from M_w 6.7· 10^3 to 1.8· 10^4 (D = M_w/M_n 1.5-1.7). Interestingly, **8-10** are not only soluble in organic solvents, but also in water. The availability of this type of polymer allows for the study of specific cation-crown ether interactions.

The successful preparation of **8-10** also prompted us to use C_8K for the preparation of non-ionic water-soluble polysilynes. In addition to linear polysilanes, in 1988 branched polysilanes, *i.e.* polysilynes -[SiR]$_n$-, were first reported (*4-6*). They form an intermediate between one-dimensional polysilanes and three-dimensional amorphous and porous silicon. The actual structure of polysilynes is a matter of controversy. Initially, it was believed that polysilynes have a two-dimensional sheet-like construction (*5*). In later studies, branched structures with interconnecting rings and hyperbranched/dendritic structures have been proposed (*25,26*). However, a predominantly one-dimensional overall appearance consisting of linear fragments with small branches and/or incorporated (branched) cyclics seems to be most likely (*vide infra*) (*27*).

Polysilynes are usually prepared by conventional Wurtz-type coupling of an organotrichlorosilane with either Na (toluene, 110 °C) or the combination of high-intensity ultrasound and NaK alloy (*4-6*). However, using these reaction conditions, it is impossible to obtain non-ionic, water-soluble polysilynes. Fortunately, these materials appear to be accessible using C_8K; poly(4,7,10-trioxaundecylsilyne) **11** and poly(4,7,10,13-tetraoxatetradecylsilyne) **12** could be isolated in 10-40% (Scheme 4) (*27*).

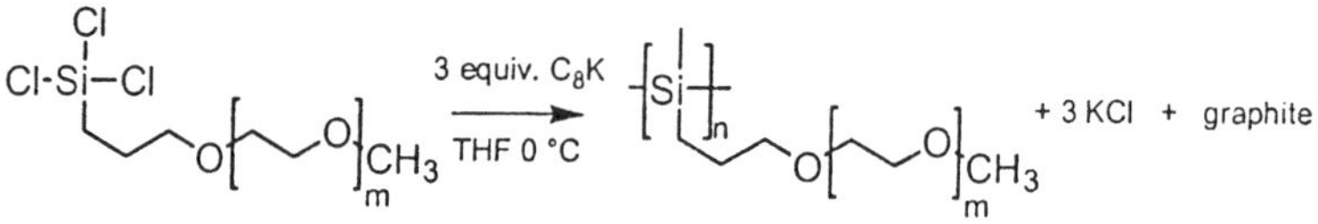

Scheme 4. *The synthesis poly(4,7,10-trioxaundecylsilyne) **11** and poly(4,7,10,13-tetraoxatetradecylsilyne) **12** using C_8K as the reducing agent.*

Thermoresponsive Behavior of Polysilanes

For asymmetrically substituted polysilanes containing aliphatic side chains, the characteristic σ-σ^* absorption maximum in regular organic solvents is positioned at *ca.* λ_{max} 305 nm (*1*). However, for an aqueous solution of **7**, λ_{max} (281 nm) is considerably blue shifted. This indicates a reduction of the degree of σ-conjugation, presumably due to the presence of a more distorted silicon backbone, *i.e.* strong solute-solvent interactions exist. Hence, in aqueous solution, **7** adopts an even more condensed random coil conformation with the polar oligo(oxyethylene) side chains extending into the water phase and the Si-CH_3 and Si-C_3H_6- groups positioned at the center of the coil, preserving the apolar character of the backbone ('unimolecular micelle' analogue) (*3,28*). Thus, at room temperature, solvent polarity dictates the backbone structure of **7**.

Upon heating an aqueous solution of **7**, the LCST is reached at 46 °C and the solution turns opaque (*3*). Figure 1 shows that the LCST can easily be determined by the onset of light scattering (monitoring wavelength λ 500 nm). According to the Schweizer theory of conformation dependent solute-solvent interactions (*29*), asymmetrically substituted polysilanes possess a small coupling constant V_D/ϵ with respect to a critical value $(V_D/\epsilon)_c$. V_D represents a measure for the solute-solvent interactions, which in the case of polysilanes primarily depends on the polarizability of the catenated silicon backbone and ϵ, the mean free energy of defect formation (*29*). When V_D/ϵ crosses this critical value $[V_D/\epsilon > (V_D/\epsilon)_c]$, abrupt thermochromism will be observed (*29*). However, in an organic solvent such as THF, $V_D/\epsilon < (V_D/\epsilon)_c$ for **7**, which therefore exhibits a (reversible) continuous shift of λ_{max} from 305 to 317 nm upon cooling to -60 °C.

UV spectroscopy at the LCST unequivocally revealed an instantaneous

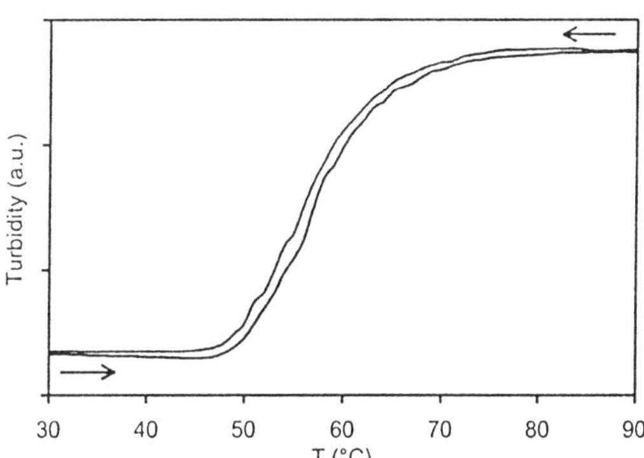

Figure 1. Turbidity vs. *temperature of 7 in aqueous solution.*

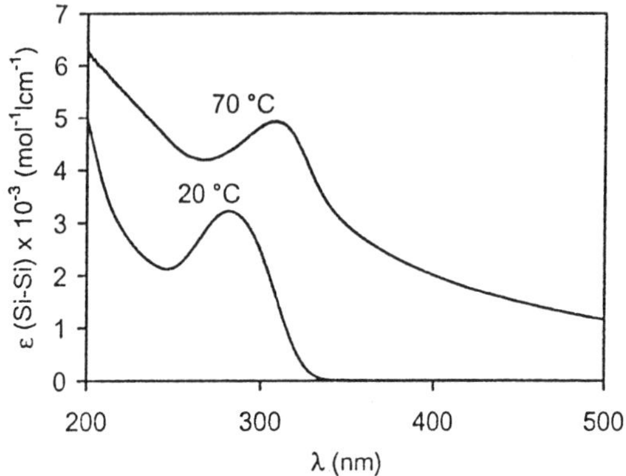

Figure 2. UV-Vis-absorption spectra of 7 in an aqueous solution at room temperature and above the LCST.

reversible bathochromic shift of λ_{max} from 281 to 320 nm (*3*). Shown in Figure 2 are the UV-Vis-absorption spectra of 7 at 20 and 70 °C. The '*inverse*' abrupt thermochromism implies that in water near/above the LCST $V_D/\epsilon > (V_D/\epsilon)_c$ as a result of an increased polarizability of the Si backbone (V_D) (*29*). The bathochromic shift of λ_{max} at the LCST further suggests an increasing degree of σ-delocalization; *i.e.* it is comparable with that of a THF solution of 7 at -60 °C (*3*). This suggests that the average silicon segmental length markedly increases at the LCST and that above the LCST, the silicon backbone structure of 7 is no longer dictated by solvent polarity (*3*). Notwithstanding, fluorescence spectroscopy indicates that the lowest energy (emitting) chromophores do not change; *i.e.* the position of the emission maximum does not shift upon heating.

It is noteworthy that the initial phase separation is *reversible* (no hysteresis), and that above the LCST, solution UV spectra can still be measured. This indicates that 7 remains partially dissolved. Only after prolonged heating (> 48 h. at 60 ± 5 °C) does complete phase separation occur. Therefore, this process can be described as the onset of full phase separation; a two-phase system containing a polymer depleted (possibly, even free) aqueous phase and a highly concentrated solution of 7 is obtained. Presumably, the polymer chains aggregate, eventually leading to complete phase separation. This was further investigated by both static and dynamic light scattering. Below the LCST, light scattering of the solution is indistinguishable from pure solvent scattering. At the LCST, the solution starts to scatter progressively in the forward direction, indicating the occurrence of aggregation. The measurements reveal that monodisperse scattering units with a radius of *ca.* 200 nm are present (*3*).

An interesting aspect of aqueous solutions of **7** is the ability to tune their LCST's by the addition of inorganic salts (*3,30*). Shown in Figure 3 is the

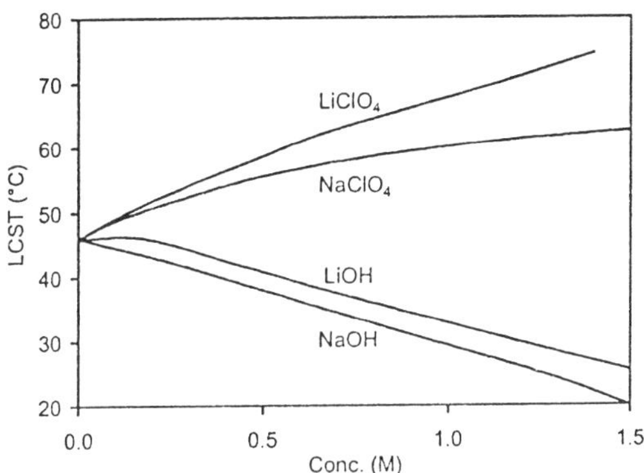

Figure 3. The LCST of 7 in an aqueous solution vs. *the concentration of added inorganic salt.*

dependence of the LCST of an aqueous solution of **7**, on the concentration of added inorganic salt. Addition of alkali metal perchlorates leads to a dramatic increase of the LCST (conc. 1.4 M LiClO$_4$, LCST 75 °C). This is not surprising, in view of the strong interaction of LiClO$_4$ with oligo(oxyethylene) fragments (*13*), which previously have been used in solid polymer electrolytes (*31*). The alkali metal cations induce physical (intramolecular) crosslinking between the oligo(oxyethylene) side chains, causing an increased stabilization of the periphery of the 'unimolecular micelle' analogue (*28*).

In marked contrast, addition of the corresponding hydroxide salts results in a decrease of the LCST. In the presence of sufficient amounts of hydroxide salts, the LCST is positioned below room temperature. Hence, the anions play a crucial role in the solubility of **7**. As a result of the incompatibility of the hydroxide ions with **7**, both cations and anions will be predominately present in the aqueous phase, instead of near the oligo(oxyethylene) fragments. This decreases the compatibility of **7** with water and hence a lowering of the LCST is observed (*3,30*). The UV behavior of aqueous solutions of **7** in the presence of inorganic salts is also quite different from that of the salt free solutions above the LCST. It can be seen in Figure 4 that above the LCST only a very weak σ–σ* transition superpositioned on a scattering background is discernible, suggesting that most of the initially absorbing chromophores are absent. Presumably, the coordinating alkali metal cations participate in the formation of a physically crosslinked network structure.

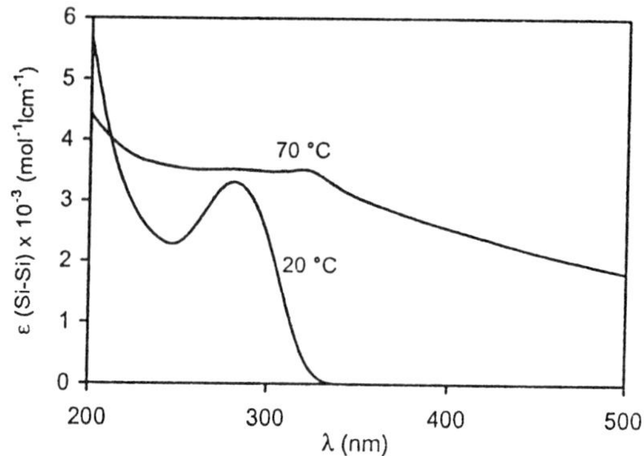

Figure 4. UV-Vis-absorption spectra of *7* in an aqueous solution containing 1.4 M LiClO₄ at room temperature and above the LCST.

The Influence of Crown Ether Substituents

For the crown ether substituted polysilanes **8-10**, the position of λ_{max} (σ-σ* transition) in THF is dependent on the size of the crown ether moiety (Table I).

Table I. Typical σ–σ* absorption characteristics for 8-10 in THF and H₂O.

Polysilane	λ_{max} (THF) (nm)	$\varepsilon \, x10^{-3}$ (THF) (mol⁻¹lcm⁻¹)	λ_{max} (H₂O) (nm)	$\varepsilon \, x10^{-3}$ (H₂O) (mol⁻¹lcm⁻¹)
8	297	5.7	283	4.1
9	299	6.7	289	6.2
10	302	5.7	294	5.1

In going from **8** to **10**, the backbone is capable of adopting a slightly more extended conformation. Apparently, the larger 19-crown-6 moieties are less sterically demanding than the small 13-crown-4 units. This can be attributed to the fact that larger crown ethers are less rigid (conformational flexibility) than the smaller representatives; *i.e.* they can fold to accommodate the silicon backbone to a greater extent (*32*).

In water the effect on the λ_{max} is even more pronounced (Table I). The substantial shift in λ_{max} of 11 nm in going from **8** to **10**, as a function of the crown ether size, is probably not only caused by steric effects, but also by the fact that **10** will be more shielded from the polar aqueous environment by the large crown ether moieties; *i.e.* **10** is "dissolved" in its substituents. Hence, **10** can adopt a slightly more extended conformation than **8** and **9**. The position of λ_{max} for **8** dissolved in water, is comparable to that of aqueous solutions of 7 (*3*). This implies that a comparable silicon backbone structure must be present, *viz.* a 'unimolecular micelle' analogue. In water, as compared to THF, the enhanced stiffening of the crown ether moieties by hydrogen bonding with water molecules is presumably responsible for a further deformation of the backbone.

In contrast to **7**, temperature dependent UV-Vis spectra of THF solutions of **8-10** do not exhibit large changes in the UV-absorption spectrum upon cooling; λ_{max} is constant. It is of considerable interest to look at these results in view of Schweizer theory (*29*). The observed temperature dependent behavior of **8-10** is a special case of Schweizer theory with $V_D/\epsilon \ll (V_D/\epsilon)_c$ (*29*). When the mean free energy of defect formation ϵ, *viz.* the energy required to create a twisted conformer ("non-*trans* defect") in a fully extended backbone, is large due to the bulky substituents, the coupling constant V_D/ϵ will be small and Schweizer theory predicts that no shift will be discernible (*29*).

The water-solubility of **8-10** instigated the investigation of their thermoresponsive behavior. Indeed, aqueous solutions of **8-10** also have an observable LCST (*24*). The observed phase separation process appeared to be reversible; no significant hysteresis was found. Interestingly, the position of the LCST changes concomitantly with the size of the crown ether moiety. In view of the λ_{max} values for **8-10** in aqueous solution (Table I), it is not surprising that the position of the LCST will be dependent on the size of the crown ether moiety. Polysilane **8** possesses at room temperature in aqueous solution the most distorted catenated silicon backbone. The LCST of **8** is 54 °C, which is only slightly higher than that of **7** (*3*). The compatibility of **9** and, especially, **10** with the aqueous phase appears to be enhanced. This is both reflected in their red shifted values of λ_{max} (Table I) as well as their substantially higher LCST values, *viz.* **9** LCST 62 °C and **10** LCST 74 °C. As is shown in Figure 5, a (linear) relationship exists between λ_{max} in water and the position of the LCST (R^2 0.99). This indicates that the position of the LCST is strongly correlated with the global conformation of polysilanes in solution.

Remarkably, at the LCST, the reversible shift in the UV-Vis spectrum (σ-σ^* absorption) of **8-10** is not abrupt, as was found for **7**. Instead, as shown in Figure 6, a continuous shift is observed. At 90 °C λ_{max} is shifted to *ca.* 300 nm, *i.e.* the same position as was found for THF solutions of **8-10**. In THF for **8-10** $V_D/\epsilon \ll (V_D/\epsilon)_c$, whereas in aqueous solution $V_D/\epsilon < (V_D/\epsilon)_c$ (*29*). Apparently, compared to the THF solution, V_D has now increased sufficiently, allowing for a continuous transition upon heating from 20 to 90 °C. However, in comparison to V_D/ϵ of **7**, which exhibits an abrupt transition around the LCST, the V_D/ϵ of **8-10** remains small.

It is noteworthy that upon addition of inorganic perchlorates for **7**, an increase in LCST was found, whereas for **9** a marked decrease is found. The effect is considerable; already at a concentration of *ca.* 0.3 M $NaClO_4$, the LCST is positioned below room temperature. Presumably, the stabilizing effect of intramolecular complexation is now absent, which is not surprising in view of

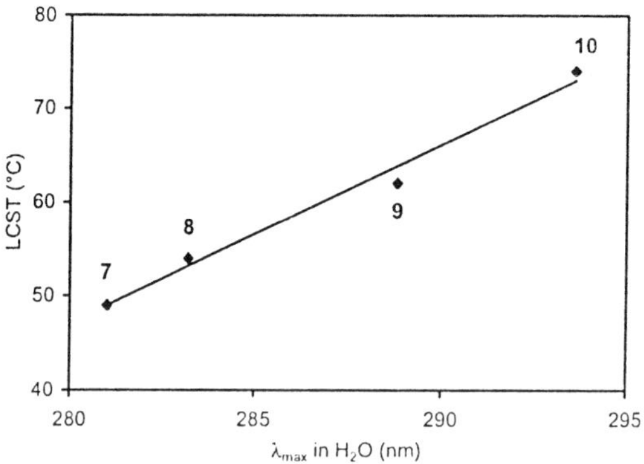

Figure 5. The position of the LCST vs. the position of λ_{max} (σ–σ* transition) in aqueous solutions of non-ionic water-soluble polysilanes *7* and *8-10*.

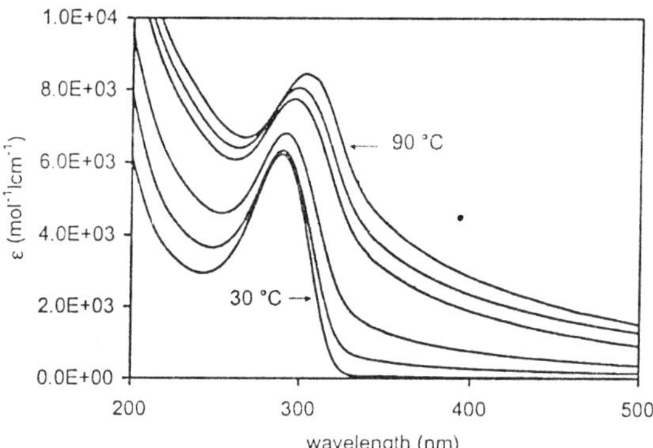

Figure 6. Temperature dependent UV-Vis-absorption spectra of an aqueous solution of *9*.

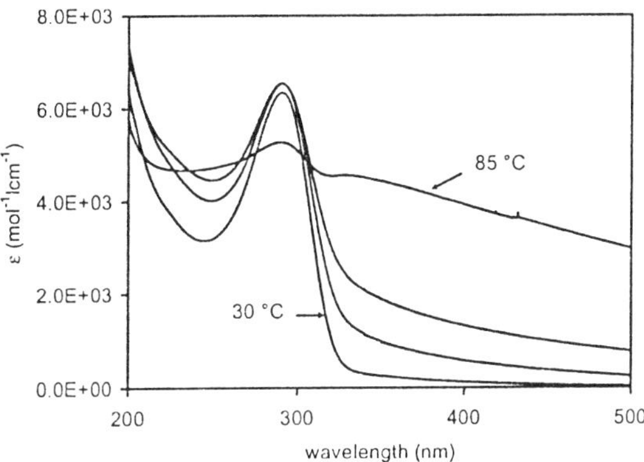

Figure 7. Temperature dependent UV-Vis-absorption spectra of an aqueous solution of 9.in the presence of NaClO₄.

the excellent Na^+ complexation by the crown ether moieties of **9** (*33*). Since complexes of **9** and $NaClO_4$ are partially soluble in water, the UV-Vis characteristics above and below the LCST can be determined and are shown in Figure 7. The increase in the bulkiness of the substituents as a result of the complexation process, *i.e.* increased ε, is reflected in the fact that no difference exists between the position of λ_{max} above and below the LCST present, *i.e.* $V_D/\in$ $<< (V_D/\in)_c$. In addition, the intensity of the σ-σ* transition has decreased, *i.e.* solid polymer electrolyte-like aggregates precipitate.

Thermoresponsive Behavior of Polysilynes

Since polysilanes **6** and **7** possess thermoresponsive behavior, the question arises whether polysilynes **11** and **12** exhibit similar behavior. Indeed, upon heating of aqueous solutions of **11** and **12**, an LCST was found at 49 °C, as can be seen by the onset of light scattering (monitoring wavelength λ 750 nm) in Figure 8 (*27*). Unexpectedly, the LCST of **11** and **12** is close to that found for the related linear **6** and **7** (*3*). In a similar fashion also upon addition of $LiClO_4$, the LCST of **11** and **12** shifts to higher temperatures (conc. 1.4 M $LiClO_4$, LCST 72 °C) (*27*). Hence, as with **11** and **12**, alkali metal cations stabilize the polymer perimeter, presumably by complexation with the oxyethylene moieties. Unfortunately, upon addition of alkali metal hydroxide salts, such as LiOH and NaOH, instead of a decrease of the LCST, degradation occurs.

The observed thermoresponsive behavior of aqueous solutions of **11** and **12**, in combination with the observation that their LCST is nearly identical to that found for **6** and **7**, suggests that the polysilyne and polysilane structures are correlated. This is an unexpected finding, since polysilynes obtained by the

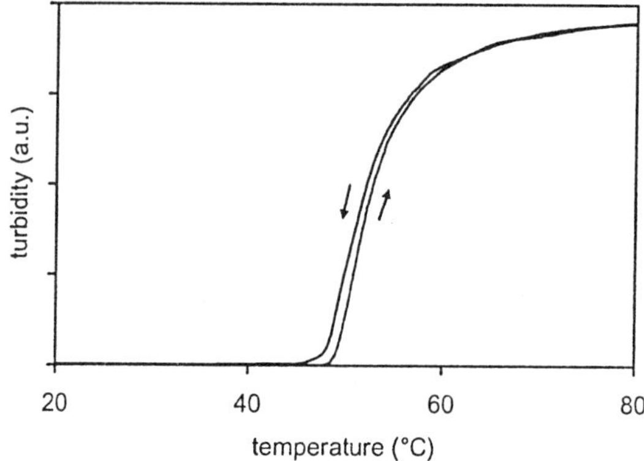

Figure 8. *Turbidity vs. temperature of an aqueous solution of 11.*

reductive coupling of organotrichlorosilanes with either Na or C_8K were proposed to possess two-dimensional, extended sheet-like structures consisting of a large number of interconnecting rings or hyperbranched/dendritic structures (*5,25,26*). However, in view of the thermoresponsive behavior of aqueous solutions of **11** and **12,** such structures appear to be implausible; *i.e.* folding/unfolding of such structures in solution is unlikely to occur. Moreover, if such a process is still feasible, a significantly different (if not absent) LCST is expected for **11** and **12** in comparison with that of their corresponding polysilane analogues. In addition, PM3 calculations give no evidence for the formation of hyperbranched/dendritic structures (*27*).

Instead, the thermoresponsive behavior, as well as the PM3 calculations, suggest that polysilynes possess a hybrid structure with a predominantly one-dimensional overall appearance consisting of linear fragments with small branches and/or incorporated (branched) cyclics (*27*). However, the overall appearance of polysilynes will be semi-linear, which reconciles their excellent solubility properties as well as the thermoresponsive behavior of **11** and **12** (*27*). Hence, the availability of these non-ionic water-soluble polysilynes finally has shed some light on the (controversial) structure of branched polysilanes.

Global Conformation

The observation that changes in the catenated silicon backbone conformation of polysilanes markedly affect their opto-electronic properties has prompted the investigation of the relationship between the polysilane optical properties and global conformation in solution (*34*). In this context it is interesting to review the observed differences in λ_{max} ($\sigma-\sigma^*$ transition) between polysilanes in aqueous and organic solutions. The main questions that are addressed is to what extent solute-solvent interactions influence their global conformation and whether in aqueous/organic binary solvent mixtures the optical properties are linearly dependent on the solvent mixture composition or if preferential solvation occurs. To address this issue, solutions of **7** in a series of binary solvent mixtures, *i.e.* H_2O-EtOH, H_2O-ACN and H_2O-THF, of different composition ratios were studied using optical spectroscopy.

As can be seen from Figure 9, already upon addition of small amounts of the organic solvent to the aqueous solution of **7**, a red-shift in the λ_{max} ($\sigma-\sigma^*$ transition) occurs. This shift continues to a mole fraction of 0.09-0.15, after which the λ_{max} value of the pure organic solvent is reached. Further addition of organic solvent (mole fractions 0.2–1.0) does not affect λ_{max}. This non-linear behavior is similar for all three organic solvents. These solvents presumably exhibit much better compatibility with the silicon backbone and the

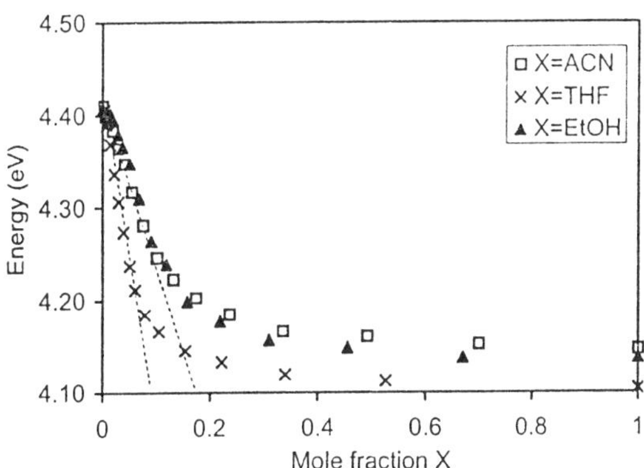

Figure 9. UV-absorption maximum, λ_{max} ($\sigma-\sigma^*$ transition), of 7 vs. the binary solvent mixture composition.

Si-CH_3 and Si-C_3H_6- groups than water. Hence, preferential solvation occurs. Consequently, upon the addition of organic solvents, the solute-solvent interactions, which in pure water prevent the silicon backbone from adopting a more extended conformation, are reduced, causing the catenated silicon backbone to become less distorted.

Although an estimate of the global conformation can be derived from λ_{max}, a better understanding of the average silicon backbone segmental length L is desirable. In this context, the extinction coefficient ε and the full width at half maximum (FWHM) of the $\sigma-\sigma^*$ absorption are important, since empirical

252

relationships have been derived which correlate L with ε, FWHM and the viscosity index α (*34*). The analysis of the σ–σ* absorption shown in Figure 10

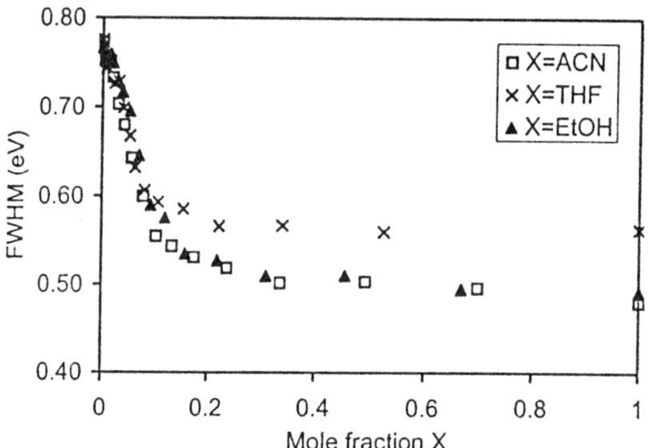

Figure 10. *Full width at half maximum, FWHM (σ–σ* transition), of 7 vs. the binary solvent mixture composition.*

reveals that not only λ_{max}, but also ε and the FWHM are markedly dependent on the binary solvent mixture composition.

Since for polar polysilanes the FWHM values are more accurate than the ε values (*23*), it is more reliable to calculate L (repeating units) using the empirical relationships FWHM = 3.15 exp (-2.77α) and ε = 1130 exp(2.90α) in combination with ε = 330L (*34*). Using this approach, in water, a value for **7** of 15 is found, which upon addition of the organic solvents increases to 21-24. This confirms that the silicon backbone is more folded in water and extends (partial unfolding) upon addition of small amounts of the organic solvents. This conjunction is corroborated by the observation that the exciton coherence length doubles upon going from aqueous to organic solutions; *i.e.* a twofold increase in the quantum yield Φ in going from polar to apolar solvents is observed (*35*).

Conclusions

Evidence is presented that the structure of non-ionic water-soluble polysilanes and polysilynes in aqueous solution is strongly dependent on the temperature and the presence of additives, *i.e.* inorganic salts. Whereas the polymer is present as a random coil at room temperature, at the LCST, depending on the bulkiness of the substituent, a more extended conformation is adopted. The interaction with inorganic salts can be further adjusted by introduction of crown-ether substituents.

References

1. Miller, R.D.; Michl, J. *Chem. Rev.* **1989**, *89*, 1359.
2. Miller, R.D. *Angew. Chem., Adv. Mater.* **1989**, *101*, 1773.
3. Cleij, T.J.; Jenneskens, L.W.; Kluijtmans, S.G.J.M. *Adv. Mater.* **1997**, *9*, 961
4. Bianconi, P.A.; Weidman, T.W. *J. Am. Chem. Soc.* **1988**, *110*, 2342.
5. Furukawa, K.; Fujino, M.; Matsumoto, N. *Macromolecules* **1990**, *23*, 3423.
6. Walree, C.A van; Cleij, T.J.; Jenneskens, L.W.; Vlietstra, E.J.; Laan, G.P. van der; Haas, M.P. de; Lutz, E.T.G. *Macromolecules* **1996**, *29*, 7362
7. West, R.; David, L.D.; Djurovich, P.I.; Stearley, K.L.; Srinivasan, K.S.V.; Yu, H. *J. Am. Chem. Soc.* **1981**, *103*, 7352.
8. Rehahn, M. *Acta Polym.* **1998**, *49*, 201.
9. Hayase, S. *Chemtech* **1994**, *October*, 19.
10. Cleij, T.J.; King, J.K.; Jenneskens, L.W.; Lubberhuizen, W.H.; van Faassen, E. *Polym. Prep.* **1998**, *39 (1)*, 94.
11. Hrkach, J.S.; Matyjaszewski, K. *J. Polym. Sci., Polym. Chem.* **1994**, *32*, 1949.
12. Yuan, C-H.; West, R. *Macromolecules* **1998**, *31*, 1087.
13. Walree, C.A. van; Cleij, T.J.; Zwikker, J.W.; Jenneskens, L.W. *Macromolecules* **1995**, *28*, 8696.
14. Oka, K.; Fujiue, N.; Nakanishi, S.; Takata, T.; Dohmaru, T.; Yuan, C-H.; West, R. *Chem. Lett.* **1995**, 875.
15. Seki, T.; Tamaki, T.; Ueno, K. *Macromolecules* **1992**, *25*, 3825.
16. Seki, T.; Tohnai, A.; Tamaki, T.; Kaito, A. *Chem. Lett.* **1996**, 361.
17. Cleij, T.J.; Tsang, S.K.Y.; Jenneskens, L.W. *Chem. Commun.* **1997**, 329.
18. Lacave-Goffin, B.; Hevesi, L.; Devaux, J. *J. Chem. Soc., Chem. Commun.,* **1995**, 769.
19. Jones, R.G.; Benfield, R.E.; Evans, P.J.; Swain, A.C., *J. Chem. Soc., Chem. Commun.,* **1995**, 1465.
20. Cleij, T.J.; King, J.K.; Jenneskens, L.W. *Macromolecules* **2000**, *33*, 89.
21. Cleij, T.J.; Jenneskens, L.W. *Macrom. Chem. Phys.* **2000** in press.
22. Walree, C.A. van *Charge Transfer and Migration in π- and σ-Conjugated Optical Molecules and Materials*, Ph.D. Thesis, Utrecht University, Utrecht, The Netherlands **1996**, Chapter 7.
23. Cleij, T.J. *Conjugated Silicon Based Polymeric Materials. Novel Strategies for Tailor-Made Polysilanes and Polysilynes*, Ph.D. Thesis, Utrecht University, Utrecht, The Netherlands **1999**, Chapter 7.
24. Cleij, T.J.; King, J.K.; Jenneskens, L.W. *Polym. Prep.* **1999**, *40 (2)*, 216
25. Sasaki, M.; Matyjaszewski, K. *J. Polym. Sci., Polym. Chem.* **1995**, *33*, 771.
26. Wilson, W.L.; Weidman, T.W. *Synth. Met.* **1992**, *49-50*, 407.

27. Cleij, T.J.; Tsang, S.K.Y.; Jenneskens, L.W. *Macromolecules* **1999**, *32*, 3286.

28. Newkome, G.R.; Moorefield, C.H.; Baker, G.R.; Saunders, M.J.; Grossman, S.H. *Angew. Chem. Int. Ed. Engl.* **1991**, *30*, 1178.

29. Schweizer, K.S.; Harrah, L.A.; Zeigler, J.M. *Silicon Based Polymer Science: A Comprehensive Resource*, Ziegler, J.M.; Gordon Fearon, F.W., eds.; Advances in Chemistry Series 224, ACS, Washington, DC **1990**, p. 379 and references cited.

30. Taylor, L.D.; Cerankowski, L.D. *J. Polym. Sci., Polym. Chem.* **1975**, *13*, 2551.

31. Cleij, T.J.; Jenneskens, L.W.; Wübbenhorst, M.; Turnhout, J. van *Macromolecules* **1999**, *32*, 8663.

32. Patai, S.; Rappoport, Z. *Crown Ethers and Analogs*, John Wiley & Sons, Chichester **1989**.

33. Izatt, R.M.; Pawlak, K.; Bradshaw, J.S.; Bruening, R.L. *Chem. Rev.* **1991**, *91*, 1721

34. Fujiki, M. *J. Am. Chem. Soc.* **1996**, *118*, 7424.

35. Cleij, T.J.; Jenneskens, L.W. *J. Phys. Chem. B* **2000**, *104*, 2237.

Chapter 15

Synthesis and Properties of Water-Soluble Thermosensitive Copolymers Having Phosphonium Groups

T. Nonaka, K. Makinose, and S. Kurihara

Department of Applied Chemistry & Biochemistry, Faculty of Engineering, Kumamoto University, Kumamoto 860-8555, Japan

Water soluble thermosensitive copolymers were synthesized by copolymerization of N-isopropylacrylamide(NIPAAm) with acryloyloxyethyl trialkyl phosphonium chlorides(AETR) having varying alkyl lengths. Terpolymers with butyl methacrylate(BMA) were also synthesized. The relative viscosities of the copolymer solutions increased with increasing content of phosphonium groups in the copolymers and decreased with increasing chain length of alkyl chains in the phosphonium groups. The relative viscosities of the copolymer solutions decreased with increasing temperature, and they decreased sharply at around the lower critical solution temperature (LCST) of poly(N-isopropylacrylamide) (PolyNIPAAm). The copolymers had high flocculating ability against kaolin or bacteria suspensions. These copolymers were also found to have antibacterial activity. The flocculating ability and antibacterial activity were considerably affected by the alkyl chain length in the phosphonium groups.

Introduction

PolyNIPAAm is a thermosensitive polymer having an LCST in aqueous solution around 32°C. The copolymers containing PolyNIPAAm have been widely studied from fundamental and practical points of views.

On the other hand, it has been reported that water-soluble polymers having phosphonium groups had high antibacterial activity against *S. aureus* or *E. coli* (*1,2*). We have also reported the preparation of several water-insoluble resins or hydrogels having phosphonium groups and that they have antibacterial activity against *S. aureus* or *E. coli*. By using these insoluble bactericides, we can prevent the residual toxicity of bactericides in water (*3,4*).

In this study, the solution properties of water-soluble thermosensitive copolymers having phosphonium groups synthesized by copolymerizing NIPAAm with AETR with and without BMA were studied.

Experimental

Synthesis of water-soluble copolymers. The structure of AETR-NIPAAm-MBAAm terpolymers is shown in Figure 1.

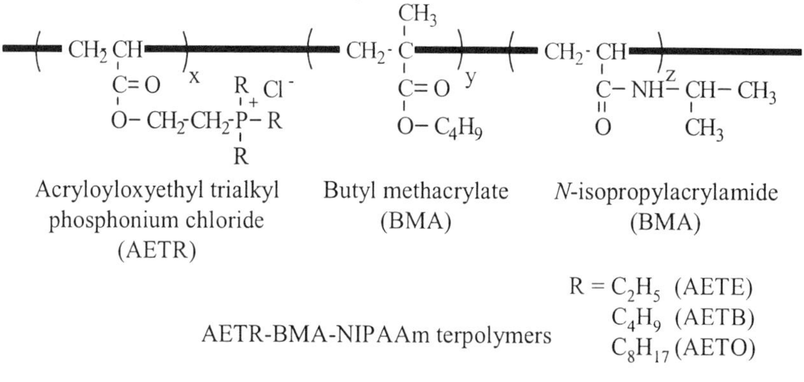

Acryloyloxyethyl trialkyl phosphonium chloride (AETR)

Butyl methacrylate (BMA)

N-isopropylacrylamide (BMA)

AETR-BMA-NIPAAm terpolymers

$R = C_2H_5$ (AETE)
C_4H_9 (AETB)
C_8H_{17} (AETO)

Figure 1. *Structure of AETR-BMA-NIPAAm terpolymers*

BMA was used to give hydrophobicity to the water-soluble polymers obtained. AETR with butyl, hexyl, and octyl group in phosphonium groups are abbreviated as AETE, AETB, and AETO, respectively.

First, AETR and NIPAAm were dissolved in 20mL of dimethyl-sulfoxide (DMSO) in a glass vessel under a nitrogen atmosphere. The copolymers were obtained by radical copolymerization using 2,2'-azobis-isobutylonitrile as a radical initiator at 50°C for 1 h.

Measurement of relative viscosity of copolymer solutions. The relative viscosity of copolymer solutions was measured with an Ostwald viscometer at various temperatures.

Measurement of flocculating ability against kaolin suspension or bacteria suspensions. The clarifying ability was studied by observing the optical density at 660nm of the supernatant after standing for 2 minutes after mixing the copolymer solutions with the 500mg/L kaolin suspensions or about 10^7 cells/mL bacteria suspensions. The flocculating ability was studied by observing the sedimentation rate and sedimentation volume of kaolin after mixing copolymer solutions with a 5wt/vol% kaolin suspension in a test tube.

Measurement of antibacterial activity. The bacteria used in this study were *E. coli* (IFO 3301) and *S. aureus* (IFO 13276), which were obtained commercially from the Institute for Fermentation, Osaka. Calculated cell suspensions containing about 10^7-10^8 cells/mL were prepared for each strain and used for antibacterial tests. About 40 mg of copolymers were dissolved in 20mL of sterile deionized water in a 50mL Erlenmeyer flask. The copolymer solution was added into 20mL of cell suspension and the flask was shaken at 30°C for a prescribed time. After mixing the copolymer solutions with a bacteria suspension for a prescribed time, 1mL of the bacteria suspension was pipetted from the flask and 9mL of sterile water were added to the bacteria suspension. The suspension was diluted several times and 0.1mL of the diluted suspension was spread on an agar plate made of nutrient agar. The plate was kept at 30°C for 15-24 h and the numbers of viable cells were calculated from those of the colonies formed on the plate.

Measurement of residual copolymers in water. The amount of residual copolymers after heating at the temperature above 35°C, followed by filtration was determined by UV spectroscopy.

Results and Discussion

Synthesis of water-soluble copolymers and terpolymers.

Table 1 shows the mole ratio of each monomer in the feed and the phosphorus contents and molecular weight of AETR-NIPAAm copolymers obtained. The phosphonium content in the copolymers was calculated from the phosphorus content in the dried copolymers, which was determined by the phosphovanadomolybdate method (5). The molecular weight of the copolymers was determined by GPC using dimethylformamide as a solvent and using polystyrene standards.

Table 1. Content of phosphorus and weight-average molecular weight of copolymers

Component	Mol ratio in feed	P content(wt%)		$Mw^{a)}$ (g/mol)
		Calculated	Observed	
AETE-NIPAAm	1:100	0.28	0.46	16700
	3:100	0.80	0.88	306000
	5:100	1.27	1.03	164000
AETB-NIPAAm	1:100	0.27	0.50	525000
	3:100	0.78	0.74	182000
	5:100	1.23	1.19	142000
AETO-NIPAAm	1:100	0.27	0.52	50700
	3:100	0.75	0.84	438000
	5:100	1.16	1.18	262000

a) These values were determined by GPC using polystyrene standards.

In the thermosensitive AETR-NIPAAm copolymers, the NIPAAm content was held constant while the AETR content was varied to give AETR:NIPAAm ratios from 1:100 to 5:100.

The phosphonium content of the copolymers varied according to the AETR:NIPAAm ratio. The AETR-NIPAAm copolymers had molecular weights of several ten thousands. The AETR-BMA-NIPAAm terpolymers were prepared in the same manner.

Temperature dependence of viscosities of copolymers and terpolymers

Figure 2 shows the relative viscosities of various copolymer solutions measured with an Ostwald viscometer at various temperatures.

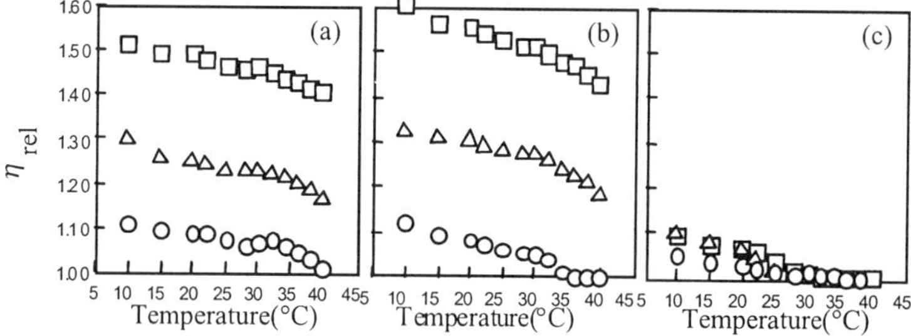

Figure 2. *Temperature dependence of relative viscosities of 0.2g/dL Copolymer Solutions Copolymers:(a)AETE-NIPAAm, (b)AETB-NIPAAm, (c)AETO-NIPAAm AETR-NIPAAm(mol ratio) : ○ : 1:100, △ : 3:100, □ : 5:100*

The relative viscosities of the aqueous solutions of 0.2g/L of AETR-NIPAAm copolymers increased with increasing contents of phosphonium contents in each copolymer.

This increase in relative viscosities is brought about by the expansion of the polymer chains in water due to repulsion of cationic charges introduced into the copolymers. It is usually known that the relative viscosity of aqueous PolyNIPAAm solutions decreases abruptly at around 33°C of its LCST. However, as shown in Figure 2, the degree of the decrease in the relative viscosity of these copolymers having phosphonium groups with increasing temperature is small. That is, the thermosensitivity of these copolymers was considerably reduced by the introduction of phosphonium groups into the copolymers.

This indicates that the association of PolyNIPAAm moieties above the LCST became difficult because of the presence of phosphonium groups.

It was also found that the relative viscosities of the aqueous solutions of AETO-NIPAAm copolymers were considerably low compared with those of other copolymers. This is because of the greater hydrophobicity of the octyl groups as compared to the shorter alkyl moieties of the other copolymers.

Figure 3 shows the temperature dependence of the relative viscosities of the aqueous solutions of AETB-BMA-NIPAAm terpolymers having different composition to investigate the effect of introduction of BMA into AETB-NIPAAm copolymers.

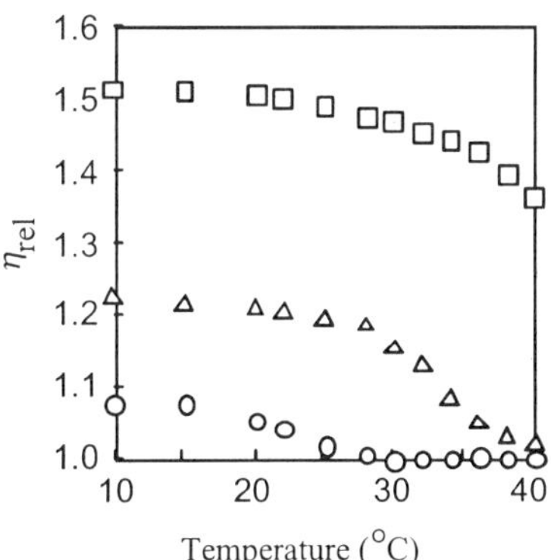

Figure 3. *Temperature dependance of relative viscosities of 0.2g/L AETB-BMA-NIPAAm terpolymer aqueous solutions AETB-BMA-NIPAAm(mol ratio) ;*
○1:4:100, △2.5:2.5:100, □4:1:100

The relative viscosities of the aqueous solutions of AETB-BMA-NIPAAm terpolymers obtained at 100mol ratio of NIPAAm increased with increasing AETB content and decreasing BMA content. The temperature at which the relative viscosity decreased sharply shifted to lower temperature with decreasing AETB content and increasing BMA content. This means that the introduction of BMA into AETB-NIPAAm copolymers increased the hydrophobicity of the copolymers.

Temperature dependence of transmittance at 660nm of copolymer solutions

Figure 4 shows the transmittance at 660 nm of the solutions of 2g/L of three kinds of AETR-NIPAAm (5:100 mol ratio) copolymers measured at various temperatures.   The transmittance at 660 nm of the solutions decreased remarkably at around 32°C.   This decrease in transmittance of light indicates that the copolymers became insoluble in water above around 32°C. Furthermore the degree of decrease in transmittance increased with decreasing chain length of alkyl chains in the phosphonium groups, although we forecast that aqueous solutions of AETR-NIPAAm copolymers having longer alkyl chains in the phosphonium groups would show a greater decrease in transmittance near the LCST. The reason for these phenomena is not clear at present.

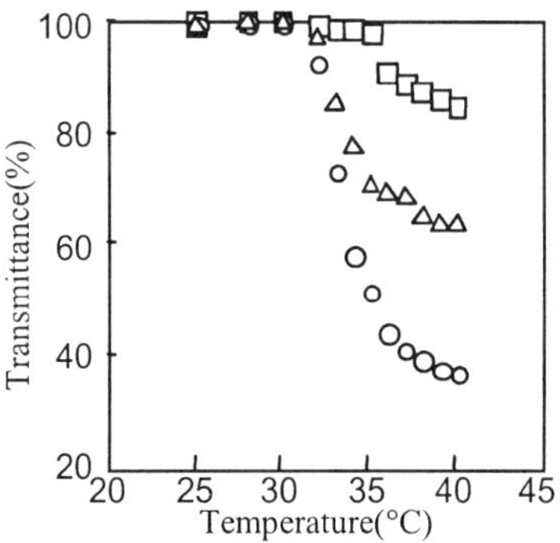

Figure 4. *Changes in transmittance at 660 nm of AETR-NIPAAm (5:100 mol ratio) copolymer solutions at various temperatures Concentration of copolymer aqueous solution = 2.0g/dL AETR : ○;AETE, △;AETB, □;AETO*

Effect of NaCl addition on the transmittance of AETR-BMA-NIPAAm terpolymer solutions

Figure 5 shows the transmittance at 660 nm of solutions of AETR-BMA-NIPAAm (4:1:100 mol ratio) terpolymers in the presence of NaCl.

The transmittance of the solutions of these copolymers decreased sharply by addition of NaCl at the concentration above 0.06wt/vol% and the degree of decrease increased with increasing chain length of alkyl groups in the phosphonium groups.

This means that the dehydration effect of NaCl became higher as the chain length of alkyl groups in the phosphonium groups increased.

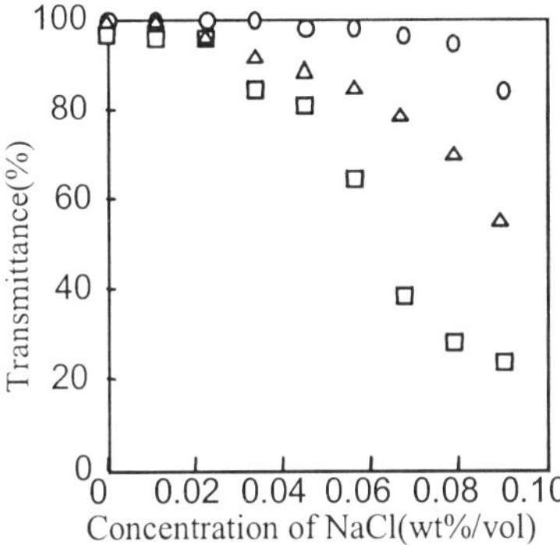

Figure 5. *Changes in transmittance at 660 nm of terpolymer solutions in the presence of NaCl at 35˚C AETR-BMA-NIPAAm (4:1:100 mol ratio), Concentration of terpolymer aqueous solution = 2.0g/dL, AETR : ○;AETE, △;AETB, □;AETO*

Flocculation of kaolin with AETR-NIPAAm copolymers

The clarifying ability of the AETR-NIPAAm copolymers for 500mg/L kaolin suspension was studied by observing transmittance at 660nm of the supernatant after standing for 2 minutes after mixing the copolymer solutions with the kaolin suspension.

Figure 6 shows the relationship between the transmittance of kaolin suspensions and the amount of copolymers added into kaolin suspensions.

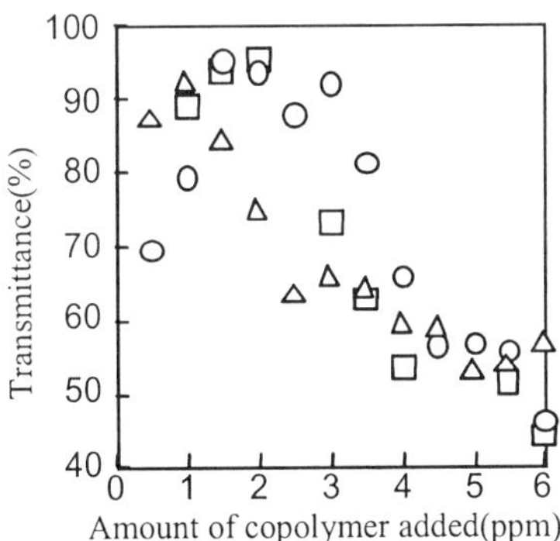

Figure 6. *Transmittance at 660nm of kaolin suspension after addition of copolymers at 22°C AETR-NIPAAm (5:100 mol ratio), AETR : ○;AETE, △;AETB, □;AETO*

The transmittance increased with increasing amounts of copolymers added, then decreased. The concentarion of copolymers added at which the highest transmittance was obtained is defined as the optimum dosage of the copolymer for the clarification of kaolin suspensions. All the AETR-NIPAAm copolymers exhibited almost the same clarifying ability against 500mg/L kaolin suspension,

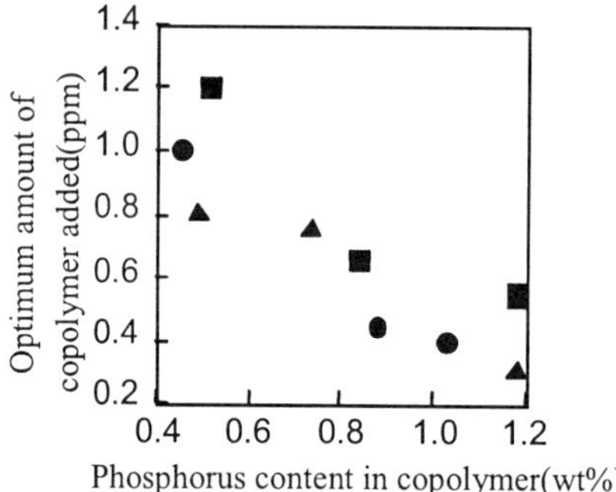

Figure 7. *Optimum amount of phosphorus added for kaolin suspention at 22°C
Copolymer : ● AETE:NIPAAm, ▲ AETB:NIPAAm, ■ AETO:NIPAAm*

although the optimum dosages of each AETR-NIPAAm (5:100 mol ratio) copolymer are a little bit different. Figure 7 shows that the optimum amount of copolymers added for flocculation of kaolin decreases with increasing content of phosphonium groups in the copolymers.

This result means that the flocculation of kaolin particles was mainly brought about by neutralization of anionic charges on the surface of kaolin particles with cationic copolymers, followed by bridging kaolin particles. The remarkable decrease in transmittance above the optimum dosage of copolymers as shown in Figure 6 is due to the dispersion of the kaolin particles because of excess adsorption of cationic polymers on the surface of kaolin particles.

The flocculating ability was also studied by observing the sedimentation rate and sedimentation volume of kaolin after mixing copolymer solutions with a 5wt/vol% kaolin suspension in a test tube. Figure 8 shows the results of AETB-NIPAAm with different phosphonium contents.

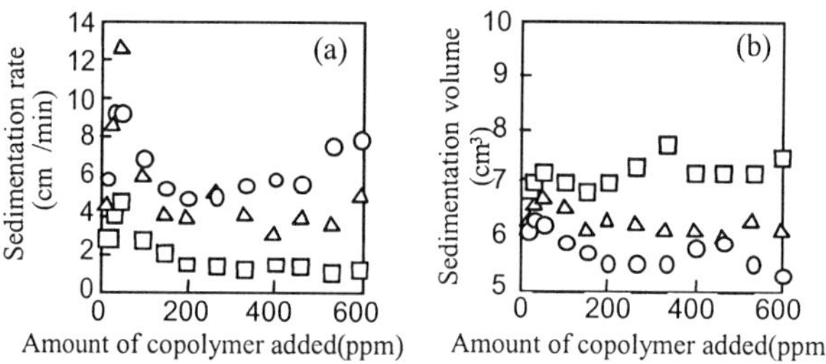

Figure 8. *Relation between sedimentation rate (a) (or sedimentation volume
(b)) and the amount of the AETB-NIPAAm copolymers added at 22°C
AETB:NIPAAm (mol ratio) ; ○ 1:100, △ 3:100, □ 5:100*

The sedimentation rate of kaolin increased with increasing amount of copolymers added up to 500ppm, then decreased. The maximum sedimentation rate of kaolin was obtained at the polymer dosage of 0.1wt% to kaolin.

In the case of AETB-NIPAAm (1:100 mol ratio) and AETB-NIPAAm (3:100 mol ratio) copolymers, the sedimentation rate increased again above the concentration of 200 and 400 ppm, respectively. The sedimentation volume of kaolin was in the order of AETB-NIPAAm (5:100) > AETB-NIPAAm (3:100) > AETB-NIPAAm(1:100) copolymer.

These results indicate that the density of flocculated kaolin particles decreased as content of the phosphonium groups in the copolymer increased.

The flocculating ability of the copolymers against bacteria (*E. coli*) suspensions was also evaluated by measuring the optical density at 660 nm of the supernatant after mixing the copolymers with a bacteria suspension (Figure 9).

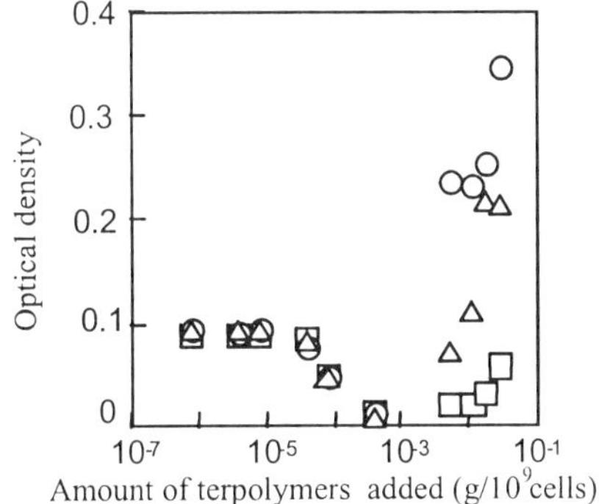

Figure 9. *Changes in optical density at 660 nm after contacting with terpolymers AETO-BMA-NIPAAm (4:1:100 mol ratio),* E.coli *suspension in deionized water :10mL, Temperature:○;22°C, △;30°C, □;35°C*

The optical density decreased with increasing amount of terpolymers added, then increased. This means that the flocculating ability of the terpolymers against the bacteria (*E. coli*) suspension increased with increasing amount of terpolymers, then decreased. This result indicates that the mechanism of flocculation of *E. coli* suspension with the terpolymers is the same as that of kaolin particles. It is known that bacteria such as *E. coli* have the same negative charges on the surface as kaolin particles in water.

All the AETO-BMA-NIPAAm terpolymers exhibited high clarifying ability against *E. coli* suspensions in the temperature range from 22 to 35°C at the optimum dosage, but the concentration range of copolymers for giving high flocculation of bacteria became a little broader as the temperature increased. This means that the flocculating ability of the terpolymers for bacteria was strengthened by increasing hydrophobicity of the copolymers and increasing temperature.

Antibacterial activity of the copolymers and terpolymers against *E. coli*

The antibacterial activity of the copolymers and terpolymers against *E. coli* was investigated. The antibacterial activity was evaluated by counting numbers of viable cells after contacting with copolymers.

Figures 10 and 11 show the results of copolymers containing no phosphonium groups and AETR-BMA-NIPAAm terpolymers, respectively. The BMA-NIPAAm (5:100 mol ratio) copolymers containing no phosphonium groups exhibited no antibacterial activity. On the other hand, the AETO-BMA-NIPAAm (4:1:100 mol ratio) terpolymer exhibited the highest antibacterial activity. The antibacterial activity of AETE- and AETB-BMA-NIPAAm terpolymers was fairly low. The AETO-NIPAAm copolymer exhibited almost the same antibacterial activity as that of the AETO-BMA-NIPAAm terpolymer.

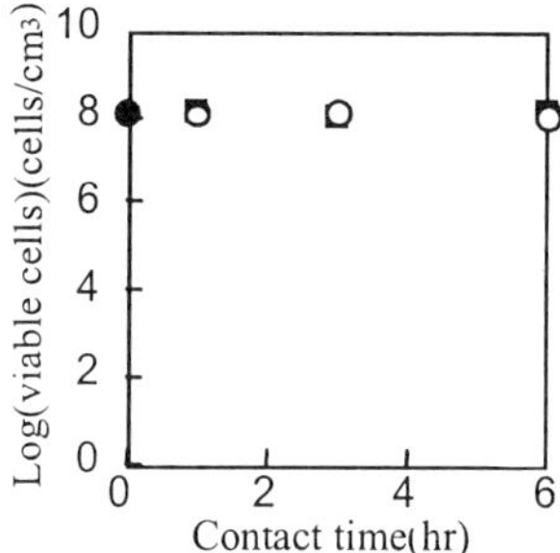

Figure 10. Changes in viable cell numbers after contacting with polymer at 30°C E.coli suspension in deionized water :40mL, Weight of polymers: 0.04g, Δ PNIPAAm; ○ BMA-NIPAAm (5:100 mol ratio); ● Blank

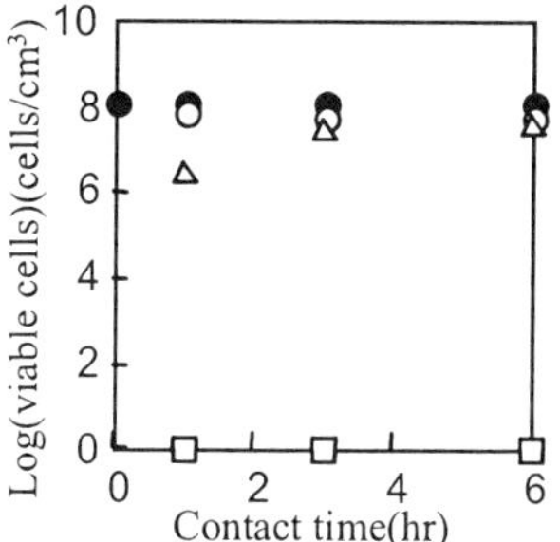

Figure 11. Changes in viable cell numbers after contacting with terpolymers at 30°C E.coli suspension in deionized water: 40mL, Weight of terpolymers: 0.04g, AETR-BMA-NIPAAm (4:1:100 mol ratio), AETR : ○ AETE; Δ AETB; □ AETO; ● Blank

In order to compare the antibacterial activity of the AETR-BMA-NIPAAm terpolymers having different phosphonium groups, the decrease coefficient of the terpolymers against *E. coli* was evaluated using equation (1).

$$D \ (mLg^{-1}h^{-1}) = (V/W \cdot t) \log (N_0/N_t) \tag{1}$$

where V is volume of cell suspension (mL), W is weight of dry copolymers (g), t is contact time (h), N_0 is initial viable cell numbers (cells/mL), N_t is viable cell numbers after contact time t (cells/mL).

The results are shown in Figure 12. The terpolymers containing AETO exhibited the highest antibacterial activity, which increased with increasing phosphonium content in the terpolymers. Thus, the antibacterial activity of the terpolymers was found to be significantly affected by both the chain length of alkyl groups in the phosphonium groups and the content of the phosphonium groups in the terpolymers.

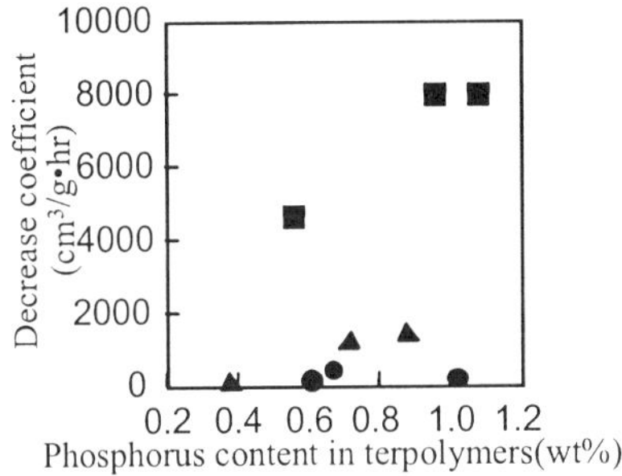

Figure12. *Decrease coefficient of terpolymers against* E.coli *at 30˚C E.coli suspention in deionized water:40mL Weight of terpolymers:0.04g AETR-BMA-NIPAAm, AETR: •;AETE, ▲;AETB, ■;AETO.*

Removal of water-soluble terpolymers from aqueous solutions

Water-soluble AETE-NIPAAm copolymers and AETR-BMA-NIPAAm terpolymers were found to have high antibacterial activity. Removal of the antibacterial polymers from solution after use as a bactericide is necessary to avoid toxicity to other organisms. Therefore the removal of AETR-BMA-NIPAAm terpolymers from its aqueous solutions was studied by changing temperature. The results are shown in Figure 13.

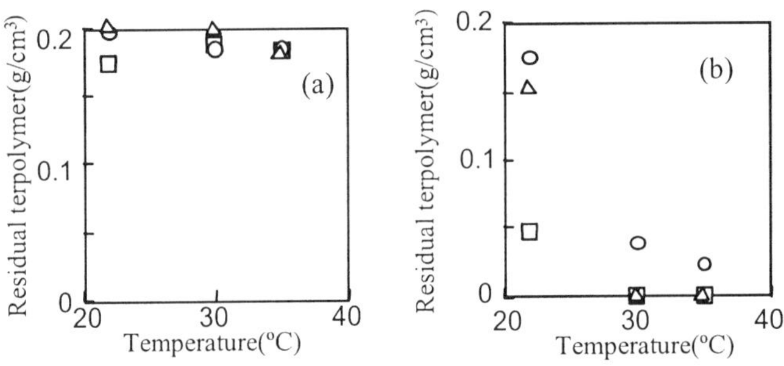

Figure 13. *Effect of temperature on the removal of terpolymers from aqueous solution(a) (or 0.9wt% NaCl solution(b)) Initial concentration of terpolymer:0.2g/L, Rate of shaking:220min^{-1}, Shaking time:2hr, SETR-BMA NIPAAm (4:1:100 mol ratio), AETR: ○;AETE, Δ;AETB, □;AETO.*

The amount of residual copolymers after heating the aqueous solutions at various temperatures, followed by filtration, was determined by UV spectroscopy.

AETR-BMA-NIPAAm terpolymers dissolved in deionized water could not be removed even when the solution was heated at the temperature above 35°C. However, AETB-BMA-NIPAAm and AETO-BMA-NIPAAm terpolymers dissolved in 0.9wt% NaCl solution were removed completely at temperature above 30°C. This result indicates that the presence of NaCl accelerates the association of terpolymers in water because of its dehydration effect.

Conclusions

1) Water-soluble polymers having phosphonium groups were prepared by copolymerization of AETR withNIPPAm with or without BMA.
2) The relative viscosity of the solutions of AETR-NIPAAm copolymers increased with increasing AETR content and decreased with increasing temperature.
3) The thermosensitivity of AETR-NIPAAm copolymers was affected by the kind of AETR hydrophobic moiety introduced into the copolymers.
4) AETR-NIPAAm copolymers and AETR-BMA-NIPAAm terpolymers had high flocculating ability for kaolin and bacteria such as *E. coli* or *S. aureus.*
5) AETO-NIPAAm copolymer and AETO-BMA-NIPAAm terpolymer had high antibacterial activity against bacteria such as *E. coli* or *S. aureus* and the antibacterial activity increased with increasing AETO content.
6) AETR-BMA-NIPAAm terpolymers dissolved in water could be removed by heating the solution above 30°C in the presence of NaCl.

References

1. A. Kanazawa, T. Ikeda, T. Endo, *J. Polym. Sci., Polym. Chem. Ed.*, <u>31</u>, 335 (1993).
2. A. Kanazawa, T. Ikeda, T. Endo, *J. Polym. Sci., Polym. Chem. Ed.*, <u>31</u>, 3003 (1993).
3. T. Nonaka, Y. Uemura, S. Kurihara, *Nippon Kagaku Kaishi*, <u>1994</u>, 1097 (1994).
4. M. Utsunomiya, S. Kurihara, T. Nonaka, *Proceeding of the Network Polymer Symposium*, Japan 1998, 17.
5. Yuki Biryo Bunseki Kenkyu Kondankai Hensyu, *Yuki Biryo Teiryo Bunseki*, Nankodo, Tokyo, 427 (1969).

Chapter 16

Proteins as Amphipathic Biopolymeric Materials

Paul A. Stroud[1], Shawn Goodwin[2], Gregory G. Martin[2], Johanna M. Kahalley[1], Mark Logan[1], Charles L. McCormick[1,2], and Gordon C. Cannon[2]

Department of Polymer Science[1], University of Southern Mississippi, Hattiesburg, Mississippi, 39406
Department of Chemistry and Biochemistry[2], University of Southern Mississippi, Hattiesburg, Mississippi, 39406

Many naturally-occurring macromolecules perform their respective native functions by changing conformation in response to local environmental changes. Of special interest are amphipathic transport proteins designed to capture and sequester hydrophobic molecules in water in a stimuli-responsive manner. Recombinant DNA methods have been utilized in our ongoing research with the eventual goal of exploiting physical and chemical attributes inherent to such native proteins. Reported in this chapter are (1) the design and synthesis of an artificial gene encoding desired microstructural sequences of a *de novo* protein, DN3L, and (2) the production of recombinantly modified proteins based on apolipophorin-III and hydrophobin.

Introduction

Many amphipathic proteins that exist in nature function as stimuli-responsive biopolymeric materials capable of interactions with hydrophobic molecules in aqueous media. Some of these proteins possess properties that make them ideal candidates for commercial applications such as drug delivery, protective encapsulation and wastewater remediation. In addition, proteins are biodegradable, renewable resources and can be produced biosynthetically in microorganisms without the use of organic solvents in a cost-effective manner.

Understanding the mechanism by which these naturally occurring amphipathic polypeptides interact with hydrophobes can lead to targeted modifications of their structure by traditional synthetic means or by using the vast tools of molecular biology in order to enhance their existing properties. Moreover, such information can provide the basis for the design and biosynthesis of *de novo* proteins with specifically desirable functions. With the development of recombinant DNA technology, it is reasonable to assume that almost any sequence modification of a known functional protein is possible and the complete design and *de novo* synthesis of novel protein-based materials is feasible.

The studies conducted by our group first focused on *de novo* synthesis of a completely artificial polypeptide and then progressed to the detailed investigation of the naturally occurring polypeptides apolipophorin-III (Apo-III) (*1, 2*) and hydrophobins (*3-8*). Native functions make these proteins excellent candidates for recombinant modification and production as stimuli-controllable carriers of hydrophobic molecules in aqueous environments.

pH Responsive Microdomain Formation in a Recombinantly Produced *De Novo* polypeptide (*9*)

Biosynthesis of proteins has emerged as a useful method of producing polypeptide-based materials (*10-12*). Proteins have been considered for many years as potentially useful materials for commercially based applications (*13*), but to date most biosynthetic efforts have been limited to synthetic homo- and random polymers from α-amino acids. The advantage of molecular biology and protein biosynthesis is the production of monodiperse proteins and peptides with exact sequences of residues based on a pool of 20 monomers naturally occurring (amino acids).

Realizing the potential of *de novo* biosynthesis, we initially designed an amphipathic polypeptide with pH triggerable functional groups with the eventual aim of producing a biopolymer capable of trapping a hydrophobic guest molecule (*9*). Solid state oligonucleotide synthesis and fragment ligation followed by transformation and then expression in *E. coli* were employed to produce an artificial gene encoding a polypeptide capable of forming pH responsive hydrophobic microdomains. The design of the polypeptide (DN3Lx1) which resulted upon expression of the gene in *E. coli* was based upon an idealized conceptual model where electrostatic, hydrophobic and hydration forces are responsible for the association of amphipathic α-helical elements. The physical properties of DN3Lx1 were examined in solution by size exclusion chromatography, circular dichroism, and fluorescence probe analysis. It was determined that the helical elements of DN3lx1 self-associated to create putative hydrophobic micro-domains in a pH- and ionic strength-dependent fashion. Fluorescence spectroscopy using the probe 1-anilino-8-napthylene (ANS) confirmed these finding by showing that the ANS probe's mobility varied as a function of pH inside the core of the DN31x1 aggregate. Gel permeation chromatography also revealed a decrease in aggregation as the pH of the solution was decreased. While the original molecular objective was to produce a polypeptide with at least 10 repetitive units that could act as centers of hydrophobic aggregation, the expression host used was not able to tolerate nucleotide repeats long enough to produce high molecular weight oligomers of DN3Lx1. This study, although falling somewhat short of its original goals, guided the development and implementation in our laboratory of recombinant biosynthetic method and analytical procedures which are now being applied to the modification and optimization of two naturally occurring proteins with characteristics similar to those of the original design goal.

Apolipophorin-III

The protein apolipophorin III (Apo-III) aids in the transport of phospholipids through the aqueous hemolymph of the insect *Manduca sexta.* The protein is thought to mediate this transfer by associating with the lipids stored in vesicles or cells and forming lipoprotein particles that are stable in aqueous solution (*14, 15*). *In vitro*, Apo-III is known to interact with phospholipid vesicles, reorganizing them into much smaller protein/lipid complexes which appear as discs when viewed by electron microscopy (*1, 2*). Apo-III consists of five amphipathic α-helices that are arranged in a bundle when not associated with lipids or other hydrophobic molecules (*16*). In the non-lipid associated form, hydrophobic elements of the amphipathic α-helices face toward the interior of the α-helical bundle allowing solubility in aqueous media (*16*). Upon contact with a phospholipid vesicle or lipoprotein particle, the α-helical bundle is thought to open, hinging at chains between the α-helices and allowing the inner hydrophobic surfaces of the amphipathic helices to interact and associate with the hydrophobic portion of the lipid molecules (*1, 17, 18*). The initiating interaction between the Apo-III and hydrophobic molecules leading to the conformational change has not yet been elucidated although one model suggests that two conserved leucine residues located at the solvent-exposed end of the helical bundle act as sensors for the surface of the storage vesicle (*19*). This has led to what is known as the "hydrophobic sensor" theory in which these leucines are thought to trigger the conformational change once they have contacted a hydrophobic patch on the surface of the phospholipid vesicle or lipoprotein particle (*20-22*).

Evidence from our laboratory has shown that the conserved leucines present in apolipophorin-III from various organisms (*21*) are not likely involved in the initiation of the conformational change in *Manduca sexta* as predicted by the model described above (*2*). The cDNA encoding Apo-III was cloned into an expression vector that allows gram quantities of the protein to be produced in standard laboratory production facilities (*2*). After determining that the recombinant molecule had properties similar to those reported for the wild type, a mutant was produced in which one of the helix-ending leucines (leucine 30) was replaced with a cysteine residue (*2*). Light scattering measurements of the mutant protein's interaction with phospholipid vesicles revealed that the leucine residues are not needed for the association of protein and lipid to take place (*2*). Weers *et al.* reported supporting evidence that the homologous conserved leucines of Apo-III in the organism *Locusta migratoria* were important for lipoprotein binding but did not inhibit the ability to clear phospholipid vesicles *in vitro* (*23*). These findings may be simply explained by the fact that Apo-III's ability to bind phospholipid vesicles may involve a different process than its ability to bind lipoprotein particles. The three dimensional structure of Apo-III for *Manduca sexta* revealed a short helix at the end of the helical bundle (*16*) and site-directed mutagenesis of a valine in this short helix results in a reduction in lipoprotein binding by Apo-III (*24*). This had led to a proposed model in which this short helix may be the trigger responsible for the conformational change observed when associated with phospholipid vesicles or lipoprotein particles (*24*).

Mutation of the conserved leucine 30, thought to be involved in triggering of the conformational change, to the reactive amino acid cysteine (*2*) was accomplished. Fluorescence probes were then covalently attached to the free sulfhydryl moiety in order to follow conformation changes in that region of the molecule when associated with lipids. Both pyrene maleimide and acrylodan were used to label the cysteine and the fluorescence of each labeled protein was measured during interaction with phospholipid vesicles. In addition to fluorescence spectroscopy, dynamic light scattering and transmission electron microscopy were used to study the labeled proteins' interactions with phospholipid vesicles.

Pyrene maleimide was chosen since the label can form eximers, indicating that two or more pyrene molecules have come into close contact. By measuring the area under the eximer and monomer peaks, the ratio of eximer to monomer can be calculated, revealing information about protein aggregation. When the pyrene maleimide labeled Apo-III (Pyr-Apo-III) was placed in solution without the DMPC vesicles, a large eximer to monomer ratio of 2.7 was observed (Figure 1). This high ratio indicates that the Pyr-Apo-III is highly aggregated in solution. This was confirmed by dynamic light scattering showing that the Pyr-Apo-III aggregates were 109 nm in size (Table 1).

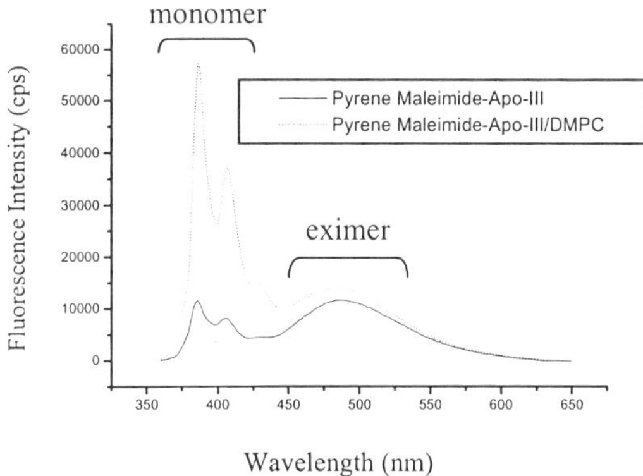

Figure 1. *The steady state fluorescence of Pyr-Apo-III with and without DMPC vesicles.*

When monodisperse unilamellar DMPC vesicles were incubated with Pyr-Apo-III, dramatic size changes are observed. The vesicles were 111 nm in size on average before the protein was added (Table 1). After addition of the protein, the aggregates were broken apart and the vesicles were reorganized into much smaller protein/lipid complexes. This observation is supported by a large drop in the eximer to monomer ratio from 2.7 without lipids to 1.0 with DMPC vesicles (Figure 1). Dynamic light scattering indicate 15 nm sized complexes in solution after the lipids and protein were incubated (Table 1). The pyrene label remains in an aggregated state in both protein aggregates and to a lesser extent in the protein/lipid complexes. This finding may indicate that the label is inside a hydrophobic core of the protein aggregate when lipid vesicles are not present, but when the phospholipid vesicles are added, the label then becomes buried within the acyl chains of the phospholipids, explaining the drop of the eximer to monomer ratio. This leads to the conclusion that the labeled region of the protein is at least involved in binding of hydrophobic portions of the protein/lipid complexes.

Research indicated that incorporation of an environmentally sensitive fluorescent label at the site of the cysteinyl mutation should provide information on the interactions of this specific site with the phospholipid vesicles. The reactive fluorophore, acrylodan, was chosen to label the unique cysteine because the emission spectrum of this molecule changes dramatically as the fluorophore experiences environments of differing polarities. Characterization of the synthetically modified protein, using prodan to construct a Beer's law plot, indicated a 1:1 molar ratio of acrylodan to protein, suggesting a single label addition to each mutant apolipophorin-III.

Table 1. Protein Aggregate, Protein/Lipid complexes and DMPC vesicle Sizes Determined by Dynamic Light Scattering

Modified Apo-III	Unilamellar DMPC vesicles	Protein no DMPC	Protein with DMPC
Pyr-Apo-III	111 nm	109 nm	15 nm
Acry-Apo-III	111 nm	N/D	20 nm

Figure 2 indicates the responsiveness of the prodan probe (acrylodan without the sulfhydryl reactive double bond) to polarity changes as observed by a shift in the wavelength maximum as well as fluorescence intensity in the presence of the DMPC vesicles. This type of behavior was expected for the acrylodan labeled apolipophorin-III (Acry-Apo-III). When the Acry-Apo-III was placed in solution without the lipid vesicles, the wavelength maximum was approximately 470 nm indicating that the acrylodan label is in a very hydrophobic environment. This may be due to the fact that the protein is aggregated in solution in a similar manner to the pyrene labeled apolipophorin-III with the label residing in the hydrophobic core of the aggregate.

When the Acry-Apo-III was incubated with the DMPC vesicles, the fluorescence wavelength maximum did not change, and only a slight increase in intensity was observed (Figure 2). This indicates that the label is again in a hydrophobic environment very similar to that observed for the protein only solution. Dynamic light scattering (Table 1) indicates a much smaller size (20 nm) once the Acry-Apo-III and the DMPC vesicles are incubated. Like the Pyr-Apo-III, the Acry-Apo-III reorganizes the DMPC vesicle into much smaller protein/lipid complexes. The fluorescence data of the Acry-Apo-III support this in both the lipid and non-lipid associated forms. In the case of the protein only solution, the protein is probably aggregated with the labels buried within the hydrophobic core of the aggregate. When the DMPC vesicles are added, the protein reorganizes these lipids and the label becomes buried among the acyl chains of the phospholipids.

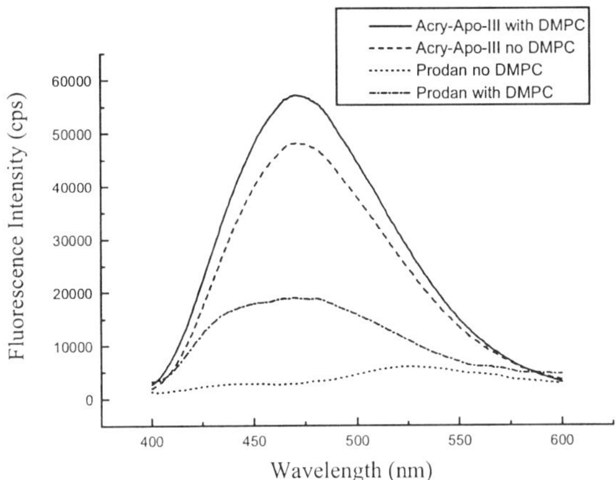

Figure 2. The fluorescence emission spectrum of the acrylodan and pyrene labeled Apolipophorin-III and the probe prodan incubated with and without DMPC.

Steady state fluorescence anisotropy of the acrylodan label was also performed with and without the DMPC vesicles present in solution. These results allowed the calculation of the rotational diffusion coefficient which elucidates how fast the label is rotating during the lifetime of the fluorophore. The values for rotational diffusion coefficient of Acry-Apo-III and Acry-Apo-III incubated with phospholipid vesicles are listed in Table 2. The Acry-Apo-III only solution had a lower rotational diffusion coefficient when compared to the protein and lipid containing sample. This could be interpreted as an indication that the label is in a more tightly packed environment, such as a protein aggregate, which supports the dynamic light scattering data showing aggregates of 109 nm in solution. When phospholipid vesicles are incubated with the Acry-Apo-III, there is a 66% increase in the rotational diffusion coefficient compared to the Acry-Apo-III without phospholipid vesicles. This indicates that the labels are in a much less rigid environment such as that provided by the hydrophobic tails of the phospholipids of the small protein/lipid complexes. Further support of the smaller Acry-Apo-III/DMPC complexes was demonstrated by transmission electron microscopy where the disc-like structures observed were identical to structures produced by the native Apo-III protein (*1*). These experiments again show that this region of the Apo-III is important for binding hydrophobic portions of the reorganized protein/lipid complexes.

Table 2. Rotational Diffusion Coefficients of Acry-Apo-III With and Without DMPC Vesicles

Acry-Apo-III only	*Acry-Apo-III/DMPC*
$1.17 \times 10^7 \text{ sec}^{-1}$	$1.94 \times 10^7 \text{ sec}^{-1}$

Understanding the mechanism of Apo-III's interaction with phospholipid vesicles and lipoprotein particles could prove to very important in the development of Apo-III as a potential candidate for applications such as drug delivery and water remediation. Knowing the triggering mechanism of the Apo-III would lead to possible synthetic or site-directed modification of important regions to incorporate a stimuli-responsive triggerability.

Hydrophobin

In the course of our biopolymer studies, we began investigating a recently discovered group of proteins, the hydrophobins, that are produced by fungi. Hydrophobins were initially discovered in the mid-1980's as the products of genes that were abundantly expressed during the emergence of fruit bodies and aerial hyphae in the wood-rotting fungus *Schizophyllum commune* (*7, 25*). Although the biological role of these proteins is still unclear, they are thought to play important roles in the emergence of fungal reproductive structures and in the attachment of parasitic fungi to plant leaves or insect cuticles of the host organism. The hydrophobins belong to a family of small, moderately hydrophobic polypeptides (about 100 amino acids) with short (about twenty amino acids) N-terminal signal sequences for secretion and eight cysteine residues located at positions that are conserved from protein to protein (*7*).

The best characterized hydrophobin to date, Sc3p, is from the fungi *Schizophyllum commune*. Sc3p is able to form a self-assembled coat on various surfaces such as Teflon and Parafilm and render these surfaces hydrophilic based on water contact angle measurements (*3, 26*). Sc3p can also coat hydrophilic surfaces such as glass or mica increasing the hydrophobic nature (*3, 26*). Treatment of the hydrophobic surfaces with hot SDS, normally destructive to the secondary and tertiary structure of proteins, is ineffective in removing the hydrophobin coating.

Sc3p also has the ability to stabilize both oil and air in aqueous media forming a self-assembled protein "brush" around these vesicles (*26-28*). Our laboratory as well as others have shown the ability of Sc3p to stabilize hydrocarbon material in water (*26, 27, 29*). These stabilized droplets remain intact for many days compared to water and protein controls (*29*). Our group has recently confirmed the location of Sc3p at the water/hydrocarbon interface utilizing epifluorescence microscopy. Fluorescently-tagged secondary antibodies were shown to interact exclusively with Sc3p-bound antibodies on the hydrocarbon vesicles. In a second experiment the hydrocarbon-soluble, nile red fluorescence probe was shown to reside exclusively in the interior of the emulsified droplet. (*29*). Sc3p/hydrocarbon mixtures have also been extruded through polycarbonate filters allowing production of monodisperse vesicles (*29*).

We have isolated a cDNA clone of the Sc3 gene and have successfully expressed the cDNA in a bacterial expression system. This expression system will allow us to directly determine the role the mannose residues covalently attached to the native Sc3p purified protein from fungal culture. Because these mannose residues should not be present in the recombinant Sc3p, the role they play in the structure of the protein and its ability to sequester hydrophobic materials can be elucidated. More importantly, we can now make specific changes in the amino acid sequence of the protein. These changes will allow us to gain and understanding of the self-assembly process and subsequently engineer the protein to provide environmental responsiveness.

Conclusions

The work reviewed here illustrates the possible use of naturally occurring and "designed" proteins for stimuli-responsive sequestration in aqueous media. It is clear that recent advances in genetic engineering and biotechnology open the door to many exciting new approaches to polymeric design and synthesis. However, as these results show, a thorough understanding of the molecular mechanisms utilized by nature is required before the full force of genetic engineering technology can be successfully brought to bear upon biopolymer design.

Acknowledgements

Support for this research by the Department of Defense, Office of Naval Research, and the U. S. Department of Agriculture are gratefully acknowledged.

References

1. Wientzek, M., Kay, C. M., Oikawa, K., and Ryan, R. O. *J Biol Chem* **1994**, *269*, 4605.
2. Kahalley, J., Stroud, P., Cannon, G., and McCormick, C. L. *Biopolymers* **1999**, *50*, 486.

3. Martin, G. G., Cannon, G. C., and McCormick, C. L. *Biopolymers* **1999**, *49*, 621.

4. Martin, G. G., Cannon, G. C., and McCormick, C. L. *Biomacromolecules* **2000**, *1*, 49.

5. de Vocht, M. L., Scholtmeijer, K., van der Vegte, E. W., de Vries, O. M., Sonveaux, N., Wosten, H. A., Ruysschaert, J. M., Hadziloannou, G., Wessels, J. G., and Robillard, G. T. *Biophys J* **1998**, *74*, 2059.

6. Wessels, J. G. *Fungal Genet Biol* **1999**, *27*, 134.

7. Wessels, J. G. *Adv Microb Physiol* **1997**, *38*, 1.

8. Wosten, H. A., de Vries, O. M., van der Mei, H. C., Busscher, H. J., and Wessels, J. G. *J Bacteriol* **1994**, *176*, 7085.

9. Logan, M., Cannon, G., and McCormick, C. L. *Biopolymers* **1997**, *41*, 521.

10. O'Brien, J. P. *Trends Polym. Sci.* **1993**, *1*, 228.

11. Tirrell, J. G., Fournier, M. J., Mason, T. L., and Tirrell, D. P. *Chem. Eng. News* **1994**, *72*, 40.

12. McMillan, R. A., Lee, T. A. T., and Conticello, V. P. *Macromolecules* **1999**, *32*, 3643.

13. Hudson, S. M. (1997) *Protein-based materials*, Birkhauser, Boson.

14. Blacklock, B. J., and Ryan, R. O. *Insect Biochem Mol Biol* **1994**, *24*, 855.

15. Soulages, J. L., and Wells, M. A. *Biochemistry* **1994**, *33*, 2356.

16. Wang, J., Gagne, S. M., Sykes, B. D., and Ryan, R. O. *J Biol Chem* **1997**, *272*, 17912.

17. Ryan, R. O. *Biochem Cell Bio* **1996**, *74*, 155.

18. Wang, J., Sahoo, D., Sykes, B. D., and Ryan, R. O. *Biochem Cell Biol* **1998**, *76*, 276.

19. Breiter, D. R., Kanost, M. R., Benning, M. M., Wesenberg, G., Law, J. H., Wells, M. A., Rayment, I., and Holden, H. M. *Biochemistry* **1991**, *30*, 603.

20. Kawooya, J. K., Meredith, S. C., Wells, M. A., Kezdy, F. J., and Law, J. H. *J Biol Chem* **1986**, *261*, 13588.

21. Smith, A. F., Owen, L. M., Strobel, L. M., Chen, H., Kanost, M. R., Hanneman, E., and Wells, M. A. *J Lipid Res* **1994**, *35*, 1976.

22. Soulages, J. L., Salamon, Z., Wells, M. A., and Tollin, G. *Proc Natl Acad Sci U S A* **1995**, *92*, 5650.

23. Weers, P. M., Narayanaswami, V., Kay, C. M., and Ryan, R. O. *J Biol Chem* **1999**, *274*, 21804.

24. Narayanaswami, V., Wang, J., Schieve, D., Kay, C. M., and Ryan, R. O. *Proc Natl Acad Sci U S A* **1999**, *96*, 4366.

25. Mulder, G. H., and Wessels, J. G. H. *Exp. Mycol* **1986**, *10*, 214.

26. Wosten, H. A., Schuren, F. H., and Wessels, J. G. *Embo J* **1994**, *13*, 5848.

27. van der Vegt, W., van der Mei, H. C., Wosten, H. A., Wessels, J. G., and Busscher, H. J. *Biophys Chem* **1996**, *57*, 253.

28. De Vries, O. M. H., Fekkes, M. P., Wosten, H. A. B., and Wessels, J. G. H. *Arch. Microbiol.* **1993**, *159*, 330.

29. Goodwin, S., Cannon, G. C., McCormick, C. L., XVI International Botanical Congress **1999**, St. Louis.

Chapter 17

Responsive Polymer/Liposome Complexes: Design, Characterization and Application

Françoise M. Winnik* and Tania Principi

Department of Chemistry, McMaster University, 1280 Main St W, Hamilton, Ontario Canada L8S 4M1

The interactions between vesicles and hydrophobically-modified copolymers of N-isopropylacrylamide (NIPAM) and N-glycine acrylamide (Gly) have been examined by fluorescence spectroscopy, dynamic light scattering, and dye-release studies. Four different polymer samples were employed: a copolymer of NIPAM and Gly (PNIPAM-Gly), a copolymer of NIPAM, Gly, and N-(1-pyrenylmethyl-acrylamide) (PNIPAM-Gly-Py), a copolymer of NIPAM, Gly, and N-(n-octadecylacrylamide) (PNIPAM-Gly-C_{18}), and a copolymer of NIPAM, Gly, and N-[4-(1-pyrenyl)butyl]-N-n-octadecylacrylamide (PNIPAM-Gly-C_{18}Py). The polymers form soluble micelles in cold water but their solutions undergo phase separation upon heating above a critical temperature. In the presence of liposomes the polymeric micelles are disrupted and the hydrophobic substituents of the polymer are incorporated within the liposome bilayer to yield polymer/liposome complexes. The binding of PNIPAM-Gly-C_{18}Py with liposomes was assessed as a function of the chemical composition of the liposome membrane, the total lipid concentration, and the incubation time. Three factors control the polymer/liposome interactions: (1) hydrophobic interactions driven by the nonpolar side groups of the polymer; (2) hydrogen-bond formation between the amide residues of the NIPAM units and the ethylene glycol head groups of the lipid; and (3) hydrogen-bond formation between the Gly residues of the polymer and those linked to the liposome bilayer via incorporation in the bilayer of the glycine-terminated lipid, 1',3'-dihexadecyl N-[1-(N-glycyl)succinyl]-L-glutamate ((C_{16})$_2$-Glu-C_2-Gly). The latter interactions are labile and can be disrupted by changes in pH. Thus, efficient release of calcein took place when a pH 7.4 suspension of calcein-loaded liposomes containing (C_{16})$_2$-Glu-C_2-Gly in their bilayer and coated with PNIPAM-Gly-C_{18}Py was acidified to pH 6.0.

Introduction

Liposomes form spontaneously when lipids are dispersed in aqueous media. They consist of one or several bilayer membranes, which capture a pool of water. The practical value of liposomes is derived from two unique properties: 1) their ability to entrap either water-soluble materials in the internal water reservoir or liposoluble compounds in the lipid bilayer; and 2) their compatibility with natural membranes making them safe for medical use and facilitating their penetration into cells. Current applications range from drug vehicles (*1,2*) and diagnostics tools to cosmetics formulations and encapsulating media in the food industry. Liposomes are fragile structures created in water as a result of a delicate balance of interacting forces, which arise when amphiphilic compounds are added to water. In most practical situations vesicles have to be stabilized so that they conserve their integrity in hostile environments. They must also deliver the entrapped materials once they have reached their target.

Among the various approaches to achieve these goals, one option explored in several laboratories seems particularly promising. It consists in adsorbing securely on the outer surface of the liposome a polymer, which responds, by some controllable physical change, to an external stimulus, such as a pulse of light or a sudden change in pH, temperature, or ionic strength of the suspending medium (*3*). Extensively studied systems include complexes between phosphatidyl choline membranes and the pH-sensitive polyelectrolyte poly-(2-ethylacrylic acid) (*4*) or the temperature-responsive poly-(*N*-isopropylacrylamide) (PNIPAM) (*5*). In most studies of liposome/PNIPAM systems, anchoring of the polymer on the bilayer is effected via a small number of long alkyl chains attached to the polymer backbone, an approach pioneered by Ringsdorf and coworkers (*6*), and employed effectively by several research teams (*7,8,9,10*). In systems using these polymers, termed 'hydrophobically-modified–PNIPAMs' (HM-PNIPAM), the alkyl substituents serve as anchors and the N-isopropylacrylamide chain provides the thermosensitivity. In water, HM-PNIPAMs, like PNIPAM itself, undergo reversible phase transition when heated above a lower critical solution temperature (32 °C in the case of PNIPAM (*11*), LCST). Detailed investigations of this phase transition have established that, at the LCST, the PNIPAM chains undergo a collapse from hydrated extended coils to hydrophobic globules, which aggregate and form a separate phase (*12,13*). It is believed that the extended hydrated form of the polymer contributes to the stabilization of the liposome bilayer. This stabilizing effect can, however, be curtailed dramatically by an increase in the temperature above the LCST, as the collapse of the chain triggers a contraction of the external lipid bilayer.

Polyelectrolyte/liposome complexes are sensitive to pH changes. In certain lipid bilayers, polyelectrolytes can promote phase changes, create local defects, or cause aggregation or fusion. Alternatively, as demonstrated in the elegant work of Tirrell and collaborators, the polyelectrolyte induces pH-dependent lateral diffusion of charged lipids incorporated in the bilayer, affecting the permeability of the membrane (*4*). Particularly effective in promoting such effects are derivatives of poly(2-ethylacrylic acid) anchored into the bilayer through phospholipid residues (*4,14*). The design of liposomes sensitive to two different stimuli under controlled conditions is a natural extension of these recent achievements. Copolymers of NIPAM and comonomers of varying hydrophilicity have been prepared (*15*) and a few recent reports describe their interactions with liposomes (*16,17*). We have designed liposome/polymer complexes based on the thermosensitivity of PNIPAM and the pH-sensitivity of the carboxylic acid substituents of glycine residues incorporated along the polymer backbone. The copolymer employed, PNIPAM-Gly-C_{18}Py (*18*), Figure 1, consists of a PNIPAM chain that carries at random approximately 20 mole percent of glycine residues as well as a small number of hydrophobic groups labeled with a fluorescent dye. This unique copolymer combines the phase transition characteristics of PNIPAM with the responsiveness to changes in pH, via protonation/deprotonation of the glycine carboxyl groups. In response to specific changes in temperature, solution pH, or ionic strength, it will change its conformation.

The thrust of this article is a demonstration of the controlled release of substances entrapped in liposome/polymer complexes using copolymers of NIPAM and glycine acrylamide anchored onto phospholipid and non-phospholipid liposomes. To appreciate the intricacies of the system, it is necessary to understand the solution properties of the polymers in the absence of liposomes. These will be reviewed briefly in the first section of this manuscript.

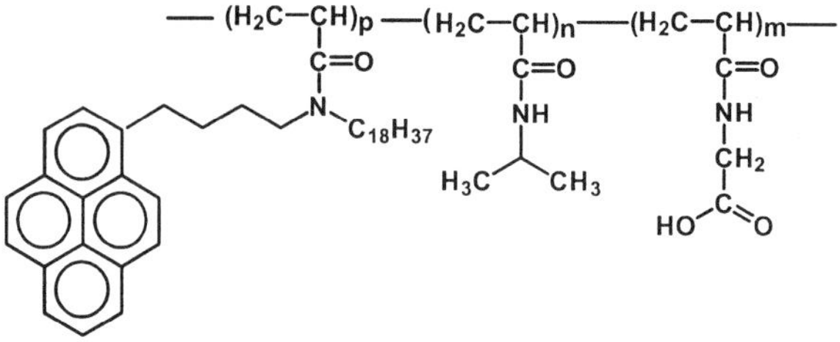

PNIPAM-Gly-Py
p = 1, n = 83, m = 16

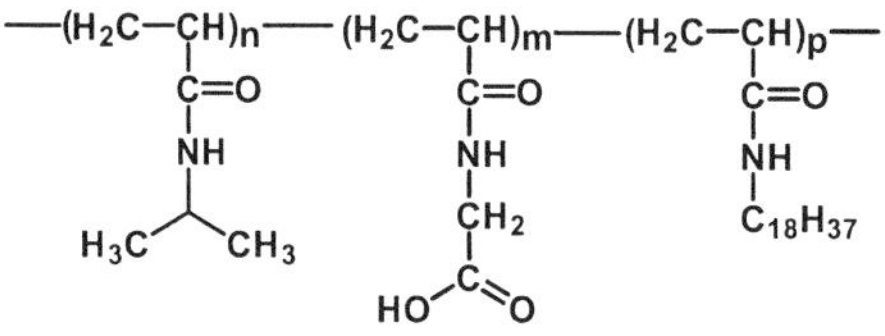

PNIPAM-Gly-C18
n = 80, m = 18, p = 2

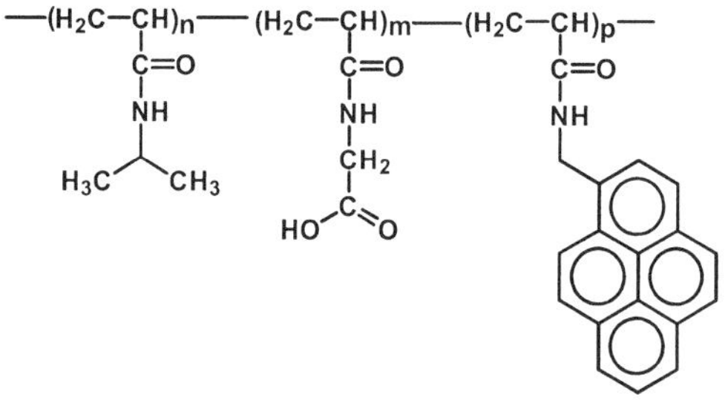

PNIPAM-Gly-Py
n = 80, m = 18, p = 2

Figure 1. *Structure of the polymers used in this study*

The physicochemical properties of polymer/liposome complexes will be discussed in a second part, starting with simple HM-PNIPAM's anchored onto phospholipid liposomes. Techniques employed to monitor the interactions include microcalorimetry, fluorescence spectroscopy and dynamic light scattering. The release properties of complexes between PNIPAM-Gly-C_{18}Py/liposomes will be described. Systems examined include complexes of PNIPAM-Gly-C_{18}Py with cationic phospholipid liposomes as well as with non-phospholipid liposomes made up of n-octadecyldiethylene oxide and cholesterol spiked with either the cationic surfactant dioctadecyldimethylammonium bromide (DDAB), the anionic amphiphile dioctadecylphosphate (DP), or a neutral lipid with a glycine-terminated head group, $(C_{16})_2$-Glu-C_2-Gly (Figure 2). Results are interpreted in terms of the relative importance of electrostatic interactions, hydrogen bond formation, and hydrophobic interactions in guiding the formation, stability, and release properties of the various liposome/polymer complexes.

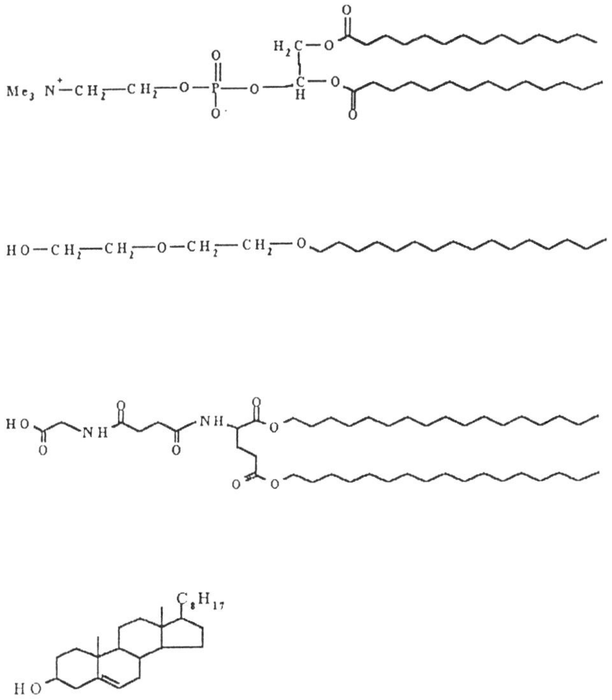

DMPC

Brij 72
(EO$_2$C$_{18}$H$_{37}$)

(C$_{16}$)$_2$-Glu-C$_2$-Gly

Cholesterol

Figure 2. *Structure of lipids used in this study*

Experimental Section

Materials

Dimethyldioctadecylammonium bromide (DDAB), palmitoleyl phosphatidyl choline (POPC), palmitoleyl phosphatidyl serine (POPS) and dimyristoyl phosphatidyl choline (DMPC) were purchased from Avanti Polar Lipids. Cholesterol, n-octadecyldiethylene oxide (Brij 72, EO$_2$-C$_{18}$H$_{37}$), and calcein were obtained from Sigma Chemical Co. 1',3'-dihexadecyl N-[1-(N-glycyl)succinyl]-L-glutamate ((C$_{16}$)$_2$-Glu-C$_2$-Gly (Figure 2)) was prepared following the published procedure (*19*). The copolymer of N-isopropylacrylamide, N-[4(1-pyrenyl)butyl)]-N-n-octadecylacrylamide, and N-glycidylacrylamide (PNIPAM-Gly-C$_{18}$Py), Mn 25,000, Mw/Mn 2.1 (from GPC data calibrated against poly(ethylene oxide) standards) was prepared as previously described (*18*). The copolymers PNIPAM-Gly, PNIPAM-Gly-Py, and PNIPAM-Gly-C_{18} (Figure 1) were obtained as described (*20*). A sample of PNIPAM-C_{18}Py was synthesized following a procedure reported earlier (*21*).

Instrumentation

Dynamic light scattering measurements were performed with a Brookhaven Instrument Corporation Particle Sizer Model BI-90. UV spectra were measured with a Hewlett Packard 8452A photodiode array spectrometer, equipped with a Hewlett Packard 89090A temperature controller. Fluorescence spectra were recorded on a SPEX Industries Fluorolog 212 spectrometer equipped with a GRAMS/32 data analysis system. Temperature control of the samples was achieved using a water-jacketed cell holder connected to a Neslab circulating bath. The temperature of the sample fluid was measured with a thermocouple immersed in a water-filled cuvette placed in one of the four cell holders in the sample compartment. The slits setting ranged from 0.5 to 2.5 mm (emission) and 1.0 to 2.0 mm (excitation) depending on the chromophore concentration. The excitation wavelength was 346 nm, unless otherwise stated. The pyrene excimer to monomer ratio (I_E/I_M) was calculated by taking the ratio of the intensity (peak height) at 480 nm to the half sum of the intensities at 378 and 397 nm. In time-dependent studies, a suspension of liposomes (15 µl, 20 g L^{-1}) was added to a solution of PNIPAM-Gly-C_{18}Py (3 mL, 0.002 g L^{-1}). Spectra were measured immediately prior to liposome addition and every 2 min thereafter.

Cloud Point Determinations

Cloud points were determined by spectrophotometric detection of the changes in turbidity (λ =600 nm) of aqueous polymer solutions (1.0 g L^{-1}) heated at a constant rate (0.5 °C min^{-1}) in a magnetically stirred UV cell. The value reported is the temperature corresponding to a decrease of 20 % of the solution transmittance.

Calorimetric Measurements

Calorimetric measurements were performed on a NANO differential scanning calorimeter N-DSC (Calorimetry Sciences Corp.) at an external pressure of 3.0 atm. The cell volume was 0.368 mL. The heating rate was 1.0 °C min^{-1}, unless other wise specified.

Sample Preparation

Polymer solutions for spectroscopic analysis were prepared from stock solutions (5.0 g L^{-1}). Ionic strength was adjusted by the addition of NaCl. Solutions in water were not degassed. Solutions in methanol were degassed by vigorous purging (1 min) with methanol saturated argon. Liposomes were prepared as follows. A solution in chloroform of the desired amounts of lipids was poured into glass test tubes. The solvent was evaporated under a stream of nitrogen. The resulting lipid film was dried under high vacuum for at least two hours. The dry lipid film was hydrated in an aqueous solution of NaCl to give a lipid suspension of concentration 20 g L^{-1}. For experiments with calcein-loaded liposomes, calcein (70 mmol L^{-1}) was added to the hydration. The lipid suspension was subjected to extrusion (60 °C in the case of non-phospholipid liposomes, 30 °C in the case of phospholipid liposomes) using a Lipofast extruder (Avestin, Canada) fitted with 100 or 200 nm polycarbonate filters obtained from Millipore. Liposome/polymer mixtures were prepared by addition of an extruded liposome suspension to polymer solutions in the desired proportions. The mixtures were allowed to equilibrate at room temperature for at least 30 min.

Calcein Release

Efflux of calcein from the liposomes was observed using fluorescence spectroscopy (22). Extruded vesicles containing calcein were purified on a Sephadex G-75 column to isolate loaded vesicles from free calcein. An aliquot of dispersion of liposomes encapsulating calcein was added to 2.5 mL of glycine buffer solutions of desired pH (50 mmol glycine, 0.1 M NaCl) in a quartz cell kept at 37 °C. The fluorescence intensity of the solution was monitored for 1100s (λ_{exc} = 450 nm, λ_{em} = 515 nm). The percent release of calcein from liposomes is defined as: % release = $(I-I_0)/(I_\infty-I_0)$ x 100, where I_0 and I are the initial (t = 100 s after addition) and final (t = 1100 s) fluorescence intensities, respectively and I_∞ is the fluorescence intensity of the liposome suspension after the addition of sodium dodecyl sulfate (SDS, final concentration 1.6 mmol L^{-1}).

Results and Discussion

Materials and Spectroscopy

Synthesis and Structure of the Copolymers

The pyrene-labeled copolymer PNIPAM-Gly-Py was prepared from a copolymer of NIPAM, N-acrylamidoglycine ethyl ester, and N-

Table I. Physical Properties and Composition of the Polymers

Polymer	Composition (mol %)[b]			Pyrene Content[c] (mol g^{-1})	M_n	M_w (M_w/M_n)
	NIPAM	Gly	C$_{18}$			
PNIPAM-Gly[a]	80	20	---	---	30,000	77,000 (2.5)
PNIPAM-Gly-Py	78	17	---	1.2 x 10^{-4}	23,000	
PNIPAM-Gly-C$_{18}$	78	19	3	---	22,000	
PNIPAM-Gly-PyC$_{18}$[a]	83	19	1	9.4 x 10^{-5}	25,000	54,000 (2.2)

[a] from ref. 18,

[b] from analysis of the[1] H NMR spectra

[c] from UV absorbance spectra

acryloxysuccinimide, by treatment first with 1-pyrenylmethylamine to introduce the pyrene chromophore through an amide link, and, second, with a mild base to hydrolyze the ester groups of the protected glycine residues (*20*). The hydrophobically-modified copolymer, PNIPAM-Gly-C$_{18}$Py, was obtained by copolymerization in dioxane of NIPAM, N- acrylamidoglycine ethyl ester, and N-[4-(1pyrenyl)-butyl]-N-n-octadecylacrylamide followed by base-catalyzed hydrolysis of the ethyl ester group (*18*). The composition and physical properties of the copolymers are listed in Table I, together with those of several HM-PNIPAM's needed to carry out control experiments. The chemical structures of the polymers are depicted in Figure 1.

Spectroscopy of the Polymers in Solution

The emission of pyrene attached to a polymer chain is sensitive to changes in the chromophore separation distance. This feature was used extensively in this study, on the basis of the principles outlined in this section. The emission of locally isolated excited pyrenes ('monomer' emission, intensity I_M) is characterized by a well-resolved spectrum with the [0,0] band at 379 nm. The emission of pyrene excimers (intensity I_E), centered at 480 nm is broad and featureless. Excimer formation requires that an excited pyrene (Py*) and a pyrene in its ground-state come in close proximity within the Py* lifetime. The process is predominant in concentrated pyrene solutions or under circumstances where microdomains of high local pyrene concentration form even though the total pyrene concentration is very low. This effect is shown for example by comparing the spectra of PNIPAM-Gly-C$_{18}$Py in water and in methanol (Figure 3). The emission of pyrene from a methanolic solution of PNIPAM-Gly-C$_{18}$Py

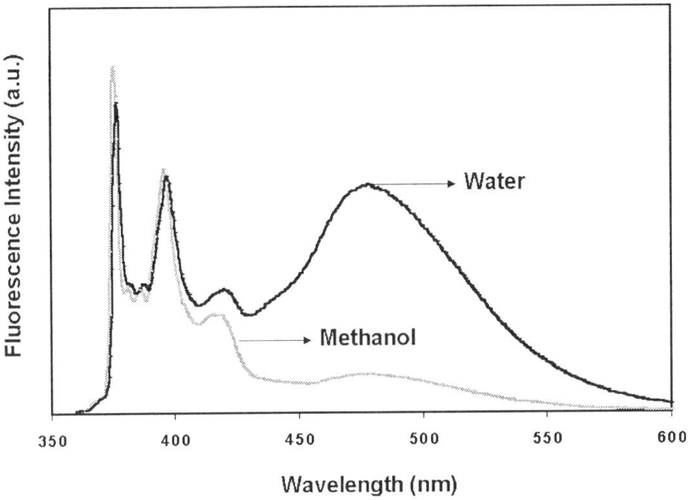

Figure 3. Fluorescence spectra of PNIPAM-Gly-C_{18}Py in water and in methanol; polymer concentration: 0.05 g L^{-1}; $\lambda_{exc} = 341$ nm.

displays mostly pyrene monomer emission, a common occurrence in dilute solutions ([Py] = 3.0 x 10^{-6} mol L^{-1}). In contrast, the emission of pyrene from an aqueous solution of the polymer in the same concentration exhibits a strong excimer emission that vouches for the presence of hydrophobic microdomains in which pyrene groups come in close contact. Disruption of the hydrophobic microdomains, by added surfactants or co-solvents for example, resulting in the local dilution of the chromophores can be detected easily by a significant decrease in excimer emission and a concomitant increase of the intensity of Py monomer emission. The fluorescence spectra of a pyrene-labeled copolymer of NIPAM and glycine acrylamide, PNIPAM-Gly-Py, which does not carry pendent octadecyl chains, are presented in Figure 4 for solutions in water and in methanol. In methanol, the emission originates almost entirely from locally isolated Py* In water, the relative contribution of excimer emission is larger. Although not apparent from Figures 3 and 4, the overall emission intensity is lower for polymers in water, indicating a large extent of pyrene self-quenching. The excimer in this case originates mostly from pre-formed pyrene dimers or higher aggregates and is not formed via the dynamic mechanism postulated by Birks (*23*) and exemplified by the pyrene emission from methanolic solutions of PNIPAM-Gly-Py or PNIPAM-Gly-C_{18}Py. Pyrene dimers have been detected in aqueous solutions of several pyrene-labeled polymers (*24*).

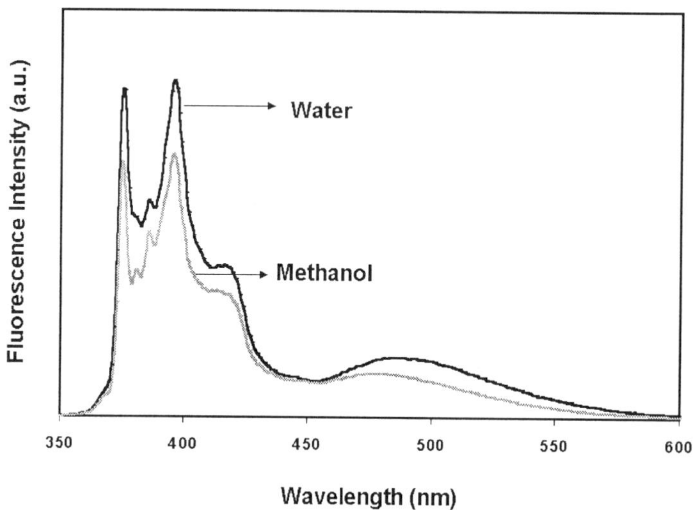

Figure 4. *Fluorescence spectra of PNIPAM-Gly-Py in water and in methanol; polymer concentration: 0.05 g L^{-1}; λ_{exc} = 342 nm, methanol; λ_{exc} = 344 nm, water*

Properties of Aqueous Solutions of Copolymers of NIPAM and N-glycine acrylamide

Even though PNIPAM exhibits many properties characteristic of amphiphilic polymers, it does not form aggregates in water below the cloud point of the solution. Similarly, PNIPAM-Gly and PNIPAM-Gly-Py do not aggregate even in the case of solutions of low pH as long as solutions remain below their cloud point, as concluded from the absence of signal in dynamic light scattering measurements (DLS). In contrast, a strong signal was detected when the DLS measurements were performed on aqueous solutions of the hydrophobically-modified copolymer (20 $^{\circ}$C, 0.1 g L^{-1}, [NaCl] = 0.1 M, pH 2.0 to 9.0). The effective hydrodynamic diameter of the PNIPAM-Gly-C$_{18}$Py aggregates was16 nm $\pm$ 2 nm (20). The pH of the solution did not affect the size of the polymeric micelles. Moreover, their size was not altered by changes in salt concentration ([NaCl] =0.1 to 2.0 M) for a PNIPAM-Py-Gly-C$_{18}$Py solution of neutral pH (7.1). Only in solutions near the conditions of ionic strength, pH, and temperature corresponding to the point of macroscopic phase separation did we observe a significant change in the size of the aggregates. Thus, the dynamic light scattering studies point to the fact that in aqueous solutions of PNIPAM-Gly-C$_{18}$Py the structure of the hydrophobic microdomains is determined by the nature of the hydrophobic group and its level of incorporation along the chain, and that it is hardly affected by changes in the degree of protonation of the glycine residues. The hydrophobic core of such micelles is composed of octadecyl groups and pyrene (Py) labels. It is surrounded by a corona consisting of hydrated poly-(N-isopropylacrylamide) moieties and partially neutralized glycine residues. The collapse of the polymeric corona can be effected either by changes in temperature, at constant pH and ionic strength, or by a pH-jump at constant temperature (20). Return to the original conditions of pH and/or temperature restores polymeric micelles in their original solvated form. This conclusion is strengthened by results gathered from a study of the pH-dependence of the photophysical properties of PNIPAM-Gly-C$_{18}$Py in aqueous solutions. The ratio I$_E$/I$_M$ remained constant, independent of pH ranging from 2.5 to 8.0 (20).

All the NIPAM/Gly copolymers were soluble in water at or below room temperature, independent of the pH of the solutions. However, depending on pH, their aqueous solutions became turbid when heated above a critical temperature, or cloud point. This temperature was determined for polymer solutions of various pH by the simple spectrophotometric method based on the detection of changes in a solution transmittance at a wavelength of light absorbed by neither the solvent nor the solute. The cloud points of aqueous solutions of all NIPAM/Gly copolymers exhibit a marked dependence on the pH of the solution (Table II). They decrease in value with decreasing pH, reflecting the increased hydrophobicity of the copolymers as they assume their fully protonated form. Microcalorimetry offers another view of the phase transition phenomenon. A decrease in pH or progressive protonation of the polymer results in a decrease in the transition temperature and a broadening of the endotherm, as illustrated in Figures 5, which displays the temperature dependence of the partial molar heat capacity for solutions of PNIPAM-Gly-Py and PNIPAM-Gly-C_{18}Py of various pHs. Each polymer solution exhibits an endotherm near the cloud point. Thus, the collapse of the copolymers is triggered primarily by the response of the NIPAM units to changes in solution temperature, and only mildly affected by the presence of hydrophobic substituents.

Table II. Cloud Points and Temperatures of Maximum Heat Capacity (T_{max}) of Aqueous Solutions of N-isopropylacrylamide-N-Glycine Acrylamide Copolymers

Polymer	pH	Cloud point (^oC)[a]	T_{max} (^oC)[b]
PNIPAM-Gly	1.98	28.3 ± 0.3	---
	2.77	30.4 ± 0.4	31.4 ± 0.4
	3.83	32.7 ± 1.7	
	4.25	34.4 ± 1.2	35.0 ± 0.4
	5.10	38.3 ± 2.3	37.4 ± 0.4
PNIPAM-Gly-Py	3.30	29.5 ± 0.2	31.3 ± 0.4
	3.70	38.5 ± 0.4	---
	4.10	43.5 ± 1.5	35.0 ± 0.4
	4.53	45.5 ± 2.4	37.4 ± 0.4
PNIPAM-Gly-C_{18}Py	1.98	26.2 ± 0.3	---
	2.77	28.6 ± 0.4	28.4 ± 0.4
	3.70	36.7 ± 1.3	36.3 ± 0.4
	4.30	49.0 ± 1.7	---
	4.53	55.6 ± 2.3	
	5.10	none	40.2 ± 0.4

[a] from turbidity measurements

[b] temperature of peak maximum of calorimetric endotherm

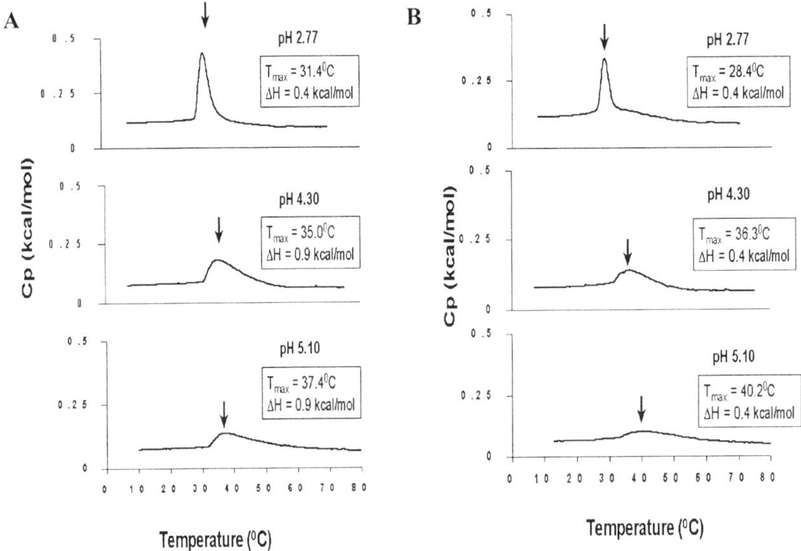

Figure 5. *Microcalorimetric endotherms for aqueous solutions of (A) PNIPAM-Gly and (B) PNIPAM-Gly-$C_{18}Py$ (1gL^{-1}) at pH 2.77, 4.30 and 5.10; citric acid buffers, [NaCl] = 0.1molL^{-1}; heating rate: 1^{0}C min^{-1}*

Polymer/Liposome Interactions

Experimental Design

All liposomes were prepared by dispersion of lipid films in water or buffer, followed by sonication of the dispersion and, finally, extrusion though a polycarbonate membrane. Under these conditions, small unilamellar vesicles (SUV) are formed predominantly. Their diameters, measured by dynamic light scattering, ranged from 160 nm to190 nm. In all cases, the lipid composition was such that the membrane had three components: (a) a neutral lipid, either dimyristoylphosphatidyl choline (DMPC) or *n*-octadecyldiethylene oxide (70 to 80 % w/w); (b) cholesterol, a 'modulator' to strengthen the membrane (20 % w/w), and (c) either an ionogen, such as dimethyldioctadecylammonium bromide

290

(DDAB) or a lipid carrying a glycine terminated head-group (from 0 to 10 %). The minor component was included in studies aimed at probing the effect of electrostatic interactions or hydrogen bonding on the polymer/liposome interactions. All membranes were in the fluid phase at the experimental temperatures selected (> 25 °C).

Another important consideration in the design of the experiments had to do with the relative concentration of polymers and liposomes in the samples under study. In fluorescence experiments using pyrene as a probe, concentrations were chosen such that there was an excess of liposomes, relative to polymer. Under these conditions one can monitor easily the changes in the properties of the polymers, but since only part of the liposomes are coated with polymer, experiments aimed at probing liposome properties, such as changes in size, will yield only average values between coated and naked liposomes. In liposome leakage tests, it is important to choose conditions where there is excess polymer, relative to liposomes, since only polymer-coated liposomes will exhibit controlled release characteristics.

A final consideration concerns the contact time between liposome and polymer that is required to reach an equilibrium situation, where the concentration of free and bound polymer remains constant in time. This equilibration time depends on temperature, the relative amounts of polymer and liposomes, and on the chemical composition of polymer and liposomes. This time is usually on the order of 3-5 hr (*25*).

Spectral Response of the Labeled Polymers during the Adsorption Process

The addition of non-phospholipid cationic liposomes to a solution of PNIPAM-Gly-C$_{18}$Py in water triggered noticeable changes in the emission of the labeled polymer: the pyrene monomer emission intensity increased significantly at the expense of the pyrene excimer emission intensity (Figure 6). This effect is a clear indication that the interchromophoric separation distance in polymer/liposome systems grows beyond the distance probed by pyrene excimer formation. It is taken as evidence that the hydrophobic microdomains formed by the polymer in water are disrupted and that the hydrophobic substituents are incorporated within the liposomes bilayer. When increasing amounts of polymer were added to a liposome suspension of constant lipid concentration, a polymer concentration was reached for which the excimer emission increased again. This signals the point at which the liposome surface becomes polymer-saturated and free polymer micelles coexist in solution with polymer-coated liposomes. Conversely, when increasing amounts of liposomes are added to a polymer solution of constant polymer concentration, a lipid concentration is reached, after which further addition of lipids has no effect on the label emission.

This concentration is best determined in plots of the changes with lipid concentration of the ratio, I_E/I_M, of pyrene excimer to monomer intensity.

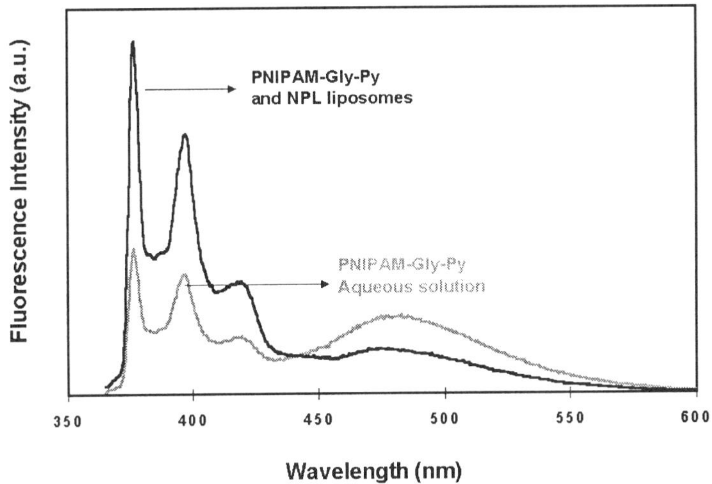

Figure 6. Fluorescence spectra of PNIPAM-Gly-C$_{18}$Py in water and in the presence of non-phospholipid liposomes; polymer concentration: 0.002 g L^{-1}; λ_{exc} = 344 nm.

The changes in the ratio I_E/I_M as a function of lipid concentration in mixtures of PNIPAM-Gly-C$_{18}$Py and various liposomes are presented in Figure 7. Overall, the value of I_E/I_M dropped from 1.1 in the absence of liposomes to 0.1 after saturation was reached. In the case of cationic NPL's a sharp decrease in excimer intensity upon addition of liposomes was observed, up to a polymer/lipid ratio of 1:5 (w/w). Cationic phospholipid liposomes exhibited the same type of interactions with PNIPAM-Gly-C$_{18}$Py. We note however that for similar membrane charge density, provided by 5-mol % DDAB, the ratio I_E/I_M dropped to a value of 0.6 in the case of the phospholipid liposomes, whereas in the case of NPL's the ratio I_E/I_M took a value of 0.1. Monitoring the ratio I_E/I_M for various polymer/liposome samples proved to be an extremely useful tool, not only to detect the occurrence of interactions, but also to unravel the mechanism of the interactions of PNIPAM derivatives and lipid bilayers of various compositions. Results reported previously uncovered major differences in polymer/liposome interactions depending on the chemical composition of the membrane (*26*). The data all point to the occurrence of strong interactions, presumably via hydrogen-bonding, between the PNIPAM backbone units and the glycol head groups of n-octadecyldiethylene oxide, a major component of the membrane of NPL's.

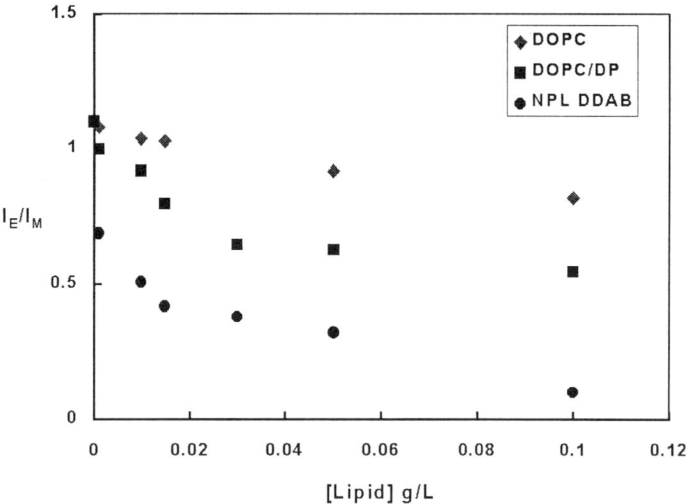

Figure 7. Changes in the ratio of excimer to monomer emission intensities I_E/I_M as a function of lipid concentration in mixtures of PNIPMA-Gly-C_{18}Py with liposomes of various types: ◆ *POPC liposomes;* ■ *POPC/DDAB (9 : 1) liposomes;* ▲ *$EO_2C_{18}H_{37}$/ cholesterol/DDAB (7.5: 2: 0.5) liposomes; polymer concentration, 0.002 g L^{-1} in Tris buffer (5 mM) containing NaCl (0.15 mol L^{-1} (from reference 26).*

Interactions of PNIPAM-Gly-C_{18}Py with Neutral Liposomes Containing Gly-terminated Lipids

If, indeed, the cooperative effect of hydrogen bonding between lipids and polymer is responsible for the strong interactions between NPL's and HM-PNIPAM, it may be possible to design responsive systems via formation of complementary H-bonds. The preliminary experiments described in this section were carried out to assess the feasibility of this design concept. We prepared liposomes containing in their bilayer a lipid carrying a glycine head group (Figure 2), able to undergo specific hydrogen bonds with the glycine residues of PNIPAM-Gly-C_{18}Py, as depicted schematically in Figure 8.

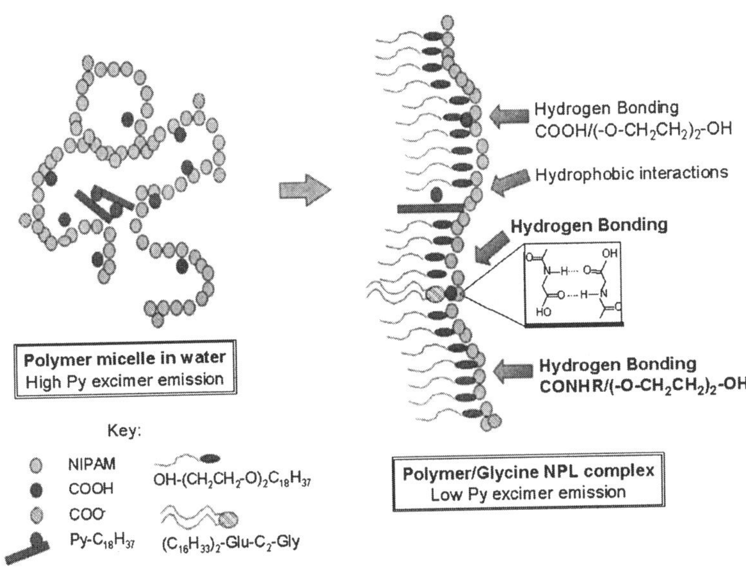

Figure 8. Idealized representation of the interactions between a Glycine containing NPL and PNIPAM-Gly-C_{18}Py (for clarity only half a bilayer is shown and cholesterol has been omitted).

The polymer PNIPAM-Gly-C_{18}Py (2 x 10^{-3} g L^{-1}) was added to liposomes containing increasing amounts of the Gly-terminated lipid, and the changes in the ratio I_E/I_M were monitored as a function of liposome composition. Both phospholipid liposomes and NPL's were employed in these measurements. In a first series of experiments we measured emission spectra of polymer/liposome systems as a function of the time of addition of liposomes to a polymer solution. Spectra were recorded every two minutes over a period of 40 min. In all systems we observed a gradual decrease in excimer emission with concomitant increase of the monomer emission intensity. However the effects vary greatly depending on the composition of the liposome bilayer: the rate of complex formation increased with increasing amounts of Gly-lipid incorporated in the liposome bilayer. The effect is illustrated in Figure 9, which presents the relative changes in the ratio I_E/I_M over 40 minutes for different lipid compositions. The percent change in I_E/I_M was calculated using the following equation: % change in I_E/I_M = 100 x [$(I_E/I_M)_{t=40\ min} - (I_E/I_M)_{t=0}$]/ $(I_E/I_M)_{t=0}$, where $(I_E/I_M)_{t=0}$ and $(I_E/I_M)_{t=40\ min}$ represent the values of the ratio in liposome-free polymer solution and in the mixed system 40 min after addition of liposomes, respectively.

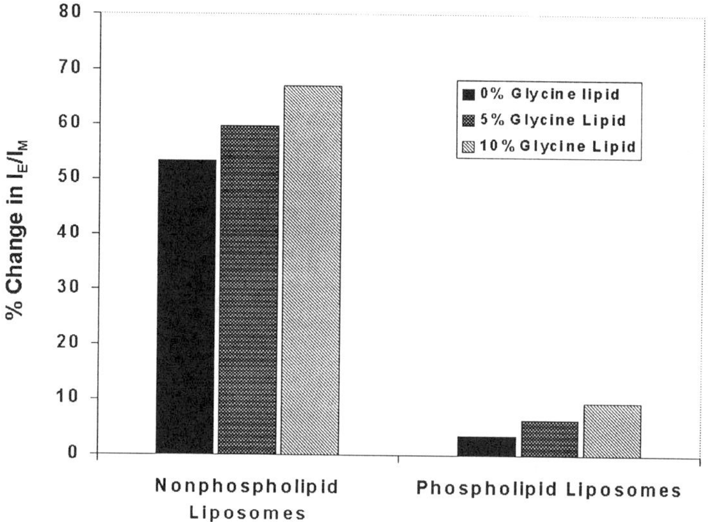

Figure 9. Relative changes in the ratio of pyrene excimer to monomer emission intensity I_E/I_M after a polymer/liposomes contact time of 40 min for various liposome composition; polymer concentration: 0.002 g L^{-1}; lipid concentration: 0.1gL^{-1} ; % change in I_E/I_M = 100 x [$(I_E/I_M)_{t=40\ min} - (I_E/I_M)_{t=0}$]/ $(I_E/I_M)_{t=0}$ (see text).

Next, we performed a fluorescence analysis of polymer/liposomes systems which had been kept at room temperature for 24 h to ensure equilibration. The ratio I_E/I_M decreased sharply upon addition of NPL's containing 0 to 10 % Gly-terminated lipid to a solution of PNIPAM-Gly-C_{18}Py (Figure 10). The value of I_E/I_M reached at saturation (0.15 ± 0.2) is nearly identical to the value reached at identical lipid concentration in mixed system of the polymer and cationic NPL's (0.1). Interactions of the polymer with DMPC-based liposomes containing the Gly-terminated lipid are much weaker than those between the polymer and

NPL's, as judged by the rather modest decrease in the ratio I_E/I_M in the mixed systems. It is important to note, however, that the ratio I_E/I_M (0.6) recorded in the case of mixed systems of polymer and DMPC liposomes containing 10 % Gly-lipid, is identical, within experimental error, to the value obtained previously in systems of PNIPAM-Gly-C_{18}Py and cationic phospholipid liposomes containing 10 % DDAB (Figure 8). A mild dependence of the interactions on the Gly-lipid content is also detected: I_E/I_M at saturation decreases from 0.8 (0 % Gly-lipid) to 0.65 (10 % Gly-lipid) (Figure 10).

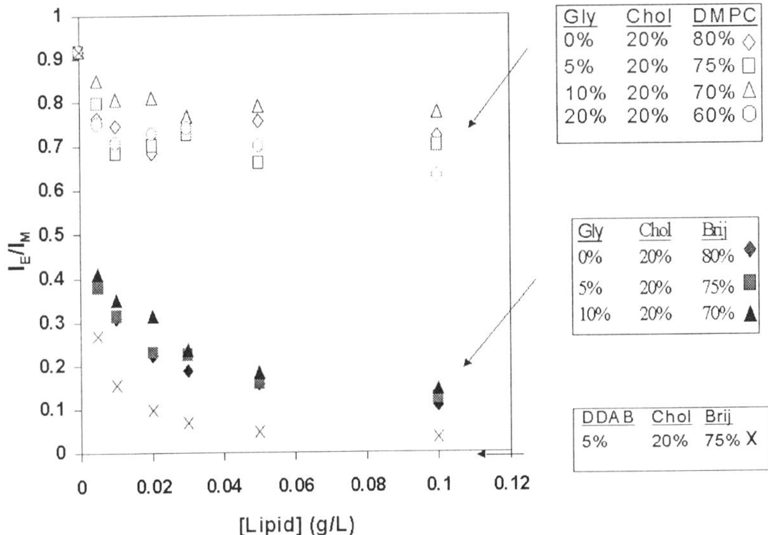

Figure 10. Changes in the ratio of excimer to monomer emission intensities I_E/I_M as a function of lipid concentration in mixtures of PNIPMA-Gly-C_{18}Py with liposomes of various compositions.

Release Properties of Polymer/Liposome Systems

When PNIPAM-Gly is mixed with liposomes at conditions far from the cloud point, the adsorption of copolymer onto liposomes surface does not disrupt the liposome integrity and, consequently, any encapsulated material will remain confined in the liposome interior. As the liposome-polymer complexes approach the conditions of cloud point, leakage of the encapsulated material may occur as a consequence of the partial or complete disruption of the liposome bilayer (*16*). We assessed the release properties of mixed PNIPAM-Gly-C_{18}Py/liposome systems with liposomes containing increasing amounts of Gly-lipid. Leakage was initiated by acidification of a solution of polymer-coated liposomes containing calcein in their aqueous pool (see experimental). The release of calcein became more efficient as the magnitude of the drop in pH increased and as the Gly-lipid content increased (Figure 11). Even a relatively mild acidification (pH 7.4 to 6.0) resulted in leakage of 40% of the vesicle content in less than 5 minutes in the case of polymer coated-NPL's containing 10 % Gly-lipid. Naked liposomes of identical composition did not respond to such a mild change in pH (Figure 11).

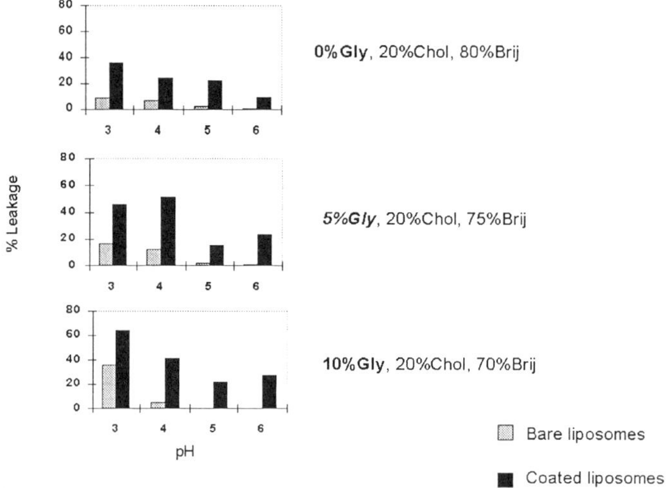

Figure 11. Plots of the percent release of calcein upon change in pH from unmodified liposomes of various compositions (light gray bars) and from liposome coated with PNIPAM-Gly-C$_{18}$Py(dark gray bars) in a buffer of pH 7.4; polymer concentration: 0.04gL^{-1}; lipid concentration: 0.1gL^{-1}; temperature: 37^{0}C.

The driving forces for the leakage are believed to be the pH-induced disruption of the polymer/lipid Gly-Gly hydrogen bonds and the conformational change of the adsorbed PNIPAM-Gly chains as they reached conditions close to the cloud point of the polymer (20). As the polymer contracts, due to the protonation of the glycine residues, it brings about severe distortions of the lipid bilayer with subsequent leakage of the liposome content. Such leakage efficiencies were not achieved in control experiments using phospholipid liposomes coated with PNIPAM-Gly-C$_{18}$Py (leakage efficiency: less then 10 %).

Acknowledgments

This work was supported by a grant from the Natural Sciences and Engineering Research Council of Canada. Ms T. Principi thanks the Natural Sciences and Engineering Research Council of Canada and McMaster University for summer research fellowships. FMW thanks her co-workers over several years who contributed to this research program: Dr. A. Polozova who carried out the key initial experiments, as well as Dr. Gangadhara, Ms. M. Spafford, and Ms E. Goh.

References

1. *Handbook of Non-Medical Applications of Liposomes;* Barenholz, Y. Lasic, D.D., Eds.; CRC Press: Boca Raton, FL, 1996.
2. Lasic, D. D. *Liposomes: From Physics to Applications;* Elsevier: Amsterdam, 1993.
3. Tirrell, D.A.; Takigawa, D. Y.; Seki, K.; Ann. N.Y. Acad. Sci. **1985,** 446, 237.
4. Thomas, J. L.; Tirrell, D. A. Acc. Chem. Res. **1992,** 25, 336.
5. Schild, H. G. *Prog. Polym. Sci.* **1992,** *17,* 163.
6. Ringsdorf, H.; Venzmer, J.; Winnik, F. M. *Angew. Chem. , Int. Ed. Engl.* **1991,** *30,* 315.
7. Kono, K.; Hayashi, H.; Takagishi, *J. Controlled Release* **1994,** *30,* 69.
8. Hayashi, H.; Kono, K.; Takagishi, T. *Biochim. Biophys. Acta* **1996,** *1280,* 127.
9. Kim, J. C.; Bae, S. K.; Kim, J. D. *J. Biochem.* **1997,** *121,* 15.
10. Meyer, O.; Papahadjopoulos, D.; Leroux, J. C. *FEBS Lett.* **1997,** 42, *61.*
11. Heskins, M; Guillet, J. E. *J. Macromol. Sci.* A2 **1968,** 1441-1455.
12. Binkert, Th.; Oberreich, J.; Meeves, M.; Nyffenegger, R.; Ricka, J. *Macromolecules* **1991,** *24,* 5806-5810.
13. Schild, H. G. Prog. Polym. Sci. **1992,** 17, 163-249.
14. Thomas, J. L.; Borden, K. A.; Tirrell, D. A. *Macromolecules* **1996,** *29,* 2570.
15. Feil, H.; Bae, y. H.; Feijen, J.; Kim, S. W. *Macromolecules,* **1993,** *26,* 2496.
16. Hayashi, H.; Kono, K.; Takagoshi, T. *Bioconugate Chem.* **1999,** *10,* 412.
17. Kono, K.; Nakai, R.; Morimoto, K. Takashita, Y.; Biochim. Biophys. Acta **1999,** *1416,* 234.
18. Spafford, M.; Polozova, A.; Winnik, F. M. *Macromolecules* **1998,** *31,* 7099.
19. Berndt, P.; Fields, G. B.; Tirrell, M. *J. Am. Chem. Soc.* **1995,** *117,* 9515.
20. Principi, T.; C. C. E. Goh; C. W. R. Liu, Winnik, F. M. *Macromolecules* (in Press, 2000).
21. Ringsdorf, H.; Venzmer, J. ; Winnik, F. M. *Macromolecules* **1991,** *24,* 1678.
22. Allen, T,. N.; Cleland, L. G.; *Biochim. Biophys. Acta* **1980,** *597,* 418.
23. Birks, J. B. *Photophysics of Aromatic Molecules* Wiley Interscience : London, 1979; chapter 7.
24. Winnik, F. M. *Chem. Rev.* **1993,** *93,* 587.
25. Winnik, F. M.; Adranov, A.; Kitano, H. *Can. J. Chem.* **1995,** *73,* 2030.
26. Polozova, A.; Winnik. F. M. *Langmuir* **1999,** *15,* 4222.

Chapter 18

Unipolymeric Micelles

Properties & Biological Significance

Jian Guo[1,3], Hongbo Liu[1,4], Stephanie Farrell[2], Leila Nikkhouy-Albers[1], Kristi Schmalenberg[1] and Kathryn E. Uhrich[1,*]

[1]Department of Chemistry, Rutgers University, Piscataway, NJ 08855
[2]Department of Chemical Engineering, Rowan University, Glassboro, NJ 08028
[3]Current Address: Merck Laboratories, Rahway, NJ 07065
[4]Current Address: Department of Surgery, Emory University, Atlanta, GA 30322

The most serious problem with formulating drugs in surfactant systems is the paucity of suitable, biodegradable materials available for drug delivery. The polymers discussed herein are directly dispersible in aqueous solution and able to water-solubilize hyrdophobic drugs. Using previously developed synthetic methods, systematic alterations in polymer structure are easily attainable. Control over the chain length, ratio of hydrophobic to hydrophilic components, and the polymer properties enables control of the size and hydrophobic characteristics of the polymer complexes in water. The nature and stability of the polymeric drug carrier are of fundamental importance in understanding their potential drug carrying capacity. Of particular interest is the physical nature of the polymer core; knowledge about the core facilitates drug solubilization and subsequent release.

Introduction

A methodology to synthesize degradable, hyperbranched, amphiphilic polymers with micellar properties was recently reported. (*1*) The polymers consist of mucic acid (a sugar), alkyl chains (fatty acid-like units) and poly(ethylene glycol) (PEG). The structures are highly branched with a hydrophobic core of alkyl chains surrounded by a hydrophilic surface layer of PEG. PEG is a nontoxic, synthetic polymer that is often adsorbed or chemically bound to surfaces of hydrophobic polymeric matrices to prolong device half-life in blood. (*2, 3*) Because the hydrophilic PEG creates a steric barrier to prevent interaction with proteins or cells (*4*), long-term blood circulation occurs due to the avoidance of renal filtration and reduction in uptake by the liver. (*5*)

Our synthetic goal was to create materials with controllable, predictable characteristics such as water-solubility and ability to encapsulate (i.e., water-solubilize) hydrophobic drugs. We hypothesized that water-solubility and encapsulation ability would be determined by the ratio of lipophilic, alkyl chains located in the polymer 'interior' (or core) relative to the hydrophilic, ethylene glycol chains at the polymer 'exterior' (or shell). The hydrophilic/lipophilic ratio may also influence the rate of hydrolysis (i.e., biodegradation), whereas the choice of the polymer components affects the polymer toxicity or biocompatibility.

Polymer Synthesis

The synthetic scheme is below. Mucic acid (MA) (**1**) was acetylated using acyl chlorides (i.e., propanoyl, hexanoyl and lauroyl chloride) to yield various mucic acid derivatives (**2-4**). The mucic acid derivatives (**2-4**) were coupled to 1,1,1-tris(hydroxyphenyl) ethane (**5**) to afford various molecules **6-8** using dicyclohexylcarbodiimide (DCC) and dimethylaminopyridine (DMAP). Tris(hydroxyphenyl)ethane (**5**) was chosen because it is trifunctional, the aromatic groups provide a spectroscopic tag for compound identification, and the aromatic moieties can potentially enhance the encapsulation of aromatic drug molecules. Finally, the core molecules (**6-8**) were condensed with monomethoxy-terminated ethylene oxide chains of various lengths using DCC/DMAP to give the unimolecular micelle structures (**9-13**). After the final coupling reaction, excess PEG is removed by a combination of fractionation and chromatographic separation techniques. All products were purified by chromatography, except the final polymers (**9-13**) were purified by multiple, fractional precipitations into ethyl ether.

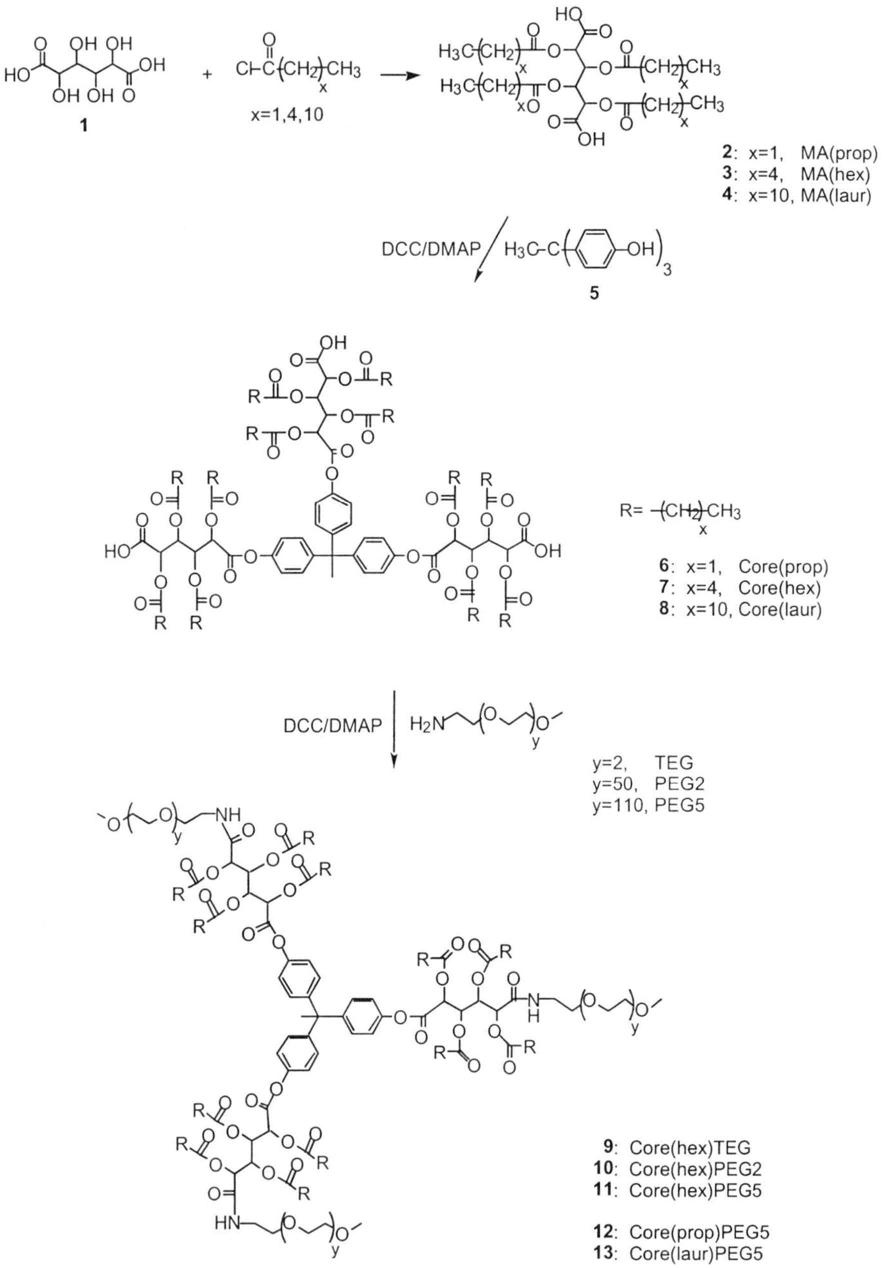

300
x=1,4,10
1
2: x=1, MA(prop)
3: x=4, MA(hex)
4: x=10, MA(laur)
DCC/DMAP
5
R= (CH2)x CH3
6: x=1, Core(prop)
7: x=4, Core(hex)
8: x=10, Core(laur)
DCC/DMAP
y=2, TEG
y=50, PEG2
y=110, PEG5
9: Core(hex)TEG
10: Core(hex)PEG2
11: Core(hex)PEG5
12: Core(prop)PEG5
13: Core(laur)PEG5

For the coupling reactions involving DCC/DMAP, the addition sequence of the reagents was crucial; DCC was added last to the reaction mixture or the reaction would not proceed. Additionally, DCC and DMAP were used in concentrations equivalent to the carboxylic acid functionality of the derivatized mucic acid.

The nomenclature used throughout this chapter is illustrated with the following examples: MA(hex) (**3**) refers to mucic acid acylated by hexanoyl chloride; Core(hex) (**7**) refers to the compound resulting from the coupling of MA(hex) with tris(hydroxyphenyl) ethane (**5**); and Core(hex)PEG5 (**11**) refers to the compound with an 'interior' of Core(hex) and an 'exterior' of PEG with molecular weight of 5000. The ethylene oxide chains were of varying lengths; triethylene glycol (TEG) consists of three ethylene oxide units, whereas PEG2 and PEG5 consists of 50 and 110 ethylene oxide units, respectively.

Hydrophilic/Lipophilic Balance (HLB)

Polymers with longer ethylene oxide chains were completely miscible with most organic solvents and water. Polymers with shorter ethylene oxide chains such as TEG were also very soluble in organic solvents but only sparingly soluble in water. Generally, all the polymeric micelles were soluble in chlorinated solvents, alcohols, aromatic hydrocarbons, and polar aprotic solvents but insoluble in diethyl ether or hexane.

For conventional micelles, hydrophilic/lipophilic balance (HLB) is a good indicator for micelle formation. The HLB values for these polymers were calculated using a modified version of Griffin's relationship for poly(ethylene oxide). (*6*) HLB is the ratio of M_H to $(M_L + M_H)$ where M_L and M_H are the molecular weights of the lipophilic and hydrophilic moieties, respectively, that can be modified in the synthetic scheme. For example, in Core(hex)TEG (**9**) there is a total of nine ethylene oxide units (@ 44 amu) in the entire macromolecule to give M_H = 9 x 44 amu = 396 amu. Similarly, in polymer **9** there is a total of twelve hexanoyl chains in the entire macromolecule each with four methylene units (@ 14 amu) to give M_L = 12 x 4 x 14 amu = 672 amu. For this molecule, the HLB value is 0.37 as taken from [396 / (396 + 672)]. The HLB values for Core(hex)PEG2 (**10**) and Core(hex)PEG5 (**11**) are 0.90 and 0.96, respectively.

As these materials were developed for *in vivo* applications, water solubility is an important parameter. Thus, only the series of materials that are fully water-soluble (HLB of 0.90 or greater) were evaluated further: Core(prop)PEG5, Core(hex)PEG5 and Core(laur)PEG5. Within this series, the differences in HLB values are strictly due to changes in the lipophilicity of the interior (i.e., core) chains and not to the PEG exterior.

Encapsulation

The ability to entrap and water-solubilize hydrophobic molecules is one of the defining characteristics of micellar systems. Encapsulation studies were performed using high pressure liquid chromatography (HPLC) (*1, 7*) to monitor lidocaine, a hydrophobic drug that is used as a local anesthetic. The encapsulation number is defined as the number of molecules that can be entrapped within the polymeric micelles. The values for Core(prop)PEG5, Core(hex)PEG5 and Core(laur)PEG5 were 0.7, 1.0 and 1.6 wt %, respectively. The encapsulation number for lidocaine in polymers **11-13** increased as the lipophilicity of the interior increased.

These encapsulation values are in agreement with entrapment capabilities of conventional polymeric micelles. For example, Barry and El Eini (*8*) investigated the solubility of several steroids in long chain poly(ethylene oxide) surfactants and calculated that two to nine molecules were associated with each micelle, representing a maximum of 3 wt % of the micelle-drug complex. In another example, Hagan et al. (*9*) reported values of 0.3-0.7 wt % of testosterone in micelles comprised of poly(lactide-*co*-ethylene glycol) surfactants.

As described above, there is a proportional relationship between the core size of these polymers and the capacity to encapsulate lidocaine, i.e., structures with larger cores can encapsulate a larger quantity of drug. (*1*) To optimize drug encapsulation efficiency, the relationship between HLB and size of the polymer as well as the drug was evaluated. At this stage, the focus is on drugs with relatively low molecular weights (less than 500 amu) over a range of partition coefficients (log P) of water/octanol. Several common drugs are shown in Table I along with their values of logP, molecular weight (MW) and common use.

Table I. The Partition Coefficients (log P) and Uses of Common Drugs

Drug	*logP*	*Use*
Tamoxifen	7.87±0.75	antitumor agent
Nicardipine	5.22±0.62	calcium blocker for hypertension
Lidocaine	2.35±0.26	local anesthetic; antiarrhythmic
Fluorouracil	-0.78±0.31	antineoplastic

Surface Tension

To determine whether the lidocaine was encapsulated by a single polymer or entrapped by an aggregation of polymers, the concentration dependence of surface tension on polymer concentration was measured. Tween®20 has a similar alkyl:ethylene oxide chain ratio as Core(hex)PEG5 (**11**) and was therefore used as the control. The polymer concentration range was the same used in the previously described encapsulation experiments (10^{-4} to 10^{-6} M). (*1, 7*) The plot of surface tension against log concentration for aqueous solutions of polymers showed no distinct change in slope, which was taken to signify the lack of aggregation. As expected for a thermodynamically stable entity, there is no indication of a critical micelle concentration (cmc) for polymer **11**. In contrast, Tween®20 under the same conditions showed a cmc of 1.8×10^{-4} M as indicated by an abrupt change in slope.

Molecular Size

To verify further the lack of aggregation, dynamic light scattering (DLS) was utilized to measure the size of polymers **11-13** in aqueous media. The diameters of Core(prop)PEG5, Core(hex)PEG5 and Core(laur)PEG5 were 31, 37 and 50 nm, respectively, at concentrations ranging from 10^{-4} to 10^{-6} M. No other populations were observed. Therefore, at the concentrations used in the surface tension studies, intermolecular aggregation was not observed; each polymer molecule behaves as an individual, unimolecular polymeric micelle.

For the water-soluble series (polymers **11-13**), there is a strong correlation between hydrodynamic size and encapsulation number (Figure 1). As predicted, increased lipophilicity of the interior (i.e., lower HLB values) leads to increased entrapment of hydrophobic molecules.

The size range for these polymeric micelles (< 100 nm) may be appropriate for drug targeting. For example, micelle-forming polymeric drugs with diameters below 100 nm were obtained from block copolymers of PEG-poly(aspartic acid). (*10*) In addition, liposomes with diameters of approximately 100 nm were reported to circulate in blood for long periods of time. (*11, 12*)

Interactions with Liposomes

Fusion of biological membranes is fundamental to a number of important physiological and pharmacological processes. Because surfactant molecules affect fusion, we expected that our polymeric micelles would similarly affect fusion processes. Liposomes (also called vesicles) of dipalmitoyl-phosphatidylcholine (DPPC) are well characterized in the literature and frequently utilized as models for cell membranes. (*13*)

Most micelle-forming surfactants are known to increase the permeability of biomembranes, yet micellar solubilization appears to reduce the rate of drug transfer across membranes. (*14*) Because these polymers cannot easily be broken down into surfactant-like moieties, they can be utilized as probes to identify potential polymer/liposome interactions.

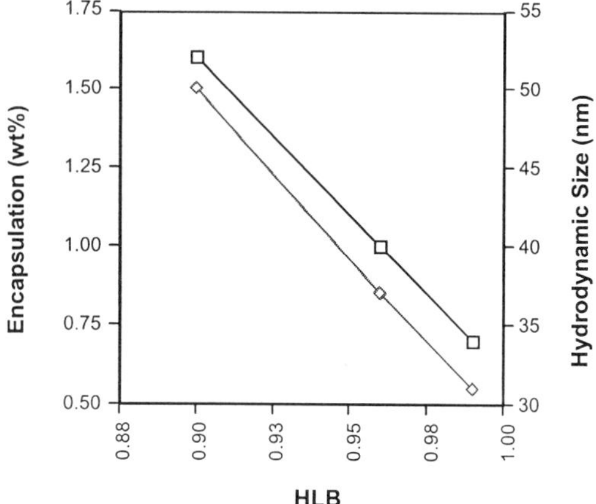

Figure 1. Relationship between encapsulation of lidocaine, hydrodynamic size and HLB values of polymer.

Differential Scanning Calorimetry Measurements

To monitor interactions of the polymer **11** with DPPC liposome structures, microdifferential scanning calorimetry was utilized to monitor thermotropic changes. After several days of stabilization, the DPPC vesicles showed two distinct transitions at 38°C and 42°C corresponding to small unilamellar vesicles (SUV) and large unilamellar vesicles (LUV), respectively. Addition of polymer **11** to the stabilized vesicles caused a rapid and complete shift in the DSC profile corresponding to the complete conversion of SUV's to LUV's. At concentrations used in the previous surface tension and fluorometric studies, polymer **11** promoted fusion processes to form more thermodynamically stable LUV's with a total enthalpy change of 12.8 cal/g. Typically, the fusion event (i.e., conversion of SUV to LUV) is promoted by lowering the solution temperature well below the phase transition to promote defects in the SUV structures. These results indicate that polymer **11** forces defects in the SUV structures that enable fusion to occur at higher temperatures.

Transmission Electron Microscopy (TEM)

Using DSC as well as confocal microscopy, it was determined that addition of polymer to aqueous solutions of liposomes stabilizes the liposome structures. Using freeze-fracture techniques, this stabilization can be visualized using TEM.

The average molecular size of liposomes, polymer **11** and polymer-liposome mixtures in aqueous media was determined by the ratio between the average diameter in microns. All the samples were prepared with a concentration of 2 mg/ml. In liposome solutions, two populations existed with average sizes of 0.10 μm and 0.50 μm. The average size of the polymeric micelles was 0.16 μm. When the polymer was added to liposome solutions, only one population was observed with an average size of 0.08 μm. These results are consistent with our DLS, DSC and confocal microscopy studies that indicate that addition of the polymer stabilizes the liposome structures.

Evaluation of Cytotoxicity

Using L-929 fibroblasts, the *in vitro* toxicity characteristics of Core(hex)PEG5 (**11**) and PEG (as control) were evaluated. The viability of the fibroblasts was monitored by counting live cells at 3 and 7 d using polymer concentrations of 1×10^{-4} M, 10^{-6} M, 10^{-7} M and 10^{-8} M. Cells, which had been incubated in solutions of Core(hex)PEG5 (**11**) at the highest concentration (10^{-4} M) did not survive at any time point. However, solutions of the control (PEG) also had the same influence on cells. The cells are likely coated by either polymer solution (**11** or PEG) at high concentrations, which prevents the ability of cells to attach. All fibroblasts maintained in polymer solutions of lowered concentrations (from 10^{-6} M to 10^{-8} M) survived and proliferated in the observed time period.

Summary

The objective of this research is to evaluate the possibility of using polymeric micelles as drug carriers. A synthetic method in which the polymer structures were systematically altered were developed to potentially achieve optimal degradation, biocompatibility and drug release characteristics. Recent enzymatic degradation studies show that the polymers break down into sub units with in 24 hours. (*15*) The excellent water-solubility of these polymers makes intravenous injection, oral administration and transdermal application of hydrophobic drug molecules possible. For controlled release applications, the small size of these polymers along with their enhanced thermodynamic stability make these desirable systems. Our polymers are fundamentally novel and are the first example of a covalently bound, "unipolymeric micelle" that is highly branched, biodegradable and biocompatible. These characteristics may make these polymer systems excellent generic drug delivery systems.

Experimental Section

Physicochemical Characterization

Chemical composition was ascertained by elemental analysis, IR, mass, ^{1}H NMR and ^{13}C NMR spectroscopies. Elemental analyses were performed by

Galbraith Laboratories (Knoxville, TN). Infrared spectroscopy will be obtained on an ATI Mattson Genesis (M100) FTIR spectrophotometer. Samples were either film cast onto KBr plates or pressed into KBr pellets. ^{1}H and ^{13}C nuclear magnetic resonance spectroscopy was obtained on a Varian 300 MHz spectrophotometer on solutions in $CDCl_3$, D_2O or DMSO-d_6 with solvent used as the internal reference.

Thermal Characterization

The glass transition temperature (T_g), crystalline melting points (T_m), heats of fusion, and heats of crystallization were determined using differential scanning calorimetry (DSC). Decomposition temperatures (T_d) were determined using thermal gravimetric analysis (TGA). Thermal analysis were performed on a Perkin-Elmer system consisting of a TGA7 thermal gravimetric analyzer equipped with PE AD-4 autobalance, and Pyris1 DSC analyzer with TAC 7/7 instrument controllers. Pyris software was used to carry out the data analysis on a Digital Venturis 5100 computer. For DSC, an average sample weight of 5-10 mg was heated at 10°C/min under a flow of N_2 (30 psi). For TGA, an average sample weight of 10 mg was heated at 20°C/min under a flow of N_2 (8 psi).

Molecular Weight Determination

Molecular weights were determined on a Perkin Elmer Series 200 LC system equipped with a PL-Gel column (5 µm, mixed bed) operated at 60°C, Series 200 refractive index detector, Series 200 LC pump and ISS 200 autosampler. A digital Celebris 466 computer was used to automate the analysis via PE Nelson 900 interface and PE Nelson 600 Link box. PE Turbochrom 4 software was used for data collection and processing. THF was the eluent for analysis at a flow rate of 0.5 ml/min. Samples (~ 5 mg/ml) were dissolved into THF and filtered using 0.45 µm PTFE syringe filters prior to column injection. Molecular weights were calibrated relative to narrow molecular weight polystyrene standards (Polysciences, Dorval, Canada).

Solubility

Polymers are placed in solutions at concentrations of 10 g/L, in water, CH_2Cl_2, DMF, DMSO, methanol, acetone, and THF.

Surface Tension

Surface tension measurements of dilute aqueous solutions of polymers were measured using a Fisher Surface Tensiomat (Model 21) at room temperature. Surface tension (dyne/cm) was plotted against the log of polymer concentration.

The required concentrations were obtained by diluting a stock solution of polymer in water such that the polymer concentrations ranged from 10^{-4} to 10^{-6} M. Surface tension measurements for each concentration were performed in triplicate.

Molecular Size

For dynamic light scattering measurements, samples were injected and illuminated by a 30 mW, 780 nm wavelength, solid state laser. The photons scattered by molecules in the flow cell (7 ml) were collected at a scattering angle of 90°. Solutions were filtered through 0.1 μm membrane filters before measurement. Ten measuring runs were performed at room temperature with the cell temperature ranging from 25°C to 27°C. Data was obtained and analyzed with the Biotage DynaPro-801 instrument and AutoPro analysis software. This system analyzes the time scale of the scattered light intensity fluctuations generated from Brownian motion of the polymers by an autocorrelation processor internal to the DynaPro-801. Using the Stokes-Einstein equation, the hydrodynamic radius (R_H) was derived. The molecular weights of molecules are estimated using the relationship between molecular weight and R_H with two assumptions. It is assumed first, that the molecules are spherical in shape, and second, that the polymers have a constant density relative to their size in order to calculate mass from the molecular volume.

Fibroblast Cell Toxicity

In vitro biocompatibility of the polymeric micelles and the breakdown products were determined by monitoring cell growth in solutions containing the polymers. Cells were cultured in media supplemented with polymer (at various concentrations) to ascertain the effects of long-term exposure on cell growth.

Cell proliferation and viability was tested using a fibroblast cell line. Fibroblasts (L-929, mouse areolar/adipose) were grown in Dubelco's modified Eagles minimal essential media supplemented with 10% fetal bovine serum, 2 mM glutamine and 50 units penicillin/streptomycin. The cells were maintained at 37°C at 5% CO_2 in air. Polymers were diluted in the culture media then sterilized by filtration (0.2 mm) to remove bacteria. The cells are plated into tissue culture plates (12-well) at a seeding density of 5×10^5 cells/well. Polymer solutions will be added to the tissue culture plates in concentrations ranging from 1×10^{-2} M to 10^{-8} M and evaluated at various time points. The cells were examined under light microscopy, then assessed for viability using cell counts (Trypan blue).

Liposome Preparation

Liposomes are prepared according to procedures provided by Avanti Polar Lipids (*16*) as briefly described herein. Dipalmitoyl-phosphatidylcholine

(DPPC) (100 mg) was dissolved into chloroform (10 ml) to obtain a clear solution. The solvent was removed by evaporation at room temperature yielding a thin lipid film. The lipid film was thoroughly dried by placing the flask on a vacuum pump overnight. Distilled water (2 ml) at 60°C was added and the flask agitated at 60°C without vacuum for one hour. The lipid suspension was sonicated for 5 minutes with a probe tip sonicator at 60°C with an output power of 75 watts. The suspension turned translucent and slightly bluish in color. The solution was then filtered through a 0.2 μm filter and stored at 60°C.

For preparing liposome/polymer solutions, the polymers were added into 1 ml of the freshly prepared liposome solution at 60°C. The solution was swirled until the polymer was completely dissolved then stored at room temperature.

Transmission Electron Microscopy

Due to the aqueous nature of our systems, samples were prepared using freeze-fracture methods. The polymer-liposome solution was dropped onto roughened copper planchets, rapidly frozen in liquid propane (-192°C for 10 sec) and fractured under high vacuum (5×10^{-8} torr) in a Balzers's BAF freeze etch system. A replica of the polymer-liposome solution was made by evaporating 2.5 μm of platinum from a 45 degree angle followed by 10 μm of carbon from a 90 degree angle. The replica was removed by immersing the planchet into distilled water. The replica was cleaned using 95% ethanol and again rinsed in distilled water.

The replicas were viewed with an JEOL 1230 transmission electron microscope operated at accelerating voltage of 80 kV. The resolution is approximately 4 Å. Images were digitized using Adobe Photoshop 4.0 and analysis accomplished using Image Pro Plus.

References

1. Liu, H.; Jiang, A.; Guo, J., Uhrich, K., *J. Polymer Sci.: Part A: Polym. Chem.* **1999**, *37*, 703-712.
2. Katre, N.; Knauf, M., Laird, W., *Proc. Natl. Acad. Sci. USA* **1987**, *84*, 1487-1491.
3. Abuchowski, A.; McCoy, J.; Palczuk, N.; van Es, T., Davis, F., *J. Biol. Chem.* **1977**, *252*, 3582-3586.
4. Dunn, S.; Brindley, A.; Davis, S.; Davies, M., Illum, L., *Pharm. Res.* **1994**, *11*, 1016-1022.
5. Gref, R.; Minamitake, Y.; Peracchia, M.; Trubetskoy, V.; Torchilin, V., Langer, R., *Science* **1994**, *263*, 1600-1603.
6. Becher, P., Schick, M. J., in *Nonionic Surfactants: Physical Chemistry* M. Schick, F. Fowkes, Eds. (Marcel Dekker, New York, 1987), vol. 23, pp. 435-491.
7. Liu, H. PhD Dissertation, Rutgers University, Piscataway, NJ, 1999.
8. Barry, B. W., El Eini, D. I. D., *J. Pharm. Pharmac.* **1976**, *28*, 210-218.

9. Hagan, S. A.; Coombes, A. G. A.; Garnett, M. C.; Dunn, S. E.; Davies, M. C.; Illum, L., Davis, S. S., *Langmuir* **1996**, *12*, 2153-2161.

10. Yokoyama, M.; Kwon, G.; Naito, M.; Okano, T.; Sakurai, Y.; Seto, T., Kataoka, K., *Bioconj. Chem.* **1992**, *3*, 295-301.

11. Allen, T. M., Chonn, A., *FEBS Lett.* **1987**, *223*, 42-86.

12. Gabizon, A., Papahadjopoulos, D., *Proc. Natl. Acad. Sci. USA* **1988**, *85*, 6949-6953.

13. Attwood, D., Florence, A. T., in *Surfactant Systems. Their Chemistry, Pharmacy and Biology* . (Chapman Hall, London, 1983) pp. 388-468.

14. Florence, A., *Techniques of Solubilization of Drugs*; Marcel Dekker: New York, 1981; , pp Pages.

15. Albers, L., Uhrich, K., Branched Polymeric Micelles: Synthesis and Degradation Studies, Middle Atlantic Regional American Chemical Society meeting, Madison, NJ (1999).

16. "Product Information" (Avanti Polar Lipids, 1997).

Chapter 19

Dilute Polymer Solutions in Extensional Flow Through Porous Media

Chad M. Garner, Robert M. Springfield, Martin E. Cowan and Roger D. Hester

Department of Polymer Science, University of Southern Mississippi, Hattiesburg, MS 39406

Extensional flow encountered by polymers used in reservoir flooding for EOR has been studied on a lab scale using flow through a series of 0.5 inch diameter woven mesh nylon screens. Dilute polymer solutions have been used as mobility control agents in EOR because of the high viscosity resulting from solutions undergoing extension. Polymer chemical composition, solvent quality, and molecular weight were found to affect extensional performance. A yield stress was measured for many of the polymers examined in this study indicating that not all polymers undergo extension at the low flow rates typical of EOR.

Polymer Flooding

Aqueous polymer solutions are often used as displacing fluids employed during reservoir flooding. To be economical, the polymers must increase the flow resistance of the displacing fluid at concentrations as low as parts per million.

During flooding, oil recovery is maximized when the mobility of the displacing fluid is equal to or less than the mobility of the residual fluids existing within the reservoir. The mobility of a polymer solution decreases as the fluid shear and extensional viscosities increase. Shear viscosities characterize the flow resistance created when a fluid experiences a shearing flow field. In shearing flow, the fluid velocities are changing with respect to a direction perpendicular to the flow direction. However, because the polymer concentrations used for displacing fluids are very low, the solution shear viscosity does not differ greatly from that of the the solvent.

In contrast, the extensional viscosity of a dilute polymer solution flowing through a reservoir can significantly increase the flow resistance of the displacing fluid. Extensional viscosity results from the polymer coils resisting deformation in an extensional flow field. In extensional flow, the fluid velocities are changing with respect to the direction parallel to the flow direction. This type of flow field exists during reservoir flooding because the displacing fluid is constantly being accelerated and decelerated as it progresses through the interstitial passages of the porous media. Thus, an understanding of the macromolecular parameters that enhance solution extensional viscosity is important in the development of more effective fluid mobility control polymers.

Dilute Polymer Solutions

High molecular weight, linear polymer molecules in solution interact with many low molecular weight solvent molecules to form random coils. Each coil is roughly spherical and contains not only the coiled polymer molecular mass, but also a much larger solvent mass. The bound solvent within a polymer coil volume has been altered because of the interaction with the macromolecule and is thermodynamically different from unbound solvent. When the coil is extended in a flow field, both the polymer macromolecule and the associated solvent are disordered from their unperturbed state. This coil deformation converts fluid kinetic energy into heat which lowers the solution's mobility. The volume fraction of polymer coils in a dilute solution is equal to the product of the intrinsic viscosity and the mass concentration of polymer in the solution, c. The solution is considered dilute when the volume fraction of polymer coils is less than unity. Therefore, the polymer solution is considered dilute when $c\eta_{intr} < 1$. Under dilute solution conditions, each polymer coil responds to the fluid flow field independently of adjacent coils.

Flow of Dilute Polymer Solutions Through Porous Media

All porous media contain many fluid passages that are composed of converging and diverging channels. As fluids travel through these channels, alternating

312

extension and compression flow fields develop. In an extensional flow field, the fluid velocities are increasing in the direction of flow, while in a compressional flow field, fluid velocity is decreasing in the flow direction. In a converging channel, the fluid surrounding a polymer coil will exert frictional drag forces on the coil that are proportional to the acceleration of the surrounding fluid. The drag forces are larger at the forward end of the coil than at the rear because fluid velocities are larger. When the difference in drag forces across the polymer coil is sufficiently large, the coil will elongate. After coil elongation, coil compression occurs due to a reversal of drag forces as the coil passes into a diverging channel. The polymer molecule and coil-bound solvent are perturbed with each cycle of extension and compression. Repeated change of molecular conformation will produce a degradation of fluid kinetic energy into heat, resulting in a decrease in the solution's mobility through porous media.

Polymer Coil as a Bingham Mechanical System

The simplest way to model the extensional flow behavior of a polymer solution is to first assume that the polymer molecules within a dilute solution act as a collection of independent and identical coils (1). For the authors' purposes, each polymer coil is considered to react to extensional fluid flow forces as a strain lock and dashpot connected in parallel. This arrangement of a dashpot and lock will be referred to as a Bingham mechanical system as shown in Figure 1. The lock prevents polymer coil deformation or strain until the fluid extensional stress is greater than a specific yield stress, σ_{yield}. Note that the lock will reset when the coil strain is returned to zero. As the dashpot is deformed or strained, fluid kinetic energy is transformed into heat providing the coil with an energy conversion property. The dashpot's viscous resistance to strain is characterized by a coil viscosity, η_c. The apparent polymer coil extensional viscosity, η_e, is the ratio of the coil extensional stress, σ_{coil}, to the coil extensional rate, $\dot{\varepsilon}_{coil} = d\varepsilon_{coil} / dt$, which is the change in the coil extensional strain, ε_{coil}, with respect to time, t. Therefore, $\eta_e = \eta_c = \sigma_{coil} / \dot{\varepsilon}$ when $\sigma_{fluid} > \sigma_{yield}$ where η_e is the coil extensional viscosity. No coil extensional strain will develop until the applied extensional stress, σ_{fluid}, equals or exceeds the coil yield extensional stress, σ_{yield}. If the yield stress is zero, coil strain will develop when any extensional stress is applied. Under these conditions, the mechanical system can be considered as only a dashpot. The coil viscosity is expected to be a function of the extensional strain developed within the polymer coil. As a first approximation, it is assumed this function is a polynomial, $\eta_c = \eta_{c0} + \eta_{c1} + \eta_{c2}\varepsilon^2$... The coefficients η_{c0}, η_{c1}, and, η_{c2} will have constant values that depend upon polymer and solvent molecular properties. However at low extensional coil strains, the higher order terms in the above equation are probably not significant. Thus, in the limit of low extensional coil strains, the coil viscosity is expected to be a constant.

The above mechanical model for polymer coil extensional flow in a porous medium suggests that for a polymer to be a good candidate as a mobility control agent in oil recovery the polymer should: 1) have a large coil volume that maximizes extensional fluid stress on the coil; 2) have a coil viscosity that is low so coil strain can be experienced at low extensional stresses; and; 3) have no significant yield stress. These properties will allow polymer coil extension and contraction at the low displacing fluid flow rates experienced during reservoir flooding. Enablement of coil extension and contraction will decrease the displacing fluid's mobility and, thereby, improve polymer flooding efficiency.

Extensional Flow of Polymer Solutions Through Screens

Several investigators have used fiber beds (*2, 3*) and arrays of cylinders (*4*) to study the extensional flow behavior of polymer solutions. Screens can also be used to simulate the extension-compression fluid flow fields found in porous media. Fine mesh screens are available and sets of these screens placed in series will produce a cyclic extension compression flow pattern similar to that found when flooding oil reservoirs. The screens have a thickness of three filament diameters, as shown in Figure 2. The acceleration of the fluid velocity across a screen can be used to estimate the average fluid extensional rate, $\dot{\varepsilon}_{fluid}$. This is the change in fluid velocity over a distance of 1.5 screen wire diameters. The fluid reaches its maximum velocity when passing through the throat formed by adjacent filaments. The average extensional rate of a fluid passing through a plain square mesh screen made using wires of diameter d_{wire} and having a fractional free projected area, f, can be estimated by Equation 1.

$$\dot{\varepsilon} = \left(\frac{16Q}{3\pi \, D_s^2 \, d_{wire}} \right) \left(\frac{1-f}{f} \right) \tag{1}$$

Q is the fluid volumetric flow rate through a circular screen of diameter D_s. The average polymer coil extensional strain can be approximated from the flow conditions and the geometry of a screen by assuming that, at low fluid extensional rates and low coil extensions, slippage between a polymer coil and its surrounding

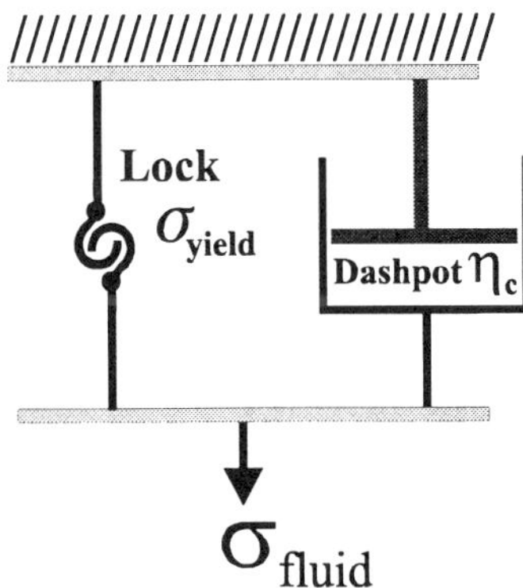

Figure 1. Bingham Model

314

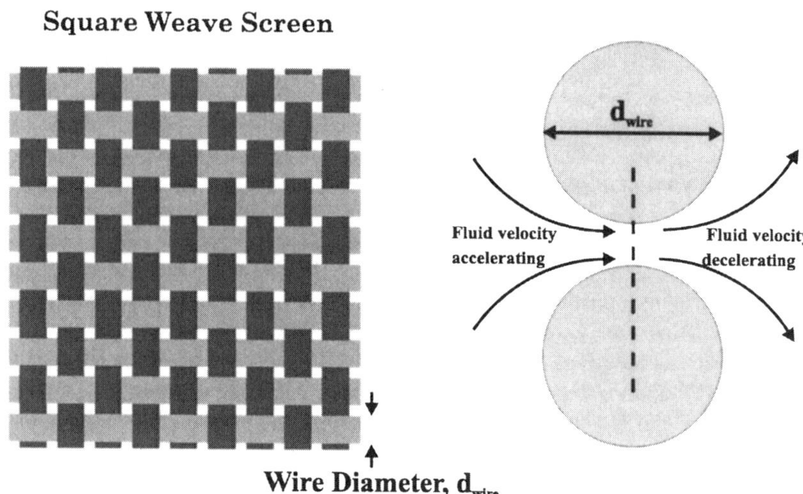

Figure 2.: Square mesh nylon screen

fluid will not develop or will be a minimum after the yield point for coil extension is exceeded. Thus, $\dot{\varepsilon}_{coil} = \dot{\varepsilon}_{fluid}\,k_s$, where k_s is the slippage factor with an assumed value of unity. A lower value for k_s would indicate slippage, i.e., the coil extensional rate would be less than the surrounding fluid extensional rate. In Equation 2, the time of coil extension was approximated as the average time for the fluid to traverse a distance equal to 1.5 screen wire diameters when $Q > Q_{yield}$. Note that when $Q < Q_{yield}$, $\dot{\varepsilon}_{coil}$ equals zero.

$$\dot{\varepsilon}_{coil} \approx \int_0^{t_{ext}} \varepsilon_{fluid}\,k_s\,dt \approx \int_0^{t_{ext}} \frac{16k_s Q}{3\pi D_s^2 d_{wire}}\frac{1-f}{f}dt \approx \frac{2k_s[1-f]}{[1+f]} \qquad (2)$$

Thus, $t_{ext} = 3\pi d_{wire}\,D_s^2\,f / [8\,Q\,(\,1 + f\,)\,]$. It was assumed that the coil extensional strain developed in passing through the first half of the screen passage of distance $1.5\,d_{wire}$ is completely recovered over the second half of the screen passage that is also equal to $1.5\,d_{wire}$. This assumption of complete coil recovery after passage through each screen is probably valid if coil extensions are small. At higher fluid volumetric flow rates, the polymer coils may retain some extensional strain after passage through each screen due to insufficient recovery time. In such a scenario, recovery to an unperturbed coil state may not occur. Coil recovery will become restrictive when the time available for recovery (equal to the time of extension, t_{ext}) is much less than the characteristic response time for the polymer coil, λ_c. The polymer molecular weight, M, solvent viscosity, μ_o, temperature, T, polymer intrinsic viscosity, η_{intr}, and gas law constant, R, can be used to estimate the Zimm response time by, $\lambda_c \approx 25M\eta_{intr}\mu_o/6\pi^2 RT$.

Working Equation to Analyze Extensional Flow Data

The equation, $\Delta P_{solution} - \Delta P_o = (k_s^2 \eta_c / \beta)Q$, has been developed (5) in order to analyze polymer solution flow data. Note that this equation is valid only when $Q > Q_{yield}$. β, a collection of constants, is defined as $\beta = \pi D^2 d_{wire} f^2/64 n \phi c \eta_{intr}(1-f)^2$. Plots with the values of the left side of above ΔP relationship, (the pressure drops across the screens due to the polymer) versus values of the fluid volumetric flow rate, Q, should yield a straight line with a zero intercept and have a slope equal to $k_s^2 \eta_c/\beta$. Note that all the parameters needed in the equation are known or can be measured. Thus, the dashpot or coil viscosity of the Bingham system can be determined from plots of flow data since the slope is equal to $k_s^2 \eta_c/\beta$ or $\eta_c = \beta b/k_s^2$, where b is the slope from the flow data.

Recall that at low coil extensional strains, it is expected that slippage is expected to be a minium and thus $k_s \approx 1$. The flow rate at which the polymer coils start to extend, Q_{yield}, can be determined by comparing plots of ΔP across the screens for both solvent and solution versus volumetric flow rate. The flow rate at which the solution pressure drop deviates upward from the solvent pressure drop is Q_{yield}. The coil yield stress, σ_{yield}, is approximated as the product of the coil extensional viscosity, η_c, and the fluid extensional rate corresponding to Q_{yield}.

Fluid Extensional Rates in Sandstone

We can use the information developed by the screen rheometer to predict polymer solution flow behavior in oil bearing reservoirs. Sandstones have porosities, ϕ, which range from 0.1 to 0.4 and permeabilities, $k_{sandstone}$, which vary from 1.0×10^{-10} to 3.0×10^{-8} cm^2 (10 to 3000 md). For many sandstone reservoirs, ϕ, can be related to $k_{sandstone}$ as $\phi = A_{sand} + B_{sand} \log(k_{sandstone})$. The parameters, A and B, vary with the particular reservoir. Woodbrine sandstone values for A_{sand} and B_{sand} are 0.465 and 0.025, respectively, where $k_{sandstone}$ is expressed in cm^2 dimensions (6).

Although a sandstone reservoir is composed of compressed or consolidated sand particles, this porous medium can be roughly characterized by an average sand sphere diameter, d. This diameter can be estimated from ϕ and $k_{sandstone}$ values of the reservoir using the Kozeny-Carman relationship (7), $d = (1 - \phi/\phi)(180 k_{sandstone}/\phi)^{0.5}$. The reservoir characteristic particle diameter, d, can be used to estimate the extensional rate, $\dot{\varepsilon}$, for a fluid forced through the medium at velocity, v, which is defined as $\dot{\varepsilon} = (2^{0.5}/\phi)(v/d)$ (8). The fluid velocity in the porous medium, v, decreases as the distance from the injection wellhead increases. At large distances from the injection well, a typical displacing fluid velocity is about 1.0 ft per day. Fluid extensional rates can be calculated using the above information. Figure 3, is a 3D plot of calculated fluid extensional rates for typical polymer flooding conditions in Woodbrine sandstone. The fluid extensional rates range from 0.5 to 5.0 sec^{-1}. Fluid extensional rates near the injection wellhead are much higher and vary from 50 to 500 sec^{-1}. Therefore, for polymers to be effective in decreasing the mobility of a displacing fluid, they must elongate at the low extensional rates of 0.5 to 5.0 sec^{-1} experienced during reservoir flooding. In addition, the polymers must not be over extended and break into smaller molecular weights by the large fluid extensional rates developed at the wellhead.

Construction of a Screen Rheometer

A screen extensional rheometer (SER), used to study the extensional flow characteristics of dilute polymer solutions, was constructed in our laboratory. Figure 4 shows a cross sectional drawing of the SER.   The rheometer uses a minimum internal fluid volume to direct fluids across a set of 0.5 inch diameter screens that are placed in the center block. Fluids are forced through the SER using an ISCO model 500D syringe pump. Two pressure transducers rated from zero to ten psi are used to measure fluid pressures before and after the screens.   A transducer calibration technique was developed so that the SER can be used to accurately measure the pressure difference across the screen set for solvent and polymer solutions.

Screen Rheometer Configuration

The 0.5 inch diameter screens used in the rheometer are made from woven nylon filaments having a wire diameter, d_{wire}, of 0.02mm.   Scanning electron micrographs of several nylon screens were used to directly measure the wire diameter and the fraction of free projected area, f, between nylon wires and was determined to be 0.16. The porosity, ϕ, is the ratio of the screen open volume to the spacial volume and was determined to be 0.515 for the screens used in this study. Table I lists the parameters defining the screens used in this study.

Polymer Solution Description

Dilute solution flow properties of several acrylamide (AM) based polymers, homo-polyacrylamide (PAM), an acrylamido terpolymer (ATABAM 5-5), and a copolymer of diacetone acrylamide and AM (DAAM 35), along with samples of poly(ethylene oxide) (PEO) of varying molecular weights were investigated using the SER. Extensional flow curves for each of the acrylamide based copolymers and PEO homopolymers were generated. Each polymer solution was prepared so that the volume fraction of polymer coils, $VF = c\eta_{intr}$, equaled 0.1. Table II shows the chemical structure of the polymers used in this study.

Analysis of Extensional Flow Behavior of Dilute Polymer Solutions

Poly(ethylene oxide) Solutions

PEO samples with molecular weights as determined from Multi Angle Laser Light Scattering (MALLS) of 0.55, 2.1 and 4.1×10^6 g/mol were supplied from Union Carbide with intrinsic viscosities of 5.5, 10.1 and 18.6 dL/g, respectively. All

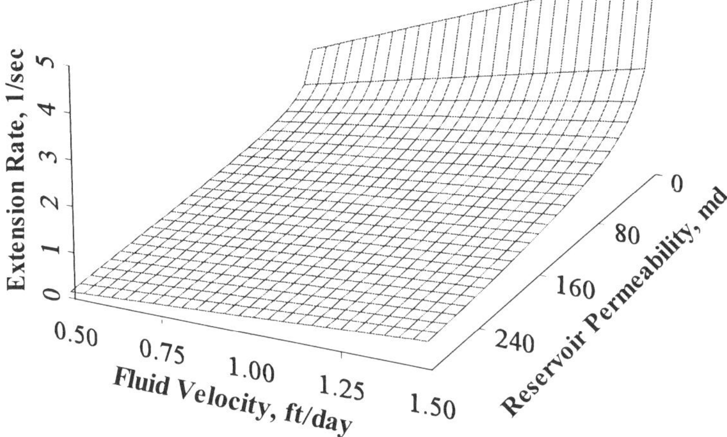

Figure 3. Fluid Extensional Rates

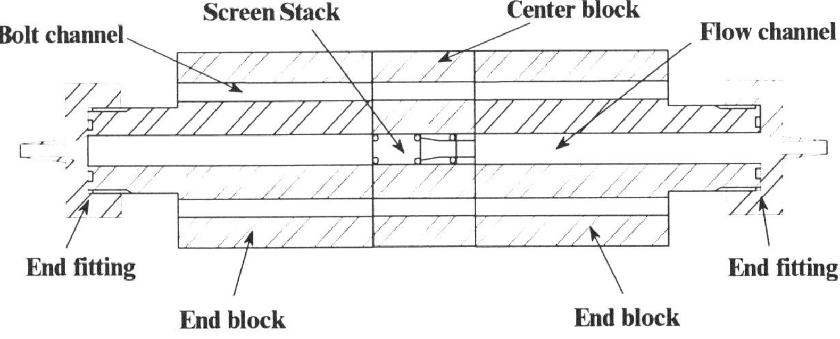

Figure 4. Screen Extensional Rheometer (SER)

Table I. Screen Rheometer Parameter Values

Screen Rheometer Parameter Description	Symbol	Value
screen diameter	D_s	1.27 cm
screen wire diameter	d_{wire}	0.020 mm
fractional free projected area of screen	f	0.16
porosity	ϕ	0.515
number of screen in series	n	30

318

Table II. Polymer Structures

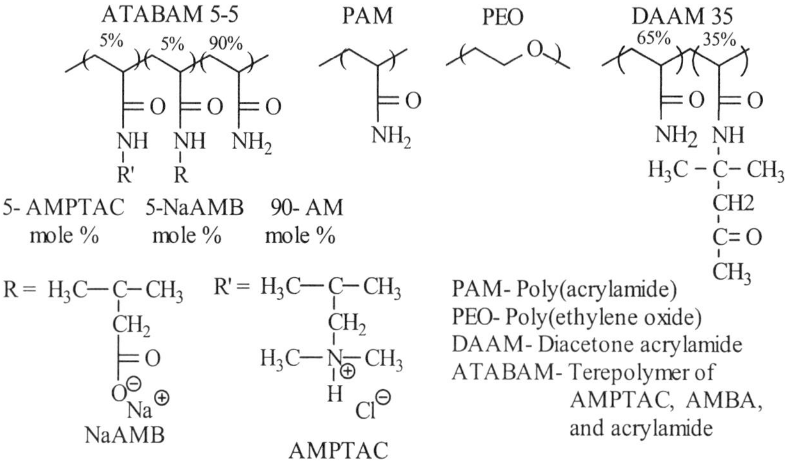

ATABAM 5-5   PAM   PEO   DAAM 35

5-AMPTAC   5-NaAMB   90-AM
mole %   mole %   mole %

$R = H_3C-\underset{CH_2}{\overset{|}{C}}-CH_3$ $R' = H_3C-\underset{CH_2}{\overset{|}{C}}-CH_3$

NaAMB   AMPTAC

PAM- Poly(acrylamide)
PEO- Poly(ethylene oxide)
DAAM- Diacetone acrylamide
ATABAM- Terepolymer of
         AMPTAC, AMBA,
         and acrylamide

Table II. Homopolymer Properties

	0.55M *PEO*	*2.1M* PEO	*4.1M* PEO	*1.89M* *PAM*
Solvent	DI Water	DI Water	DI Water	DI Water
$[\eta]$ (dL/g)	5.5	10.1	18.6	15.2
M_w (g/mol)	0.55M	2.1M	4.1M	1.89M
λ_c (msec)	0.052	0.37	1.32	0.50
Q_{yield} (mL/min)	100	36	12	16
$\dot{\varepsilon}_{yield}$ (1/msec)	4.61	1.66	0.55	0.46
η_c (dyne sec/ cm^2)	0.11	1.74	2.65	1.34
σ_{yield} (dyne/cm^2)	507	2885	1464	617

intrinsic viscosity data in the study were obtained using a Contraves, model 30, rheometer operating at a steady shear rate of 5.96 sec^{-1}. Therefore, the solution concentrations equivalent to a VF of 0.1 are 182, 99 and 54 ppm for the 0.55, 2.1 and 4.1 million molecular weight PEO polymers, respectively.

Solvent and all polymer solutions were studied using the SER with 30 screens at several volumetric flow rates. The pressure drop across the screens was measured at each flow rate. Figure 5 shows the flow data plotted as, $\Delta P_{solution}$ - ΔP_o versus Q. The value of $\Delta P_{solution}$ - ΔP_o is referred to as the normalized solution pressure (NSP) and is the pressure drop across the screens that is due only to the presence of polymer in the solvent.

Note in Figure 5, the slope of the NSP vs. Q curve for the 0.55M Mol. Wt. PEO solution abruptly changes from a value of zero at a flow rate of about 100 mL/min. This is the fluid flow rate, Q_{yield}, at which these polymer coils start to extend. This coil extension converts some of the fluid kinetic energy into heat. This energy transformation results in a solution pressure drop across the screens that is greater than that for the solvent flowing through the screens at the same volumetric rate.

Q_{yield} of the 2.1M Mol. Wt. PEO solution deviates upward at a flow rate of about 36 mL/min. The 4.1M Mol. Wt. PEO solution has a Q_{yield} that equals 12 mL/min. After identifying the Q_{yield} values for each polymer solution, the polymer coil extensional yield stresses, σ_{yield} , associated with each polymer can be determined. See Table II for a summary of homopolymer properties.

At flow rates greater than the yield point, the initial NSP versus flow rate data for each polymer solution have been fit to a straight line. As explained earlier, the slope of each line has been used to calculate the coil extensional viscosity, η_c , of each polymer and the results are listed in Table II.

Acrylamido Polymer Solutions

The flow properties of a homo-polyacrylamide sample with a MW of 1.89×10^6 g/mol, as determined by MALLS, and a low shear [η] of 15.2 dL/g were determined with the SER. Figure 6 shows the flow data plotted as $\Delta P_{solution}$ - ΔP_o versus Q. The PAM coil parameters, listed in Table II, were calculated from the flow data as described for the PEO solutions.

The flow properties of dilute aqueous solutions of an ATABAM5-5 polymer, a NaAMB25 copolymers, and a DAAM35 polymer were also evaluated using the SER. The DAAM35 terpolymer was placed in three solvents, deionized water, NaCl (0.514M), and urea (6.0M). These solvents were chosen to study the effects of environment on polymer extensional flow behavior of the DAAM35 terpolymer solutions.

The flow data for the acrylamide based copolymer solutions are plotted in Figures 7 and 8. The data for all solutions were fit to straight lines. The slopes of

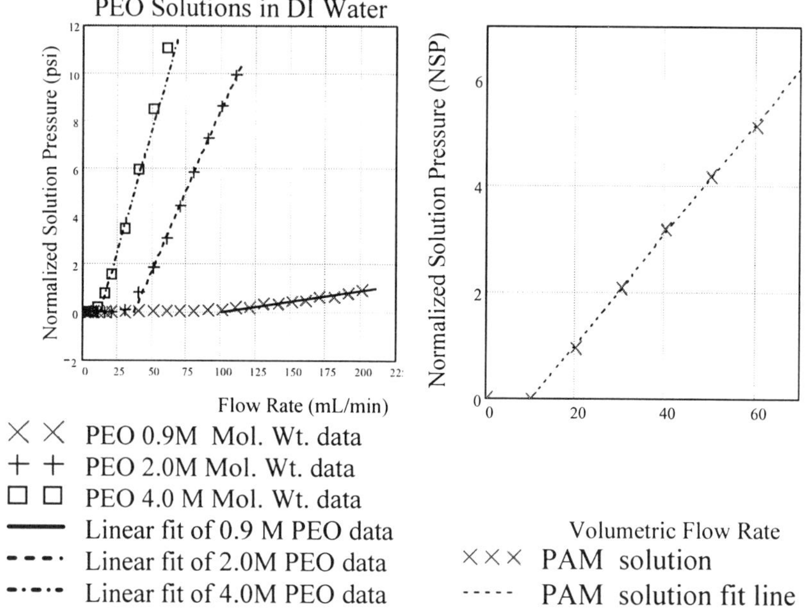

× ×   PEO 0.9M Mol. Wt. data
+ +   PEO 2.0M Mol. Wt. data
□ □   PEO 4.0 M Mol. Wt. data
─── Linear fit of 0.9 M PEO data
─ ─ ─· Linear fit of 2.0M PEO data
─·─·· Linear fit of 4.0M PEO data

Volumetric Flow Rate
××× PAM solution
----- PAM solution fit line

Figure 5. PEO Solutions   **Figure 6.** PAM Solution

the lines were used to calculate the coil viscosity of each polymer and the results are listed in Table III.

Unlike the other polymer solutions examined in this study, the three DAAM35 solutions, regardless of solvent, had no measurable yield flow rate. Yield flow rates for these solutions must be less than 0.5 mL/min, presently the lower limit of the SER measurement resolution.

Recall that for a polymer solution to be effective in decreasing fluid mobility at large distances from the injection well, it must be able to extend at fluid extensional rates between 0.5 to 5 sec^{-1}. As shown by Figure 8, DAAM35 solutions have low coil extensional yield rates and would be more efficient in flooding than the other polymers examined in this study. DAAM 35 polymers appear to expand regardless of the displacing fluid flow rates. In contrast, the PAM and PEO polymers extend only at fluid flow rates that are much higher than those typically found during reservoir flooding. Intra-molecular H-bonding between DAAM monomer units with adjacent AM units may be an explanation for the low coil extensional yield rate. The intra-molecular H-bonding is thought to increase polymer contour length resulting in a larger hydrodynamic volume than expected (10). In these systems, the SER has provided some evidence of the effect of chemical composition on decreasing fluid mobility. Polymers that posses intra-molecular H-bonding ability appear to be better suited for polymer flooding as compared to relatively linear polymers.

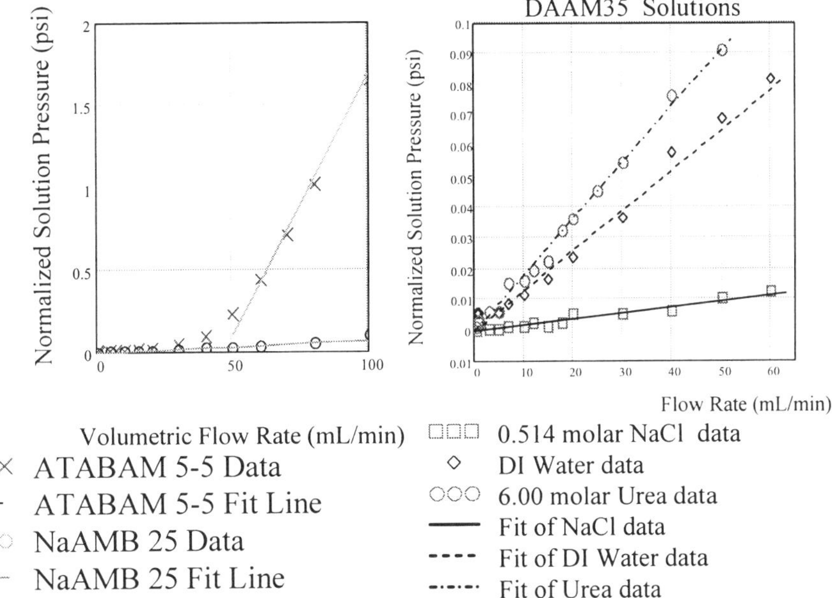

Figure 7. ATABAM and NaAMB   **Figure 8.** DAAM Environment Data

Discussion of Results

A more effective comparison of the flow resistance between polymer solutions can be made by defining the normalized solution flow resistance, NSFR (*8,9*). The NSFR is the increase in solution flow resistance, $\Delta P_{solution} - \Delta P_{solvent}$, as compared to solvent flow resistance, $\Delta P_{solvent}$, per unit volume fraction of polymer coils in solution, $c\eta_{intr}$.

$$NSFR = \frac{\Delta P_{solution} - \Delta P_{solvent}}{\Delta P_{solvent}(c\eta_{int\,r})}$$

A polymer solution is more effective at reducing fluid mobility when it has a high NSFR value and a large volume fraction of polymer coils in the solution. The NSFR is related to the ratio of coil extensional viscosity, η_c, to solvent viscosity, μ_o. In general, a polymer solution should have large NSFR values at the fluid extensional rates experienced when flooding the reservoir, typically 0.5 to 5 sec^{-1}. This condition should maximize oil displacement from the porous medium.

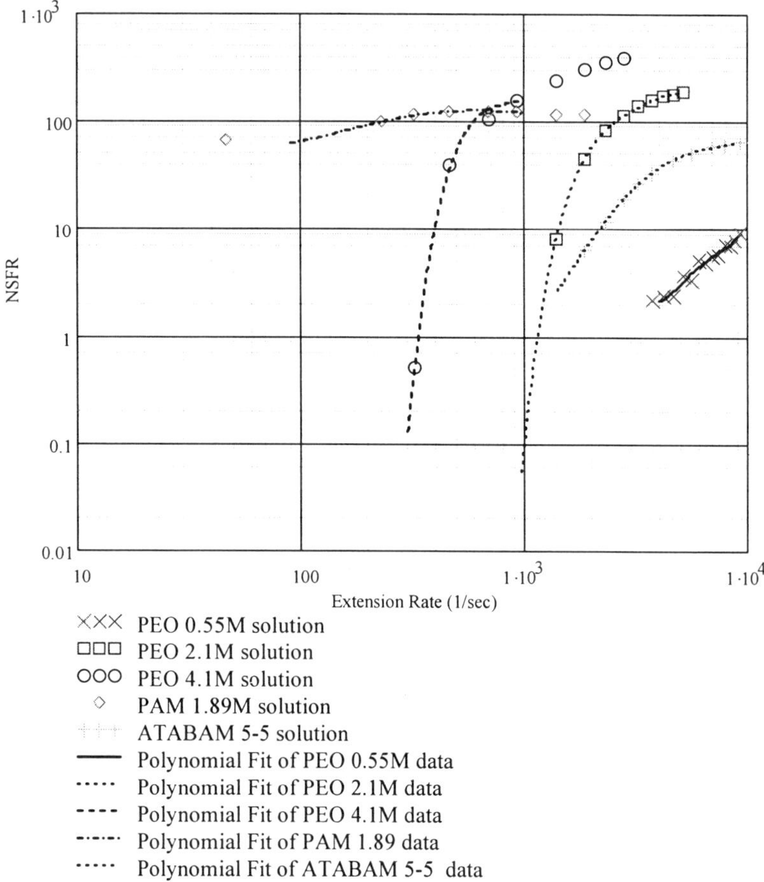

Figure 9. NSFR vs. Extension Rate

Conclusions

The effect of molecular weight, architecture, and solvent on a polymer's ability to reduce fluid mobility were studied using a screen extensional rheometer. This study suggests that for a polymer to be an acceptable mobility control agent the polymer should 1) have a large intrinsic viscosity, which minimizes the quantity of polymer needed for a given level of mobility control; 2) have a high coil viscosity, which maximizes the NSFR and thus minimizes solution mobility; 3) have a coil extensional yield stress that is zero or very low, which insures that polymer coils

As shown in Figure 8, the normalized solution pressures of the DAAM35 terpolymer solutions are solvent dependent. Table III reveals that the coil viscosity was a minimum in 0.514 molar NaCl, increased 700% in deionized water, and then increased 40% more in 6 molar urea.   Thus for this polymer, coil extensional viscosities are solvent dependent.

Table II shows that the polymer coil extensional yield rate is much lower for the higher molecular weight PEO than the lower molecular weight PEO. This suggests that lower molecular weight polymers having smaller coil diameters do not extend as easily as higher molecular weight polymers.

As shown by Figure 9, the NSFR varies with the fluid extensional rate as seen for solutions of ATABAM 5-5, NaAMB, and PEO. These solution maintain relatively constant NSFR values above a critical lower extension rate. Below this extension rate, the NSFR value for each solution drastically decreases. These solutions would not be an acceptable displacing fluid for oil reservoirs because NSFR values near zero are expected at the fluid extensional rates experienced during flooding.

In contrast, the PAM solution NSFR values remain high to much lower fluid extensional rates. Of the polymers examined, PAM appears to be the best candidate for having an appreciable NSFR value at the low extension rates experienced during application Therefore, the PAM solution would have good performance during flooding because the polymer coils of this solution extend even at very low fluid extensional rates.

Table III. Acrylamide Based Copolymer Properties

	ATABAM 5-5	*AMBA 4015-84*	*DAAM 35*	*DAAM 35*	*DAAM 35*
Solvent	DI Water	0.514M NaCl	DI Water	0.514M NaCl	6.0M Urea
$[\eta]$ (dL/g)	11.6	13.4	8.4	6.4	7.6
M_w (g/mol)	0.65M	N/A	N/A	N/A	N/A
λ_c (msec)	0.13	N/A	N/A	N/A	N/A
Q_{yield} (mL/min)	47	12	<0.5	<0.5	<0.5
$\dot{\varepsilon}_{yield}$ (1/msec)	2.16	0.055	<0.02	<0.02	<0.02
η_c (dyne sec/ cm^2)	0.41	0.009	0.017	0.0025	0.024
σ_{yield} (dyne/cm^2)	887	4.97	<0.3	<0.05	<0.5

324

will extend at the low fluid flow rates experienced during reservoir flooding. Future copolymers designed for use in reservoir flooding should be synthesized with these macromolecular properties in mind.

Acknowledgments

This work is funded by the U.S. Department of Energy (DE-AC26-98BC15111). The authors wish to acknowledge Dr. Charles McCormick's research group for providing all the polymers used in this study.

References

1. Hester, R. D.; Flesher, L. M. *Proceedings of the American Chemical Society, Div. of Polymer Materials: Science & Engineering*, 73, 1995.
2. Skartsis, L., B.; Khomami,; and Kardos, J. L. *J. Rheol.*, 36(4), 1992.
3. Skartsis, L. *Polym. Eng. Sci.*, 32(4), 1992.
4. Jones, D. M.; Walters, K. Rheol. Acta, 28, 1989.
5. McCormick, C. L., Hester, R. D. "Innovative Copolymers Systems for In Situ Rheology Control in Advanced Oil Recovery". Janurary 2000
6. Collins, R. E., *Flow of Fluids Through Porous Materials*, Petroleum Pub. Co., Tulsa, OK (1976).
7. Scheidegger, A. E., *The Physics of Flow Through Porous Media, 3rd Ed.*, University of Toronto Press, Toronto, Canada (1974).
8. Durst, F.; Haas, R., *Rheol. Acta 20*, 1981.
9. Hester, R. D.; McCormick, C. L. SPE/DOE 27823, *Proceedings of the Ninth Symposium on Improved Oil Recovery*, Tulsa, OK, 447, 1994.
10. McCormick, C. L. "Studies of the behavior of acrylamide/N-(1,1-dimethyl-3-oxobutyl) acrylamide copolymers in aqueous salt solutions" *Makroml. Chem.*, 188, 357, (1987).

Nomenclature

Symbol	Description
b	slope from NSP vs. Q curves
C^*	dimensionless concentration
c	mass concentration of polymer coils in solution
d	average sand diameter
d_{wire}	screen wire diameter
$k_{polymer}$	permeability of polymer coils
$k_{sandstone}$	permeability of sandstone
$k_{solvent}$	permeability of a Newtonian solvent
M	viscous average polymer molecular weight
N_A	Avogadro's number
Q	fluid volumetric flow rate
Q_{yield}	volumetric flow rate at which polymer coils start to extend
R	gas law constant
$\Delta P_{solution}$	fluid pressure drop across screens
ΔP_o	solution pressure drop across screens
T	absolute temperature
β	grouping of parameters
ε	coil strain
$\dot{\varepsilon}_{coil}$	polymer coil extensional strain rate
$\dot{\varepsilon}_{fluid}$	fluid extensional strain rate
η_c	polymer coil extensional viscosity
η_{intr}	polymer intrinsic viscosity
λ_c	polymer coil recovery time
μ_o	Newtonian solvent shear viscosity
σ_{coil}	fluid extensional stress
σ_{fluid}	coil extensional stress

Indexes

Subject Index

Bestsellers from ACS Books

The ACS Style Guide: A Manual for Authors and Editors (2nd Edition)
Edited by Janet S. Dodd
470 pp; clothbound ISBN 0–8412–3461–2; paperback ISBN 0–8412–3462–0

Writing the Laboratory Notebook
By Howard M. Kanare
145 pp; clothbound ISBN 0–8412–0906–5; paperback ISBN 0–8412–0933–2

Career Transitions for Chemists
By Dorothy P. Rodmann, Donald D. Bly, Frederick H. Owens, and Anne-Claire Anderson
240 pp; clothbound ISBN 0–8412–3052–8; paperback ISBN 0–8412–3038–2

Chemical Activities (student and teacher editions)
By Christie L. Borgford and Lee R. Summerlin
330 pp; spiralbound ISBN 0–8412–1417–4; teacher edition, ISBN 0–8412–1416–6

Chemical Demonstrations: A Sourcebook for Teachers, Volumes 1 and 2, Second Edition
Volume 1 by Lee R. Summerlin and James L. Ealy, Jr.
198 pp; spiralbound ISBN 0–8412–1481–6
Volume 2 by Lee R. Summerlin, Christie L. Borgford, and Julie B. Ealy
234 pp; spiralbound ISBN 0–8412–1535–9

The Internet: A Guide for Chemists
Edited by Steven M. Bachrach
360 pp; clothbound ISBN 0–8412–3223–7; paperback ISBN 0–8412–3224–5

Laboratory Waste Management: A Guidebook
ACS Task Force on Laboratory Waste Management
250 pp; clothbound ISBN 0–8412–2735–7; paperback ISBN 0–8412–2849–3

Reagent Chemicals, Ninth Edition
768 pp; clothbound ISBN 0–8412–3671–2

Good Laboratory Practice Standards: Applications for Field and Laboratory Studies
Edited by Willa Y. Garner, Maureen S. Barge, and James P. Ussary
571 pp; clothbound ISBN 0–8412–2192–8

For further information contact:
Order Department
Oxford University Press
2001 Evans Road
Cary, NC 27513
Phone: 1-800-445-9714 or 919-677-0977